Markus Ruchter

AF726901

A New Concept for Mobile Environmental Education

Schriftenreihe des

Instituts für Angewandte Informatik / Automatisierungstechnik

an der Universität Karlsruhe (TH)

Band 19

A New Concept for Mobile Environmental Education

von
Markus Ruchter

universitätsverlag karlsruhe

Dissertation, Universität Karlsruhe (TH)
Fakultät für Maschinenbau, 2007

Impressum

Universitätsverlag Karlsruhe
c/o Universitätsbibliothek
Straße am Forum 2
D-76131 Karlsruhe
www.uvka.de

SOME RIGHTS RESERVED

Dieses Werk ist unter folgender Creative Commons-Lizenz
lizenziert: http://creativecommons.org/licenses/by-nc-nd/2.0/de/

Universitätsverlag Karlsruhe 2007
Print on Demand

ISSN: 1614-5267
ISBN: 978-3-86644-182-8

A New Concept for Mobile Environmental Education

Zur Erlangung des akademischen Grades eines

Doktors der Ingenieurwissenschaften

von der Fakultät für Maschinenbau der
Universität Karlsruhe

eingereichte

DISSERTATION

von

Dipl.-Umweltwiss. Markus Ruchter

08. Oktober 1974 in Ludwigshafen/Rh.

Hauptreferent:	Prof. Dr.-Ing. habil. Georg Bretthauer
Korreferent:	Prof. Dr.-Ing. habil. Jürgen Wernstedt

Tag der Einreichung:	06.06.2007
Tag der mündlichen Prüfung:	24.07.2007

Danksagung

Die vorliegende Dissertation entstand im Rahmen meiner Tätigkeit als Doktorand am Institut für Angewandte Informatik des Forschungszentrum Karlsruhe.

Mein ganz besonderer Dank gilt Herrn Prof. Dr.-Ing. habil. Georg Bretthauer, der die Arbeit ermöglichte und mich durch Diskussionen und kritische Hinweise unterstützte. Ebenfalls danke ich Herrn Prof. Dr.-Ing. habil. Jürgen Wernstedt für die Übernahme des Korreferats und die wertvollen fachlichen Hinweise.

Besonders danke ich Herrn Dr. Clemens Düpmeier für die zahlreichen fachlichen Diskussionen und wichtigen Anregungen sowie seine tatkräftige Unterstützung bei der Entwicklung des Prototyps für den Mobilen Naturführer (MobiNaF). Weiterhin danke ich meinen Kollegen, insbesondere den Mitgliedern der Projektgruppe „Web-basierte Umweltinformationssysteme", für die gute Zusammenarbeit. Stellvertretend möchte ich vor allem Herrn Dr. Werner Geiger für seine organisatorische Unterstützung danken.

Die Erstellung dieser Dissertation wäre zudem nicht ohne die enge Zusammenarbeit mit dem Naturschutzzentrum Karlsruhe-Rappenwört im Rahmen des MobiNaf Projektes möglich gewesen. Hier sei stellvertretend Frau Heidrun Michel, Herrn Harald Dannenmayer und Herrn Klaus Hofmann für ihre Beratung bei umweltpädagogischen Fragen und ihre Hilfe bei der Evaluation gedankt.
Die Evaluation des MobiNaf Prototyps konnte nur durch die Mitwirkung eines Teams an Helfern durchgeführt werden. Für ihre engagierte Hilfe bedanke ich mich herzlich bei Andreas Gommlich, Nadine Haus, Christoph Jahn, Vilma Patalaviciute, Peter Real, Doris Simon, Christoph Singer und Rafael Sobek. Im Zusammenhang mit der Feldstudie bedanke ich mich bei Herrn Prof. Dr. Hans-Joachim Lehnert und Herrn Dr. Karlheinz Köhler von der Pädagogischen Hochschule Karlsruhe sowie Herrn Dr. Bernhard Klar von der Universität Karlsruhe für ihre fachliche Beratung insbesondere im Bereich der Umweltdidaktik sowie der statistischen Analyse.

Besonders möchte ich auch den tapferen Lesern der Vorversionen dieser Arbeit für ihre konstruktive Kritik und Korrekturen danken. Dies gilt insbesondere für Frau Nadine Haus, Herrn Gerd Zilly und Herrn Peter Brodhag.

Meiner Familie und Freunden, auf beiden Seiten des Atlantiks, möchte ich ganz herzlich für ihr Verständnis, ihre Geduld und Hilfe auf meinem Weg und während der Zeit der Erstellung der Arbeit danken.

Schließlich bedanke ich mich für die Finanzierung dieser Arbeit bei der Helmholtz-Gemeinschaft.

Karlsruhe, Juni 2007 Markus Ruchter

Ein neues Konzept zur computer-gestützten mobilen Umweltbildung

(Deutsche Kurzfassung)

Ziel dieser Arbeit ist die Entwicklung eines neuen Konzeptes zur computer-gestützten mobilen Umweltbildung. Darüber hinaus soll ein Prototyp eines „mobilen Naturführers" erstellt werden, anhand dessen die Umsetzbarkeit des Konzeptes überprüft werden kann. Schließlich soll der Prototyp in einer Feldstudie getestet werden, um seine effektive Anwendbarkeit für das Erreichen von Umweltbildungszielen zu evaluieren.

In Kapitel 1 wird, basierend auf einer Literaturanalyse, die Notwendigkeit für neue Ansätze der computer-gestützten Umweltbildung aufgezeigt. Trotz wachsender Bedeutung für eine nachhaltige Entwicklung, hat die Umweltbildung Schwierigkeiten wichtige Zielgruppen zu erreichen. Gerade Kinder und Jugendliche in urbanisierten Gesellschaften leiden mangels direkter Naturerfahrungen zunehmend an einer Entfremdung von der Natur. Damit einher geht ein mangelndes Interesse an den zurzeit angebotenen Umweltbildungsaktivitäten. Neue Medien stoßen hingegen bei der jungen Generation meist auf Faszination und Begeisterung. Existierende Konzepte zur computer-gestützten Umweltbildung werden von Umweltpädagogen und Eltern jedoch häufig mit Skepsis betrachtet. Es ist unter anderem die in der Regel gegebene räumliche Trennung von Computernutzung und direkter Naturerfahrung, welche die Sorge vor weiterer Naturentfremdung hervorruft.

Darauf aufbauend wird in der vorliegenden Arbeit die These vertreten, dass durch die zunehmende Verfügbarkeit mobiler Computer, in der Form von leichten und tragbaren Handgeräten, die Möglichkeit eröffnet wird, diese räumliche Barriere zu überwinden. Gleichzeitig wird davon ausgegangen, dass Kinder und Jugendliche, angetrieben durch die Faszination für mobile Technologien, wieder Begeisterung für die direkte Naturerfahrung gewinnen können. Kontextbezogene Informationssysteme, die als mobile Führungssysteme („Mobile Guides") in verwandten Bereichen wie z. B. dem Naturtourismus eingesetzt werden, bieten eine mögliche technische Basis für neue computer-gestützte Angebote in der informellen Umweltbildung. Eine umfassende Untersuchung bestehender mobiler Führungssysteme und verwandter Projekte zeigt, dass sich die zugehörigen Forschungsarbeiten in der Regel auf die Realisierbarkeit neuer informationstechnischer Konzepte insbesondere zur Navigationsunterstützung und der ortsbezogenen Inhaltspräsentation konzentrieren. Hingegen fehlt bisher ein umfassendes Konzept für die Anwendung umweltpädagogischer Instrumente mittels solcher mobiler Technologien. Ebenso mangelt es an empirischen Studien, die überprüfen, ob sich mit computer-gestützten, mobilen Medien die Ziele der Umweltbildung effektiv erreichen lassen.

Kapitel 2 stellt ein neues Modell für die computer-gestützte Umweltbildung unter Nutzung eines „mobilen Naturführers" vor. Hierfür werden in einem ersten Schritt, basierend auf umweltpsychologischen Erkenntnissen, Instrumente identifiziert, mit denen die Umweltbildung einen Einfluss auf das Umweltbewusstsein ausüben kann. Das in diesem Kapitel entwickelte Modell baut auf Konzepten der Umweltinterpretation auf und zeigt, wie mit Hilfe eines mobilen Führers die herausgearbeiteten Instrumente umgesetzt werden können. Es wird dabei deutlich, dass es gerade die Möglichkeit der Anpassung an den spezifische Nutzungskontext ist, die einen mobilen Führer zu einem geeigneten Medium für die Umweltinterpretation machen kann.

Kapitel 3 beschreibt den Entwurf für die Komponenten eines mobilen Naturführers auf Basis des zuvor entwickelten Modells. Auf der Grundlage einer Befragung von Repräsentanten verschiedener Zielgruppen werden die Nutzeranforderungen an einen mobilen Naturführer erfasst. Daraus werden schließlich die für ein solches System erforderlichen Komponenten abgeleitet.
Im vorgestellten Entwurf werden bestehende Ansätze zu den beiden zentralen Diensten - Navigationsunterstützung und ortsbezogene Informationspräsentation - erweitert, um sie für die Nutzung in naturnahen Gebieten sowie für die Umsetzung der vorgestellten Instrumente zu optimieren. Dies umfasst unter anderem die Nutzung zusätzlicher Kontextinformationen über die Umweltbedingungen und den Nutzer, die dem System eine Anpassung an die spezifische Nutzungssituation und das Umweltbewusstseinsprofil des Besuchers ermöglichen. Überdies wurden Konzepte für weitere Dienste ausgearbeitet. Sie umfassen einen Enzyklopädie-Dienst als Nachschlagewerk für Detailfragen, die sich vor Ort ergeben. Außerdem werden Dienste konzipiert, die ein direktes Erlebnis der Natur anregen und fördern können. Hierzu gehört insbesondere eine Reihe von „Erkundungswerkzeugen". Mit diesen Anwendungen soll ein mobiler Naturführer die Stärken eines sensor-gestützten Führungssystems nutzen, um zu einem Instrument für die Naturerkundung zu werden. Unter Anwendung von Technologien aus dem Bereich der erweiterten Realität („Augmented Reality") können „Wahrnehmungsinstrumente" eingesetzt werden, die es durch zusätzliche Sensoren (z. B. Infrarotkamera) ermöglichen, vom Menschen sonst nicht wahrnehmbare Naturphänomene zu visualisieren. Zur Unterstützung des selbstbestimmten Lernens während der Erkundung der Natur wird ein Konzept für ein elektronisches Bestimmungsbuch vorgestellt. Dieses soll auch Laien ermöglichen, die entdeckten Naturphänomene (z. B. Tier- und Pflanzenarten) zu identifizieren und mehr über sie zu erfahren. Durch ein Tourtagebuch kann dem Wunsch der Besucher Rechnung getragen werden, ihre Naturerlebnisse zu dokumentieren, um sie später mit Freunden, Mitschülern oder anderen Besuchern zu teilen. Mit Hilfe eines Manager-Dienstes soll es den Verantwortlichen von Schutzgebieten ermöglicht werden, eine effiziente Besucherlenkung zu betreiben und auch Notfalldienste anzubieten. Gleichzeitig dient der Manager-Dienst auch zur Evaluation des Umweltinterpretationsangebotes.
Alle Dienste basieren auf der Nutzung von Kontextinformationen aus verschiedenen Kategorien. Für die bessere Verwaltung dieser Daten wurde in der vorliegenden Arbeit ein spezielles Kontextmodell entwickelt, welches die Umweltbedingungen, den Nutzer, das mobile System sowie den zeitlichen Kontext repräsentiert und miteinander in Beziehung setzt.

Außerdem werden Lösungsansätze für weitere wichtige Komponenten eines mobilen Naturführers vorgestellt. Hierzu zählen die Interaktionsmechanismen und -modalitäten des Benutzerinterfaces, insbesondere für ein Gerät mit kleinem Bildschirm, das im Freien eingesetzt wird. Darüber hinaus werden Entwürfe für die Wissensbasis und die zentrale Kontrolleinheit beschrieben und Hardwareoptionen basierend auf dem Stand der Technik diskutiert.

In Kapitel 4 wird das informationstechnische Konzept für die Entwicklung eines mobilen Naturführers präsentiert. Als Grundlage für die Implementierung eines solchen Systems wird eine Drei-Schichten-Architektur vorgestellt, die durch einen modularen Ansatz die Einbettung zusätzlicher Dienste ermöglicht. Die Präsentationsschicht bildet mit den Nutzerschnittstellen für die verschiedenen Dienste die oberste Ebene. An der Basis befindet sich die Daten- und Diensteschicht, in die, je nach Verfügbarkeit, Kontext- und Datendienste eingebunden werden können. Dazwischen liegt die Kontrollschicht, die zwischen den anderen Schichten vermittelt. Des Weiteren werden die Implementierung eines ersten mobilen Naturführer-Prototyps beschrieben und Erfahrungen aus der Umsetzung des informationstechnischen Konzeptes diskutiert.

Kapitel 5 ist der praktischen Erprobung des in der Arbeit entwickelten Konzeptes gewidmet. Im Rahmen einer Feldstudie, in Kooperation mit dem Naturschutzzentrum Karlsruhe-Rappenwört, wurde der mobile Naturführer-Prototyp durch Familien und Schulklassen während einer Führung durch die Rheinauen getestet. Dabei wurde neben der generellen Gebrauchstauglichkeit des Systems auch die Wirkung einer Führung mit dem mobilen System auf die zuvor identifizierten Umweltbewusstseinsgrößen (Wissen, Einstellung, Werte, Verhalten) überprüft. Zusätzlich wurden die Ergebnisse des computer-gestützten Mediums mit den Resultaten von Testgruppen mit traditionellen Medien (Broschüre, „menschlischer Naturführer") verglichen.
Bezüglich der Gebrauchstauglichkeit hat die Untersuchung gezeigt, dass die Testnutzer noch einige Schwierigkeiten bei der Interaktion mit dem Prototyp hatten. Diese lassen sich unter anderem auf noch bestehende Limitierungen in der Leistungsfähigkeit und Stabilität des Systems als auch auf Schwierigkeiten mit den im Prototyp implementierten Interaktionsmechanismen zurückführen.
Trotz dieser Einschränkungen zeigen die Ergebnisse der Evaluation, dass durch Führungen mit dem mobilen System ähnliche Wirkungen auf die Umweltbewusstseinsgrößen erreicht werden können wie mit den traditionellen Umweltinterpretationsmedien. Testpersonen mit dem mobilen Naturführer erzielten einen vergleichbaren Zuwachs in ihrem Wissen über die Umwelt wie Teilnehmer mit der Broschüre oder dem „menschlichen Naturführer". Unabhängig von dem verwendeten Medium wurde durch die Führung keine der anderen Umweltbewusstseinsgrößen signifikant beeinflusst.
Durch die Evaluation konnte somit belegt werden, dass ein mobiler Naturführer prinzipiell als Umweltinterpretationsmedium einsetzbar ist. Darüber hinaus haben empirische Beobachtungen während der Feldstudie ergeben, dass vor allem Schüler, die den mobilen Naturführer benutzt haben, sich intensiver mit den präsentierten Inhalten auseinandersetzten als ihre Mitschüler wäh-

end Führungen mit einer Broschüre. Dies deutet darauf hin, dass mit mobilen Naturführern die Faszination für neue Technologien genutzt werden kann, um bei Kindern und Jugendlichen die Motivation zur Teilnahme an Umweltbildungsaktivitäten zu steigern. Des Weiteren kann davon ausgegangen werden, dass durch eine Verbesserung der Gebrauchstauglichkeit, einhergehend mit der weiter zunehmenden Verbreitung mobiler Geräte, sich auch bei Erwachsenen die Akzeptanz für einen solchen mobilen Naturführer weiter verbessern wird.

Zusammenfassend ergeben sich aus dieser Arbeit die folgenden neuen Beiträge und wichtigsten Ergebnisse:

1. Identifikation von Instrumenten zur Beeinflussung des Umweltbewusstseins, die sich mit Hilfe von computer-gestützten Medien anwenden lassen: a) Vermittlung von Wissen, b) Visualisierung von verborgenen Naturphänomenen, c) Motivation direkter Erkundung, d) Repräsentation von Rollenmodellen, e) Anbieten von Feedback-Mechanismen.

2. Entwicklung eines neuen Modells für mobile Naturführer, das die Verbindung von computer-gestützter Umweltbildung und direktem Naturerlebnis - basierend auf der Kombination von Umweltinterpretation und kontextbezogenen Führungssystemen - ermöglicht.

3. Erfassung von Zielgruppen: a) Schüler, b) Familien, c) Naturliebhaber. Aufstellung eines Anforderungskatalogs für mobile Naturführer basierend auf einer Nutzerbefragung. Erstellung eines detaillierten Nutzungsszenarios aufbauend auf der Analyse einer Umweltinterpretationseinheit.

4. Spezifikation der Komponenten eines mobilen Naturführers: a) kontextbezogene Dienste, b) Kontextmodell, c) Wissensbasis, d) Nutzerschnittstelle und Interaktionsparadigma, e) zentrale Kontrolleinheit, f) mobile Hardwareplattform und Technologien.

5. Entwicklung eines detaillierten Entwurfs für die kontextbezogenen Dienste unter besonderer Berücksichtigung der Einsatzbedingungen in naturnahen Gebieten. Neben den Kerndiensten für mobile Führungssysteme, dem Navigationsdienst und dem ortsbezogenen Informationsdienst, sollte ein mobiler Naturführer weitere Dienste zur Unterstützung der direkten Erkundung der Natur anbieten. Dazu gehören: a) Umwelt-Enzyklopädie, b) Wahrnehmungsinstrumente, c) elektronisches Bestimmungsbuch, d) Tourtagebuch, e) Manager-Dienst zur Unterstützung der Administratoren von Schutzgebieten.

6. Entwicklung eines informationstechnischen Konzeptes, auf dessen Basis sich ein Prototyp des Systems implementieren lässt. Entwurf einer modularen, service-orientierten Rahmenapplikation, die durch weitere Module flexibel erweiterbar ist.

7. Design der Nutzerschnittstellen basierend auf den Metaphern der Informationsbroschüre und des Naturführers. Zur Umsetzung wird die Anwendung eines animierten pädagogischen

Agenten propagiert, der eine intuitive Interaktion ermöglicht und gleichzeitig als Rollenmodell dienen kann.

8. Bestätigung der technischen Realisierbarkeit durch die Implementierung eines mobilen Naturführer-Prototyps im Rahmen des MobiNaf-Projektes. Bestimmte Probleme bei der Umsetzung des Prototyps ergaben sich vor allem durch die Verwendung von „Scalable Vector Graphics". Diese Technologie eignet sich für die Darstellung der Karten im Navigationsdienst, zeigt aber noch starke Einschränkungen bei der Repräsentation anderer Nutzerinterface-Module.

9. Erfolgreicher Einsatz des Prototyps in einer Feldstudie. Durch die Evaluation wurde belegt, dass ein mobiler Naturführer als Umweltinterpretationsmedium nutzbar ist. Trotz Einschränkungen in der Gebrauchstauglichkeit, kann mit einem mobilen Führer ein Zuwachs im Umweltwissen erreicht werden, der den Wirkungen traditioneller Umweltinterpretationsmedien entspricht. Die Ergebnisse zeigen, dass mit mobilen Naturführern die Motivation von Kindern und Jugendlichen zur intensiveren Teilnahme an Umweltbildungsaktivitäten gesteigert werden kann. Es ist anzunehmen, dass durch eine Verbesserung der Gebrauchstauglichkeit sich auch bei Erwachsenen die Akzeptanz weiter verbessern lässt.

Im Rahmen zukünftiger Arbeiten sollten, über den Navigationsdienst und den kontextbezogenen Informationsdienst hinaus, weitere der in dieser Arbeit konzeptionell entwickelten Dienste für mobile Naturführer implementiert und evaluiert werden. Vor allem durch die beschriebenen „Erkundungswerkzeuge" könnte das Potential des mobilen Naturführers als Instrument zur selbstbestimmten Naturerkundung weiter gesteigert werden.

In der Arbeit wird gezeigt, dass mobile Naturführer, die auf dem neuen Konzept für mobile Umweltbildung aufbauen, eine effiziente Alternative zu traditionellen Umweltinterpretationsmedien bieten können. Damit eröffnen sich für Bildungseinrichtungen - auch unter dem Gesichtspunkt knapper Ressourcen - neue, wirtschaftliche Möglichkeiten, um durch den Einsatz computergestützter Instrumente mehr Kinder und Jugendliche für Umweltbildungsaktivitäten zu begeistern und bei der direkten Erkundung der Natur zu unterstützen.

Contents

List of Figures

List of Tables

Abbreviations

ANOVA	Analysis of Variance
AOI	Area of Interest
APA	Animated Pedagogical Agent
API	Application Programming Interface
AR	Augmented Reality
AV	Augmented Virtuality
BT	Bluetooth
CCD	Charge-coupled device
CF	Compact Flash
CPU	Central Processing Unit
CSCL	Computer Supported Collaborative Learning
DESD	Decade of Education for Sustainable Development
DOM	Document Object Model
EC	Environmental Communication
ECA	Embodied Conversational Agent
EE	Environmental Education
EIS	Environmental Information System
ESD	Education for Sustainable Development
GIS	Geographic Information System
GPRS	General Packet Radio Service
GPS	Global Positioning System
GUI	Graphical User Interface
HMD	Head-mounted Display
ICT	Information and Communication Technologies
IT	Information Technology
JSF	Java Server Faces
LBS	Location-based Services
LCD	Liquid Crystal Display
MEI	Mobile Environmental Interpretation
MEIS	Mobile Environmental Information System
MobiNaG	Mobile Nature Guide System
MR	Mixed Reality
MVC	Model-View-Controller software design pattern
NAZKA	Naturschutzzentrum Karlsruhe Rappenwört (Nature Center Karlsruhe/Germany)
NGO	Non-Governmental Organization
PBC	Perceived Behavioral Control
PBL	Problem-based Learning
PDA	Personal Digital Assistant

POI	Point of Interest
OOI	Object of Interest
Qt	Quasar toolkit
QVGA	Quarter Video Graphics Array
RF	Radio Frequency
RFID	Radio Frequency Identification
SMS	Short Message Service
SVG	Scalable Vector Graphics
TPB	Theory of Planned Behavior
TTS	Text-To-Speech
UI	User-Interface
UMTS	Universal Mobile Telecommunications System
URL	Universal Resource Locator
USB	Universal Serial Bus
VBN	Value-Belief-Norm theory
VGA	Video Graphics Array
VR	Virtual Reality
WAP	Wireless Application Protocol
WLAN	Wireless Local Area Network
ZOI	Zone of Interest

Chapter 1

Introduction

1.1 Motivation

Despite the efforts of the world community for environmental integrity, progress has been slow, and the global environment continues to deteriorate [343]. According to the Millennium Ecosystem Assessment [134] human actions have led to unprecedent alternations of the natural environment. Sixty percent of the world's ecosystems are in decline or even degraded to an extent that we can no longer rely on their services. These services include climate regulation, clean air and water, fertile land and productive fisheries. Thus the deterioration of the natural environment remains one of the major challenges for all nations around the world.

Still environmental problems are not solely caused by governments and organizations or companies but are also a result of the life-style and behavior of individual citizens. Besides policy-based and regulatory instruments, environmental communication and in particular environmental education are considered to play a fundamental role in achieving environmental protection goals [55, 97, 221, 257]. The importance of environmental education (EE) for the process of sustainable development has been stressed by the declaration of the United Nations Decade of Education for Sustainable Development (DESD) in 2005 [345]. This educational effort intends to encourage changes in behavior that will create a more sustainable future in terms of environmental integrity, economic viability, and a just society for present and future generations.

In the face of this decade, environmental educators are confronted with major challenges. One of the fundamental obstacles they have to deal with, especially in industrialized societies, is an increasing alienation from nature, which has been observed in particular among younger generations living in more and more urbanized societies [25,45]. Empirical studies have shown that even though the young people tend to be aware of environmental issues and value conservation, they

increasingly lack direct experiences of the natural environment. Furthermore they seem to lack the interest and motivation to engage in direct experiences with nature [44, 45, 262]. Regular contact with and play in the natural world during childhood is commonly seen as an important basis for a lasting affinity to nature, along with positive environmental ethics [25, 33, 36, 163]. Consequently, these observations are particularly alarming, since the children's loss of regular contact with nature can result in a bio-phobic future generation that is not interested in preserving nature and its diversity [25, 44].
In addition, environmental education institutions commonly have to operate on increasingly tight budgets along with limited personnel resources [27, 129, 215], restricting the amount of activities that can be offered and the number of participants that can be served.

In the information and communication age an apparent solution for environmental education institutions could be to embrace contemporary media-based instruments. This implies the advanced employment of information and communication technologies (ICT) in EE as has been proposed by Siebert [315] as well as Rohwedder and Alm [277]. Such an intensified application of ICT is also encouraged by the implementation scheme for the DESD [345]. However, many educators consider the application of new media in EE as ambiguous [13, 169, 254, 313]. The use of the computer is frequently seen as an antagonism to the experience of nature, since it may keep the participant from directly experiencing his natural environment and as such acts as another potential source of alienation [13, 254, 313].
Nonetheless, it is generally accepted that, in today's information and communication societies, young people as well as other target groups tend to be interested in new technologies [13, 147]. Thus it can be assumed that a new approach in EE employing modern mobile computing technologies can help to motivate citizens to engage in EE activities. Mobile technologies further appear promising for this endeavor since they may help to overcome the antagonism between media usage and the direct experience of natural environments.
This thesis will make a contribution to this novel approach, by developing a new concept for environmental education activities employing context-aware applications. These new mobile EE instruments should assist participants in the direct exploration of their natural environment in order to increase their environmental literacy.

The Mobile Nature Guide project

The work documented in this thesis has been conducted within the scope of the Mobile Nature Guide (MobiNaG) project that is carried out as a joint project of the Institute for Applied Computer Science/Forschungszentrum Karlsruhe and the Naturschutzzentrum Rappenwört, an environmental education institution in Karlsruhe, Germany. The objective of the project is the development of new location-based mobile applications, as an addition to public environmental information systems of the State of Baden-Württemberg, that can be deployed on site to support the user during

his exploration of the natural environment [115]. As a visitor center to a floodplain conservation area along the Rhine river, the Naturschutzzentrum Rappenwört is an ideal test bed for the prototyping and evaluation of MobiNaG.
On the basis of this system, a general concept for mobile nature guides for natural areas in Baden-Württemberg is developed together with the State Institute for Environmental Protection Baden-Württemberg [283]. An exemplary realization of a MobiNaG system is intended for selected nature exploration areas [284, 285]. MobiNaG is integrated into the KEWA project of the Ministry for Environment and Traffic Baden-Württemberg [213, 284]. Additional support is granted by the Ministry of Nutrition and Rural Areas of the State of Baden-Württemberg and the City of Karlsruhe.

1.2 State of the art

1.2.1 An introduction to environmental communication and education

In order to design an effective computer-mediated instrument which can contribute to an increase in environmental awareness and responsible environmental behavior, it is necessary to come to an understanding of the fundamental principles and methods of the fields of environmental communication and education.

Environmental communication

Environmental communication (EC) is commonly acknowledged in the literature as the foundation for establishing relationships between people and the environment. It is characterized as a means for enhancing environmental literacy and sustainable environmental practices, making it a prerequisite for environmental protection [55, 97, 257]. Generally, EC can be determined as any kind of environmentally relevant information flow that implies the use of communication processes [242, 257].
Organizations working in the field of economical and environmental development, define EC as the planned and strategic use of communication processes and media products to support effective policy making, public participation and project implementation geared towards environmental sustainability [131, 242].
Pillmann [257] describes EC from an environmental informatics perspective as the sharing of environmental data and information between various audiences using different media. This environmental information flows in myriad ways, such as face to face communication, publications, mass media broadcasts, and more recently digital communication through the Web and mobile devices [248, 257]. It was acknowledged by Dickinger et al. [87] and Jelitto [153] that these digital

media have an increasing importance in EC.
Thus EC is of paramount importance in generating public awareness for environmental issues. This is especially significant with regard to environmental problems and their potential impacts, which are frequently not subject to direct experience and thus must be translated into a message that the general public can comprehend and relate to [234]. EC is considered a soft, informatory instrument that, by achieving a public understanding and participation, provides the basis for hard, legal instruments such as mandates and incentives [234]. The importance of EC for environmental policy is further stressed by the OECD and the European Environment Agency (EEA). Furthermore, EC is decisive for building the social, economic, and political action networks needed to reverse the present unsustainable and negative environmental trends [97, 242].
A number of recommendations for an effective EC campaign have been identified in the literature [55, 131, 234] and are listed in Appendix A.1.

Environmental education

Education and communication are closely related in the sense that education includes the provision of information and thus commonly relies to a certain extent on communication processes. But even though environmental education (EE) is in some cases referred to as an instrument of EC [9, 315] it is by itself a field of research of crucial significance for the development of a computer-based instrument to influence environmental awareness and behavior.

History and definitions EE emerged as an international discipline in the early 1970s as a result of the modern environmental movement [76, 221]. EE has different meanings for people from different schools of thought [76], resulting in a large spectrum of synonymous terms as well as closely related fields like outdoor and wilderness education and education for sustainable development [137]. Despite the variety in terms and definitions, much of the work in EE has been guided by international conventions. The Belgrade Charter (UNESCO-UNEP, 1976) and the Tbilisi Declaration (UNESCO, 1977) furnish a commonly acknowledged blueprint for EE [336]. The Belgrade Charter was adopted by a United Nations conference in 1976 and provides a general goal statement for environmental education:

> The goal of environmental education is to develop a world population that is aware of, and concerned about, the environment and its associated problems, and which has the knowledge, skills, attitudes, motivations, and commitment to work individually and collectively toward solutions of current problems and the prevention of new ones [346].

A comprehensive review of EE is provided by Beyersdorf [27], Thomson [336], Scoullos and Malotidi [309] and Ramsey and Hungerford [263].

Despite the international consensus on EE, achieved by the above declarations, it should be noted that due to differences in natural and social systems as well as educational traditions and schools of thought, there is of course still some variation between the EE philosophies practiced in different countries. This thesis will focus on EE in North America, where a long standing tradition of EE can be observed, and Germany, where the project is being conducted.

According to the North American Association for Environmental Education (NAAEE) [230] EE should result in the development of an environmentally literate citizenry. Environmentally literate individuals understand environmental issues and how environmental quality is impacted by human decisions. In addition, they use this knowledge to make informed, well-reasoned choices that also take social and political considerations into account [230].

Concepts for curricular of EE programs and their intended outcomes generally correspond to the overall goals of EE proposed by the Tbilisi convention. Reviews by Thomson [336] and Archie and McCrea [14] as well as Ramsey and Hungerford [263] discuss a variety of significant concepts for EE. A summary of these concepts is provided in Appendix A.2.1.

Different environmental education approaches A number of authors offer a variety of recommendations for educators who intend to support their learners in achieving environmental literacy [27,230,324,362]:

Targeting: The educators should focus on specific target groups and consider their relevant concerns. This includes captive audiences in formal settings as well as noncaptive audiences in informal settings, whose characteristics are specified in Table 1.1. Educators should create a learning environment suitable for the learner and participate with the student in the learning process. Environmental education activities at each grade level should focus on the feeling (affective), knowing (cognitive) and skill-behavior domains. Still different emphasis should be set according to age level (i.e. early years: awareness and feelings; later years: knowledge and skill-behavior).

Personalize communication: The instructors ensure ongoing communication with learner as an active participant, play close attention to level of credibility and trust, use most effective channel, encourage info exchange after initial awareness-raising stage.

Assisted access to information: The educators should assure access to a variety of information sources based on an assortment of approaches, repeating and reinforcing information flow and tell where to get more information.

Foster curiosity and enthusiasm: Instructors should foster the learners' innate curiosity and enthusiasm. They should provide positive learning experiences as well as encourage and acknowledge any behavioral changes.

Encourage exploration: The instructors should provide learners with early and continuing opportunities to explore their environment, involving all senses.

Provide Links: Educators should link science with other areas of intellectual and emotional activity.

Display responsible action: Educators should motivate independent thinking and effective, responsible action by presenting sustainable ways of life practices. They should foster a growing sense of confidence that groups and individuals can positively affect the environment.

Siebert [315] notes that the change in social, cultural, technological circumstance also leads to a shift in learning. In the scientific information society the plain acquisition of knowledge is loosing weight compared to the filtering/selection and evaluation of information. Cognitive learning (i.e. info processing, abstraction, synthesis, linking) and constructive learning (i.e. structured learning, self-directed, change in perspective, evaluation), as well as reflexive and affective learning increase in significance. This is also accompanied by a change in the quality of some learning forms. Learning by experience is still important but most experiences are only made indirectly through mass media and computational devices [315]. As a consequence also EE has to embrace a combination of traditional and modern forms of learning. Self-directed learning, based on a constructivist model of EE [315], is generally increasing. This is also due to the rise in individualization and an improved educational background as well as the availability of learning resources, frequently provided as multi-media material. Beyersdorf [27] provides a more extensive review of methods applied in the German EE domain.

Two basic forms of EE are differentiated in the literature, formal EE and informal/non-formal EE [14, 364]. Formal EE is generally characterized by taking place in a formal setting such as schools. Its participants are usually referred to as 'captive audience' since they are required to attend and are usually motivated by an external reward (e.g. grades, certificate). Informal EE typically takes place in a setting, like parks or nature centers, related to leisure activities. Participants commonly attend voluntarily, based on intrinsic motivations (e.g. interest in nature, entertainment) and are thus called a 'non-captive audience' [14, 129, 364]. While formal EE is frequently dominated by formal instruction, informal EE providers need to apply other means of communication in order to hold the attention of their participants, like for example environmental interpretation [129]. A comprehensive comparison of formal and informal EE is provided by Ham [129] as well as Wohlers [364] and summarized in Table 1.1.

Since the new EE instrument should support direct exploration of the natural environment, this thesis will focus on informal EE methods. According to Wohlers [364], the central informal EE approach is environmental interpretation. Due to its importance as an informal EE method for visitors to natural areas it is considered as particularly relevant approach for this thesis and will hereafter be discussed in more detail.

Table 1.1: Characteristics of formal and informal environmental education adapted based on [129, 364].

Trait	Formal EE	Informal EE
Target group	Captive audience	Noncaptive Audience
	previous knowledge required	no previous knowledge required
Motivation	Extrinsic	Intrinsic

Environmental interpretation

The discipline of Environmental Interpretation as a method of informal EE is rooted in the work of Freeman Tilden, who in the 1930s was asked by the U.S. National Park Service to evaluate the visitor and environmental communication services of the parks. As a result of his assignment he established a formal definition and the basic principles of Environmental Interpretation. Tilden [337] defined environmental interpretation as:

> "...an educational activity which aims to reveal meanings and relationships through the use of original objects, by firsthand experience, and by illustrative media rather than simply to communicate factual information."

As a consequence environmental interpretation distinguishes itself from traditional forms of formal education, which are frequently based primarily on facts. The goal of interpretation in contrast is to scaffold first hand experience and communicate a message. Thus the interpretor usually tells a story that generally includes a moral [129].
Based on the work of Tilden and many years of interpretive experience, Ham suggests that environmental interpretation needs to be (adapted based on Ham [129]):

Pleasurable: Since interpretation is an informal EE activity targeted at a non-captive, media spoiled audience, it needs to be entertaining in order to hold the audience's attention and make the experience memorable. This requires the use of a number of methods to make technical information more entertaining (e.g. visual metaphors, overriding analogy, personification [129]) but also the application of multi-media elements.

Relevant: Interpretation can only be effective when it is adapted to the personality, experience, and previous knowledge of the participant. So it should be meaningful (i.e. understandable language and connected to something we already know) as well as personal (i.e. relating to the participant's personal life, something he cares about) [129].

Organized: The interpretive message should be presented in a way that is easy to follow, requiring little cognitive effort. This implies that the information is well structured and easy to memorize, which can be facilitated by keeping number of main ideas less or equal to five.

Nonetheless good interpretation should capitalize on the unexpected, flexibly connecting upcoming phenomena which are surprising but obvious to the visitor [129].

Explorative and constructive: Tilden [337] describes interpretation as a form of exploration since it is not intended to instruct people but to motivate a self-directed learning experience involving direct interaction with the presented phenomena.

Thematic: The interpreter should tell an entertaining story which, however, relies on facts. It is important that this story is based on a central theme that conveys the crucial message, since the audience is in the end most likely to remember this theme.

Methods of environmental interpretation Environmental interpretation is based on the model of the interpretation triangle (see Fig. 1.1). This triangle consists of the phenomenon, the interpreter and the visitor/participant.

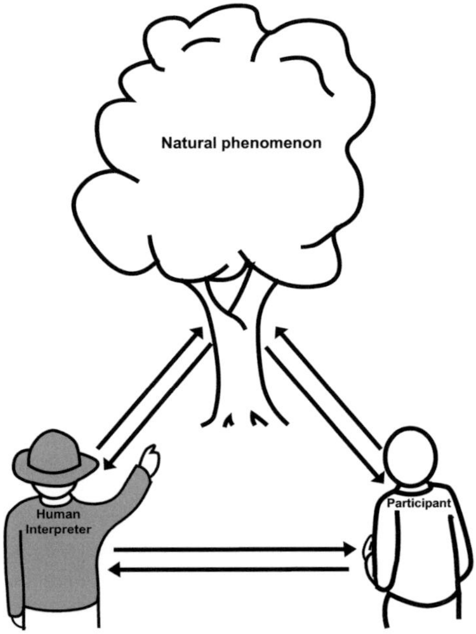

Figure 1.1: Interpretation triangle adapted from Lewis, 1993 cited in [199].

According to Carter [57] a *phenomenon* can be an object or place, a natural or cultural heritage feature, which the interpreter thinks is special (e.g. a tree, a rock, a landscape, a historical object or activity). The phenomenon should be an adequate representative of a certain type of object related to the theme. It is important though that the specific phenomenon has a "personal" story [129,199].

The *participant* is the central figure of the interpretation triangle, since the activity should be target towards him. He enters into the interpretive activity driven by certain needs, like for instance a genuine interest in certain natural phenomena or just the wish to be entertained or to relax. Since good interpretation needs to be relevant and therefore personalized, it is beneficial to learn more about participant's personal characteristics as a foundation for helping him to interact with the phenomenon in order to gain an understanding of and the linkage to the phenomenon [129].
The *interpreter* is the component of the triangle responsible for facilitating this interaction. He has to fulfill the function of a mediator between the participant and the phenomenon. His message is supposed to build a bridge between these two components of the triangle [199]. Hence he has to be able to speak the "language" of the phenomenon as well as of the participants. In order to fulfill this task, the interpreter can apply a number of methods and tools. First of all the personality that the guide projects to the audience has a crucial impact on how it experiences the tour. Ham [129] points out that interpreters acting like hosts who try to establish a friendly atmosphere making the visitor feel comfortable and promoting two-way communication are most successful. In addition the interpreter should posses certain communication qualities including a clear verbal as well as non-verbal expression. It is, however, of paramount importance that the interpreter is perceived as authentic by the visitors. This requires that the guide is enthusiastic about the natural phenomena as well as the interpretation activity and is able to convey this enthusiasm. Further methods include the provision of certain stepping stones that grant access to the interpreter's message such as analogies, metaphors, examples etc. [129, 199]. Active involvement of the participant can be achieved for example by asking the participants for his assistance during the interpretation and giving them assignments. These mechanisms should make an involvement more attractive for the participants since they should experience the possibility to help others, feel the challenge of finding something and revealing a solution or finish something incomplete. Eventually the participant should express preferences and voluntarily engaging in exploration [199].

Ham and Weiler [130] as well as Ludwig [199] differentiate between two basic forms of interpretation. Next to the traditional personal-based interpretation conducted by a human as interpreter as described by the interpretation triangle (see Fig. 1.1), there is the media-based interpretation using different EC tools such as signs, activity elements, brochures etc. to interpret the phenomena.
Ludwig [199] further describes different approaches to conducting these forms of interpretation, including short-interpretation, guided-walks or -tours, and free-interpretation. Comprehensive reviews of environmental interpretation methods are provided by Ham [129], Carter [57] and Ludwig [199].

Since this thesis will deal with a new approach for the employment of ICT in informal EE, the review of traditional approaches to EE should be completed by a brief examination of the state of computer-mediated EE.

1.2.2 Computer-mediated environmental education

Computer-mediated education has been practised for about two decades now and there have been several studies which have confirmed an overall positive effect of multimedia technologies on learning [135]. Also the EE sector has not been spared by the increasing computerization of all facets of life and is compelled to find new forms of learning applying new media [315]. This task is to some extent impeded by the, previously mentioned, reluctance of educators in the EE domain to use new ICT [13, 254]. Next to the concern of EE instructors as well as parents that a further increase in media consumption can also enhance the alienation from nature, it is assumed that a general lack of skills and knowledge related to multimedia computer applications lead to a widespread prejudice against ICT [13]. Nonetheless, there are also notions in the community to embrace new technologies [12, 153]. Apel [13] stresses that there are a variety of EE institutions as well as non-governmental organizations (NGOs) who are aware of the potential of new media and do employ ICT, which shows that there are no consistent ideological concerns throughout the community.

In the face of this ambiguity among the stakeholder in EE about the application of ICT there has previously been a certain amount of debate about the advantages and disadvantages of using multi-media instruments in EE. Payne [254] for instance put forward, that the so-called efficiency of electronic versions of EE may eventually lead to the loss of direct learning experiences in the field. Next to the alleged antagonism between experiencing nature and the use of computers, organizational and resource related issues as well as didactic challenges have been identified. Another issue of concern for EE providers includes the potential environmental impacts associated with the production and disposal of ICT hardware. On the other hand it has been acknowledged, that EE providers compete with the modern media for attention and have to keep up with time. In this respect the potential of ICT as new tool for learners and educators has been recognized. Rohwedder and Alm [277] as well as Apel [13] analyzed the challenges and also the promises of extended application of ICT in EE. Their finding are summarized in Table A.1 (see Appendix A.2.2).

Different types of digital media in EE Apel [12] argues that ICT should primarily be employed where deficits can currently be found in EE. Accordingly he differentiates four types of ICT employment:

Compensatory employment: For the things that cannot be done just as easy without electronic instruments.

Efficiency employment: In cases where a more intensive and effective learning experience needs to be provided, which can be done faster and easier with electronic media.

Motivational employment: Multimedia applications can have a strong motivational and experience mediating value. As a result, they are a suitable tool for virtual experiences of new perspectives and new encounter of phenomena (e.g. using a virtual pocket lens to stand "eye-to-eye" with a bug).

Connectional employment: Participatory potential of open-networks can be used to connect learners world wide.

Osborne and Hennessy [246] provide a comprehensive review of the role of ICT in science education relating also to environmental issues. Jelitto [153] surveyed examples of the application of digital media in the EE domain in Germany. He differentiates three types of applications: Tools, information systems, and games. The category tools comprises software, which can be used by educators as well as participants to develop EE materials or to solve tasks as part of the learning process. The category of information systems refers to a wide variety of applications including databases as well as CD-ROM based presentations or collections of materials and web-based systems. Each can include resources for educators as well as learners and the focus of these systems may range from education and information to "Edutainment" or "Infotainment" passing over into the third category of games. While in the past most applications in this domain were implemented as stand-alone systems for desktop PCs, more and more web-based services have been established in recent years. Frequently, digital media have also been employed as exhibits or as part of a displays in EE institutions.

New opportunities based on mobile technologies Computer-aided EE has already been employed for over a decade now and based on the gained experiences combined with the continuous technological innovations as well as improvements in general computer literacy of citizens, several of the initial challenges and concerns could be overcome. However, Gough [119] explicitly calls for more EE research that investigates the actual effects of ICT concerning EE objectives.
Recent research on environmental learning confirms the potentiality of information technologies for satisfaction of various essential requirements in environmental learning [243]. Still, one of the core challenges, the antagonism between experiencing nature and using a computer, has not yet been overcome by the current computer-mediated approaches. Learners have so far been trapped in front of the computer. They have learned about ecological concepts, build hypothesis, and tested them in simulations, but all in virtual learning environments. The challenge for computer-aided EE remains to relate virtual learning environment with the real environment.
New technologies in the form of mobile devices may offer mutual supplementation of direct and indirect nature experiences and thus would provide a solution to this problem [243]. Klopfer et al. [174] suggest that handheld computers promote collaboration and real world investigation techniques. Next to extensive research on mobile learning in general, there has also been some initial work with respect to employing mobile devices in the EE domain, drawing on the fields of Computer Supported Collaborative Learning (CSCL), augmented and virtual reality, and mobile computing [105, 174, 243], which will be discussed in the following sections.

1.2.3 Mobile environmental information systems

Most of the computer systems that are applied for EC, can be attributed to the category of Environmental Information Systems (EIS). EIS in general comprise a wide spectrum of different information systems [281]. Haklay [126] gives a broad definition of EIS describing them as a collection of data and information, which are of relevance for studying, monitoring and investigating the environment. Extensive reviews of EIS definitions as well as of existing EIS can be found in Günther [124], Haklay [126, 128], Ruchter [281] and Fischer-Stabel [104].
Haklay [127] stresses the driving force of ICT for the development of new types of EIS. Hence, a crucial aspect, which reinforced the development of Public EIS was especially the rapid growth of a new technological platform, the internet [87, 281, 296].
Since the beginning of this decade another ICT boost can be observed with the emergence of new mobile technologies, which again is followed by the evolution of a new generation of EIS. The rapid dispersion of mobile networks in Asia, Europe and North America enables the broad distribution of cheap devices for data delivery to a range of personal mobile terminals [87].
The fusion of EIS and mobile computing thus results in Mobile Environmental Information Systems (MEIS), defined by Antikainen et al. [10] as wireless information systems used to study, monitor, and exploit the natural environment and its interaction with human activities. The key advantage that these systems can bring to the EIS domain is, that environmental information can be accessed, added, and maintained from anywhere and anytime [87].
Considering the high public usage of mobile services such as SMS and WAP based internet access, Dickinger et al. as well as Westbomke et al. [87, 359] propose, that mobile media provide an appealing platform for public access to environmental information.

This thesis will only give a brief overview on MEIS, whereas a more extensive review of MEIS or m-environment applications is given by Antikainen et al. [10], Scharl [296] and Dickinger et al. [87].
Similar to web-based EIS, there are a number of MEIS, which primarily provide services for experts and professionals. MEIS like for instance the PAULA system [353] are designed to support environmental agency personnel in the field, by providing mobile access to expert resources contained in the EIS of the State of Baden-Wüerttemberg. The WirelessInfo project [58] does not only provide mobile information access to professionals in the field of agriculture and forestry but the application also enables the expert to collect data in the field and transfer it to the EIS. Additional mobile systems which support scientists or experts during data collection in the field are described by Pascoe [250], Fritz [111], Antikainen et al., [10], and Dickinger et al. [87].
Apart from these expert systems there are also MEIS that provide services for the public. There are a number of different Public MEIS, which can be assigned to the following categories:

Mobile Environmental Information/Data Services: These are MEIS which primarily provide mobile access to environmental pollution data such as APNEE [165], O3-WAP service and the KFUE-WAP service offered by the State Institute for Environmental Protection Baden-Württemberg [359].

Community-based Mobile Environmental Monitoring Systems: Similar to the citizen-science approach [328, 339], Storch [330] envisions a MEIS, where citizens can employ their mobile device to participate in monitoring their local environment. They send location and time dependent observations to a Weblog and can also access the environmental information generated by their virtual community with their mobile device [331]. Furthermore Storch [330] proposes the development of context-sensitive assistants for personal environmental decision making.

Mobile Clients to Public EIS: This category includes web-based Public EIS offering comprehensive environmental information specifically designed for the public, such as the "Themepark Soil System" [115,282]. These Public EIS may offer an access to the environmental information via a web-client on a mobile device.

Mobile Environmental Guide Systems: These are Public MEIS that pertain to the class of mobile guide systems. They offer context-sensitive environmental information to the user while guiding him through natural environments ranging from city parks and zoos to remote wilderness and other conservation areas. Generally these systems focus on providing services for the eco-tourism domain or the field of EE.

Since this thesis aims at providing a concept for an instrument for environmental education, it will concentrate on the category of Mobile Environmental Guide Systems.

1.2.4 Mobile guide systems

The following section will provide an analysis of key ICT concepts relevant for the design of Mobile Environmental Guide Systems and it will review a selection of existing mobile guide systems.

Mobile guide systems constitute a special category of context-aware systems. Next to their internal state, context-aware applications utilize contextual information, such as location, display medium and user profile, in order to provide tailored functionality [62, 191]. The key context element utilized by most mobile guides to offer specific services is that of location and as such mobile guidance systems belong to a class of systems commonly referred to as location-based services (LBS) [22, 172, 191]. Such services are, for instance, the provision of interactive maps, or recommendation of tours [303]. Consequently, mobile guides are defined by Baus et al. [22] as systems providing their users with location-based services such as navigational assistance where and when they need them most.

Mobile guide systems can further be characterized by the following common features that have been identified based on a reviewed literature [22, 168, 201]:

- Mobile device
- Positioning technologies providing spatial data
- Context-aware services and adaptation capabilities
- Interface and user interaction
- Control center based on a specific system architecture.

The following section will describe these features in more detail. For different states of a feature, sample mobile guide systems will be referenced. A complete list of these systems is given in Appendix A.3, Table A.2.

Mobile device

Obviously, all systems have been designed for some kind of mobile device, granting the promised mobility to the user to move around freely. A comprehensive review of current mobile computational devices is given by Sharples and Beale [311] as well as Mountain and Raper [229].
Some systems have been developed to run on just one specific type of device, while others can be operated on a variety of different mobile platforms. The types of mobile devices range from smart phones to TabletPCs or Laptops. The majority of the systems considered here, can be used on a Personal Digital Assistant (PDA) (see Appendix A.3.1, Table A.3).

Positioning technologies

Location information is used by location-aware city guides for two main purposes: Presenting information relevant to a user's location and providing route guidance [77]. In order to determine the location of the user, as a prerequisite for offering location-based services, mobile guides need to make use of some type of positioning technology. There is, however, no ultimate solution to the task of measuring the current location of the user precisely at all times [22, 77]. Positioning solutions for mobile devices can be divided into satellite-based solutions and terrestrial (i.e. network-based) solutions [229]. Location systems vary among other things in the model of location computation, the accuracy and precision of location determination, scale, recognition, and cost of the system [142]. A detailed survey of location systems for mobile computing applications is provided by Hightower and Borriello [142] as well as Kray et al. [183]. A brief account of the major positioning methods used in mobile guides follows.

Manual positioning The most basic form of determining the user location is manual positioning. It requires interaction by the user and at the same time a good sense of orientation as well as a high level of map reading skills, on the user's part. In the Sotto Voce system [366] a specific variation of self-positioning, termed "visual selection", is applied. It lets the user choose his object of interest from photographs representing the rooms of an historic house, whereas the location in the room is determined by built-in infra red transceiver. Other projects make use of manual positioning as a fall back solution if automatic solutions fail (e.g. ActiveCampus [122]), while some projects like REAL [23], utilize user interaction, including manual positioning, as part of a combination of positioning methods to achieve a higher accuracy.

Terrestrial location systems Terrestrial solutions relying upon the wireless network infrastructure, can employ a number of techniques of varying granularity to position users. The calculation of the location can take place either on the handset or the network, which forms a distinction between different approaches [229]. Such wireless positioning technologies include light-based technologies, mainly Infra Red (IR), and approaches based on Radio Frequency (RF), such as Radio Frequency Identification (RFID), Wireless Local Area Network (WLAN), Bluetooth (BT) [99].
Light-based systems such as IR-emitters, work much better inside buildings but they usually only have a limited range and thus require a tight infrastructure of beacons or tags. Radio-frequency-based network technologies such as Wireless Local Area Network (WLAN) or Bluetooth constitute another alternative for indoor as well as outdoor positioning but they also rely on a fairly dense infrastructure of access points especially due to the limited range of the mobile units with regard to sending a signal [77].
Another RF-based technology, used primarily for object identification in indoor as well as outdoor settings, is Radio Frequency Identification (RFID). RFID-tags have a unique ID, which they either passively yield to an RFID reader over short range (e.g. Imogl project [200]) or actively transmit over a longer range. An additional option are bar code scanners (e.g. Momuna [289] and the Electronic Guidebook [310] projects).
Further network-based positioning technologies include cellular location systems such as Global System for Mobile Communication (GSM) usually applied for mobile phones and employing the cell area concept and network triangulation [227] to determine the cell of origin of a caller (e.g. the Lol@ project [260]). A shortcoming is the dependency on a GSM network infrastructure and that even in urban areas the accuracy is limited to approximately 50 to 150 meters [201].

Satellite-based location systems Upon the currently most widely used methods are those applying satellite-based navigation using the Global Positioning System (GPS) [22, 77, 142, 183]. Still, employing commercially available GPS-receiver units, this positioning technology is at the time mostly restricted to outdoor settings [183]. There, the usage of GPS has crucial advantages

including its ubiquitous coverage. Furthermore, the receiver units are fairly compact and no additional infrastructure needs to be installed [77]. But even in outdoor settings the quality of the location data provided by the GPS unit may vary considerably depending on a variety of factors, such as the visibility of satellites (i.e. satellite signals may be shaded by buildings or natural features, which is referred to as the "canyon effect") or the receiver's velocity as well as certain environmental conditions, especially the prevailing atmospheric conditions [22, 77].
The majority of the surveyed systems employs GPS to some degree for the determination of the user's location. There are systems designed mostly for outdoor usage, which use only GPS (e.g. MobiDenk [187]) and there are also projects that make use of GPS in combination with other positioning technologies (e.g. CRUMPET [302]).
While the GPS uses a location computation model with the device actually computing its own position, other approaches rely on a model in which either objects like Points of Interest (POIs) periodically broadcast their location or devices transmit a signal to allow the external infrastructure to locate it [142].

Multimodal positioning and alternative solutions Kray et al. [183] point out that none of the currently available devices for position determination can supply the adequate position information parameters under all conditions. Consequently they propose, that more complex applications like mobile tourist guides need to rely on multiple sensors to obtain comprehensive position data (e.g. SaiMotion [140] and REAL [23]). Sensor fusion or multi-modal sensing seeks to improve accuracy and precision by integrating many positioning systems to form hierarchical and overlapping levels of resolution [142].
In case the determination by means of sensors entirely fails or the accuracy is not perceived as sufficient, Kray et al. [183] suggest a number of strategies to deal with the lack of positioning information such that the mobile guide system can still assist the user in completing his tasks. These strategies include:

Inference: Extrapolation of the user's position based on the history of the position information (i.e. navigation history), which can be further refined by using additional contextual information like the geographic model and/or the user model.

Interaction: Direct interaction with the user, for instance via a dialogue, to confirm or disambiguate an uncertain position information or explore position information in general.

Compensation: If sensors fail and/or interaction with the user is not possible, the system has to compensate for imprecise position information or the complete lack of information. This can be achieved by reducing the granularity of the geographic information or by extending the contextual information by offering alternative reference systems to the user. This could for example include references to prominent landmarks. Further the system can assist the user, by a more complete verbal description of navigational instructions.

Such alternative strategies for positioning are for example applied in the Deep Map [181]. The Ambient Woods system employs a specific dead reckoning system, since the GPS positioning signal was frequently degraded by the tree canopy. The dead reckoning system was devised using an accelerometer to detect movement, and a two-axis electronic compass to sense heading [264]. Electromagnetic devices such as the accelerometer or electronic compasses can assist with the positioning task but on the other hand are influenced by interference from electromagnetic fields [22].

Services

Being a subclass of LBS systems, mobile guides should naturally offer a variety of services based on the user's location or other situational factors. Among the core services of mobile guide systems are those offering navigational assistance as well as location-based information.

Navigational assistance Navigational assistance as the core feature of mobile guides [22], implies the representation of geospatial information. According to Meng and Reichenbacher [218], maps remain the most popular communication language of spatial information also for mobile applications. For a general review of map-based means of spatial communication see Appendix B.2 and Tversky [341]. The visualization of spatial information in mobile applications commonly includes maps and Points-of-Interest (POIs) [187]. A mobile map is first of all a cartographic visualization designed for the display on a mobile device [269]. In contrast to more traditional maps it should constitute a snapshot of an environment around a certain location and time with highly selective information and integrated intelligence (i.e. the capacity to dynamically adapt to the users situation) [218]. Meng and Reichenbacher [218] give an extensive review of LBS utilizing maps to assist the user, which they refer to as "map-based mobile services". At this point an overview will be given on the means of presenting spatial information and forms of navigational assistance in the surveyed mobile guide applications.
For a few projects there is no reference what so ever in the literature concerning the use of maps. It is assumed that for these systems, which are frequently designed to operate indoors, no maps are used but spatial information is conveyed by other means. This is for instance true for the Sotto Voce project. The collection of photographs of the respective rooms for visual selection also serves as means of orientation and access to location-based information for the user [366]. Such an approach of representing spatial information by using photographs, however, seems only applicable in a fairly static indoor environment, where the user can look at the POI or target only from a limited amount of perspectives. Nevertheless, it has the advantage that the user actually sees a realistic representation of the real world instead of an abstract one when using a map. This may reduce the cognitive load of the user. A similar approach is taken by the PEACH project [329],

were two-dimensional or three-dimensional images can be used to support the linguistic reference to physical objects (compare Appendix A.3, Table A.3).

The majority of mobile guidance systems though makes use of maps for the graphical representation of geographical information (compare Appendix A.3, Table A.3). According to Baus et al. [22], one reason for this is the pervasive use of maps in paper guides, the paper-based ancestors of mobile guides. But mobile guidance systems can go beyond the traditional use of paper maps in guide books [22]. They can dynamically select which section of a map to display, at what scale, and what to depict on the map as well as personalizing it to the current task and user. A variety of different types of maps are used for various services in mobile guide systems [22]. A discussion of the features that characterize the different types of maps is given in Appendix A.3.2.

Information services Next to providing navigational assistance the other core feature of most mobile guide systems is the provision of information (compare Appendix A.3, Table A.3). Already the early mobile guide system, Cyberguide [2], contained an information component providing access to all information about sights that tourists might encounter (e.g. details about buildings and associated people). There are a number of options with regard to how information services have been realized on mobile guides. The supplied information may be either general (i.e. independent of context) or context-sensitive.
Context-independent information usually has to be queried by the user and thus can be characterized as a pull-service. As one option for information retrieval, a number of systems provide such a context-independent pull-service, allowing their users to query general catalogue data including lists of all events or locations (e.g. SaiMotion [140]) or access to webpages containing general and overview information (e.g. Electronic Guidebook [310]).
In accordance with the philosophy of LBS most mobile guide systems offer context-sensitive information services. Context-sensitive information presentations can be implemented either as a pull- or push-service, whereas a push-service is characterized by automatic information presentation without the user actively requesting it. It is most frequently provided based on the user's location. A few systems offer information only through a location-based push service (e.g. Cyberguide [2]). A variety of other mobile guides offer the users a combination of pull- and push-services to access context-sensitive information (e.g. WebPark [52]). Context-sensitivity frequently implies the user's location as sole context information, but some systems also take further situational factors into account. For example CRUMPET [302], GUIDE [62], HIPS [245] and WebPark adapt context based on user context. Oppermann and Specht [245] refer to this type of information adaptation as contextualized information space, which is defined by an information repository adapted to the location, the user and the task. A key benefit for the user is that adaptive information selection reduces redundancy and information overload [245].
Eventually, there are also mobile guidance systems that only offer information via a location-based pull-service. Here the users are not automatically provided with the relevant information

according to their context but upon reaching a POI they actively have to retrieve the information by either sending a request to a server with the POI coordinates (e.g. CHIMER [355]), interacting with the POI symbol on a map or an image of the POI (e.g. Lol@ [260]) or by sensing/reading the ID of the respective artifact (e.g. Electronic Guidebook).
Further, some systems such as GUIDE [62] offer, in addition to the context-sensitive, but generally static information, also dynamic information, notifying the user about the latest changes in events that might affect their schedule.

Next to simply offering context-independent or -sensitive information services, a selection of systems also includes a functionality that combines selections of POIs and adequate route segments to a tour. Some guide applications just make suggestions on how to get to other POIs that may be of interest, either based on the current POI (e.g. Electronic Guidebook), or on the entire history of usage (e.g. HIPS). Other guides propose a number of predefined tours (e.g. Lol@). The most sophisticated functionality is offered by systems like GUIDE, Momuna [289] and SaiMotion [140], which make it possible for the user to select predefined tours and also allows them to built their own tours or schedule (e.g. SaiMotion's intelligent scheduler). DeepMap also includes a web-based pre-trip planning component [203].

Communication service Beyond these basic services that provide navigational assistance and information some mobile guides offer additional services, such as communication services (compare Appendix A.3, Table A.4).
Several of the surveyed applications account for the urge of users to communicate with others while on the move. The requirements analyses conducted for some of the projects have shown that users are interested in communicating with their peers while they engage in a guided activity [62]. There is again a variety of solutions, which have been implemented by different mobile guide services to support communication and social interaction:

- Communication with peers that are members of the same group (e.g. students of the same class etc.), facilitating social interaction as well as coordination or collaboration (e.g. Cyberguide [2])
- Communication with peers that are not members of the same group (e.g. other visitors present at the same sight or engaged in virtual visit via internet) (e.g. HIPS [245])
- Friend- or Buddy-Finders let users localize the geographic position of peers (e.g. ActiveCampus [122])
- Graffiti, tags or bookmarks allow users to leave location-based messages (e.g. warnings, comments, recommendations) for peers or other users (e.g. WebPark [52])

Emergency services Additionally, some applications offer a special location-based communication and/or information service related to safety issues. A number of systems offer specific emergency- and safety-services. Systems designed for the use in remote areas, such as PARAMOUNT [198] and WebPark [52] allow the users to send emergency calls and receive safety warnings like avalanche prediction, weather changes, warning in critical areas and information on the state of the trails. Furthermore, to facilitate visitor management and the work of search and rescue teams in case of an emergency, the users can register with a tracking-service that sends a log of their position back to the operator.

Diary service Besides services related to communication and the presentation of information, several mobile guide systems (compare Appendix A.3, Table A.4) also provide the user with an opportunity to collect information in order to document their experiences throughout the guided activity. Already the creators of the CyberGuide system proposed a travel diary service with a summary of a day's visit, including a log of time and place as well as photographs, videos and audio comments [2]. Some guide systems (e.g. National Park Information System/ReGeo [8]) have implemented such a travel log as a kind of personalized multimedia document containing a log of the geographic position as well as a list of the visited POIs and combined with the presented multimedia contents. Many of the other systems (e.g. Electronic Guidebook [310]) allow the user to further personalize this diary by adding annotations and photographs, video and audio files captured during the activity. Frequently such mobile guides provide a copy of this digital souvenir to the user or an opportunity to download it from the web. Mobile guides employed in exhibition settings (e.g. Momuna [289]) want to promote successive elaboration, reflection and discussion about the experienced and visited objects. In addition to annotations HIPS [245] also allows users to bookmark certain exhibits for an easier orientation during a return visit.

Commercial services The hype of LBS was not only indebted to the evolution of mobile technologies but also to economic opportunities that are anticipated with the successful application of LBS. Of course mobile guides can only become a success if their users are willing to pay for the offered services. As a consequence several of the surveyed projects explored options for commercial services. These include "pay-per-use" or mobile-commerce (i.e. m-commerce) services such as updates on weather info (e.g. WebPark [188]), context sensitive shopping (e.g. Momuna [289]) and push-advertisement (e.g. GUIDE [62]).

Other services which do not necessarily require payment but entail an increase in convenience for the users are booking services. Several mobile guide applications provide an opportunity for booking or reservations for accommodations, tickets or restaurants (e.g. Cyberguide [2], Lol@ [260]). Other systems allow the user to order personalized information materials based on the usage pattern (e.g. PEACH [329]).

Search and monitoring services Some systems also furnish a few out of the ordinary services. This includes a variety of search services for textual information, specific POIs or events (e.g. ActiveCampus [122] and WebPark [84]).
A number of mobile guide projects further provide a functionality not directly aimed at the user but at the service provider or operator of the system, such as administrators of museums or parks as well as instructors of school classes or groups of visitors. For example Momuna [289] and PARAMOUNT [198] provide monitoring and evaluation services, intended to monitor the usage and visit patterns and evaluate user experiences in order to facilitate the services and management of visitor groups. Finally the HIPS system [245] also offers a glossary of terms to its users

User interface and modalities

The user-interface (UI) and the available means of interaction are the parts of the system, which are most apparent to the user and thus have crucial influence on the user's perception of the system [182]. When designing user interfaces for mobile guides a few particular issues have to be taken into consideration. Mobile devices have small displays and suffer from bad contrast in direct sunlight. Further, users may be distracted by a possibly uncomfortable environment, and cannot be expected to carry a manual with them [260]. Many of the surveyed systems have been designed for non-expert users who commonly employ the guide to enhance their leisure time activity. These users (e.g. tourists) do not want to spend their time configuring or learning an application. Rather, they want a tool that works "out of the box" and supports them with their task of exploring a foreign city [260].

Interface metaphor Thus the virtue of an intuitive interaction is of particular importance in mobile guide systems. One crucial aspect in order to achieve an intuitive interaction is the selection of an appropriate interface metaphor. It should combine familiar knowledge with new knowledge in a way that will help the user understand the system [261]. The common desktop paradigm established on PCs naturally does not carry over to mobile devices [206]. Still a variety of systems have based their user interface on widely distributed metaphor, the browser. It is familiar to most users from the software they access the internet with. These systems either present the content as a hyperlink document in a common web-browser software (e.g. HIPS/Hippie [245]) or they use a modified version of a web-browser (e.g. Electronic Guidebook [310]). Other systems such as CRUMPET [302] and GUIDE [62] built their own UI based on the web-browser metaphor. Common interface elements include the use of hyperlinks to interrelate content parts, "back" and "forward" browsing, "home" or return to starting- or overview part. Generally all systems also provide an option to switch between a tour- or content-mode and a navigation-mode of the system.
Several of the reviewed mobile guides stress the aspect of location-awareness and navigational

assistance and employ a map-metaphor for their UI. In most of these systems (e.g. ActiveCampus [122]) the map is generally used as the main interface to the system. Similar to a paper map, with notes attached to important locations, the information pages are hyperlinked with the map or presented automatically upon arrival at a POI. Most of the interface elements are frequently dedicated to the manipulation of the map such as zooming and panning. In some cases a direct option to change into the content mode is provided. Some mobile guide applications also use the map as the main metaphor but at the same time offer an extensive information browser (e.g. Lol@ [260], WebPark [52]).

A few projects have taken a completely different approach. The PEACH system draws upon the metaphor of a TV-like presentation, with a presentation agent acting as one of the main interaction elements. Common multimedia control elements are provided to control multimedia presentations [190]. The interface of the Sotto Voce system on the other hand is mainly based on an "image-map" metaphor [365]. In order to reduce the cognitive load of the user, he is presented with a photograph of what he sees in the room and can access audio information clips by tabbing on the image-map of the respective object. ImogI [200] again tries to build on a metaphor that most users should be familiar with from their PC environment. By using a folder/taps metaphor the system allows the users to easily switch between the different modes of the application.

Interaction paradigm The interaction paradigm has also changed in mobile applications, since the user cannot focus entirely on the interaction with the computer, but has to dedicate his attention on a primary task (e.g. driving a car or sight seeing). This calls for an effortless interaction with the device, also termed as "natural interaction" by Abowd and Mynatt [3].

Natural interaction implies taking advantage of different interaction modalities. Based on Hyde [149] interaction modalities refer to the different input and output options of computational devices. The three main categories of modalities are related to the human sensory channels visual, auditory and haptic [149]. Bartneck and Forlizzi [21] further include the olfactory and even gustatory channels. Based on the number of communication (or sensory) channels engaged, the applications can be described as unimodal or multimodal [21].

According to the results of the literature survey, only a few of the respective guide systems can be characterized as unimodal. These systems appear to rely on a traditional single channel graphical UI (i.e. visual modality) as it is still common on desktop PCs [28] and employ text and images as output and pen-based input (e.g. ActiveCampus [122]). The majority of the examined guide applications, though, seem to take advantage of multiple modalities. Many of these multimodal systems make use of a combination of visual and acoustic modalities (e.g. AccessSight/MobiDenk [187], Electronic Guidebook [310,366]). Typically text, images, video and audio are employed as output next to a pen-based graphical input. Some of the systems such as eTour and TellMaris apply a combination of visual and haptic modalities by enabling the user to operate the device via hardware buttons. Other systems (e.g. Lol@) take advantage of all common modalities by providing a graphical user-interface and audio output as well as speech input and output and in some cases even gesture input (e.g. DeepMap [203]).

Context

Being a subcategory of context-aware systems, context plays a significant role in all mobile guide systems. This is stressed by Malaka and Zipf [203], who underline that a system, designed for tourists in a leisure setting, needs to be comfortable to carry and to use. According to Eisenhauer et al. [100], this calls for an adaptation of information presentation and other services to situation, location, task, and user.
All surveyed mobile guides make use of some category of context (see Appendix A.3, Table A.4). The following categories could be identified during the review:

User context: This involves the user's identity (e.g. PalmGuide), interests, knowledge, skills, demographics, language, behavior, social situation, preferences etc. Some systems use a user model either as a static model (e.g. GUIDE) or as a sophisticated model that is dynamically updated throughout the usage of the system.

Geographic context: This category of context corresponds essentially to the location of the user (i.e. the mobile device), which is applied by almost all systems.

Environmental context: Fewer systems take this category into consideration, which comprises mostly the state of the physical environment of the user (e.g. changes in weather or lighting conditions).

System context: Only a few of the mobile guides take the system context into account, which includes the type of the mobile device used along with the system resources as well as the available network infrastructure.

System architecture

A variety of different approaches have been taken by the surveyed projects with regard to the chosen system architecture (see Appendix A.3, Table A.4). Each approach constitutes a specific compromise between the complexity of the system and the related infrastructure, adaptability, extendability, maintainability, and scalability of the system.

Stand-alone guide system The most basic solution is a stand-alone guide system with the entire application-logic residing on the device and all information stored locally (e.g. Sotto Voce). A stand-alone solution offers a very fast response rate to information request and a high performance of the system in general. Even though the complexity of the architecture is low, the extensibility and adaptability of the system is limited. A lot of aspects in an environment are subject to change and a local, static database is, according to Abowd et al. [2], slightly more useful than a book. Furthermore, updates and maintenance of a stand-alone version are very resource intensive. A stand-alone system may frequently require less infrastructure but at the same time the mobile device has to supply a large storage capacity as well as enough computational power on the mobile device [23].

Client-server architecture Some of these issues can be resolved by a more complex architecture. The majority of the examined systems is based on some kind of client-server architecture (e.g. ActiveCampus [122], AccessSight [187]). This approach follows a well established paradigm and frequently uses a web-server accessed by a web-browser on the mobile client. It is easily extendable, allowing for a ready addition of multiple client devices [182]. Furthermore, the data is stored on a large central database, which can be easily maintained and the computational burden is generally shifted from the client to the server [122]. Still, in order for the client to access the server the system depends on a reliable network connectivity, which can not always be guaranteed. Moreover, the server being a single point of failure, the systems are less robust than decentralized approaches such as interacting applications [182].

Interacting applications Cyberguide [2] and TellMaris [66] have been implemented using such interacting applications. Even though these systems are more decentralized, they are limited with regard to the extensibility and adaptability of the system, since the applications have frequently been designed for a specific device or platform. In addition, the applications may be based on different programming languages and there is a lack of standards for the communication between these [182].

Hybrid architecture and multi-agent systems There are a few systems using a hybrid architecture like for example REAL [23], combining the interacting applications with a client-server approach and PalmGuide [332] utilizing a combination of client-server and multi-agent system. Finally, there are also systems taking the approach of a multi-agent system, such as CRUMPET [302] and DeepMap [367]. These applications allow for easy extensions by adding and removing components while at the same time being able to compensate for failures of certain components. Furthermore the communication between components is based on a standard language [182].

1.2.5 A taxonomy of mobile guide systems

The reviews by Broadbent and Marti [48] as well as Baus et al. [22] showed that corresponding to the continuous increase in the number of mobile devices, the number and variety of mobile guide systems continues to grow. This is accompanied by a rise in the number of new application domains that these systems are being designed for. While Baus et al. [22] presented an extensive review of mobile guide systems focusing on their common components and functionalities, a comprehensive survey of mobile guides with respect to different application domains is still lacking. Such a survey has been performed for this thesis as part of the literature review for the state of the art, which included the analysis and comparison of 30 mobile guide systems (see Appendix A.3). As a result a taxonomy of mobile guide systems by application domains is proposed by this thesis.

This taxonomy for mobile guides, which is visualized in Fig. 1.2, suggests five main application domains: Tourism, museums/exhibits, community events, mobile learning, and personal assistance. Each domain comprises a number of different types corresponding to the main geographic setting they are used in (e.g. city, museum, nature) or the main task (e.g. personal organization) they are used for.

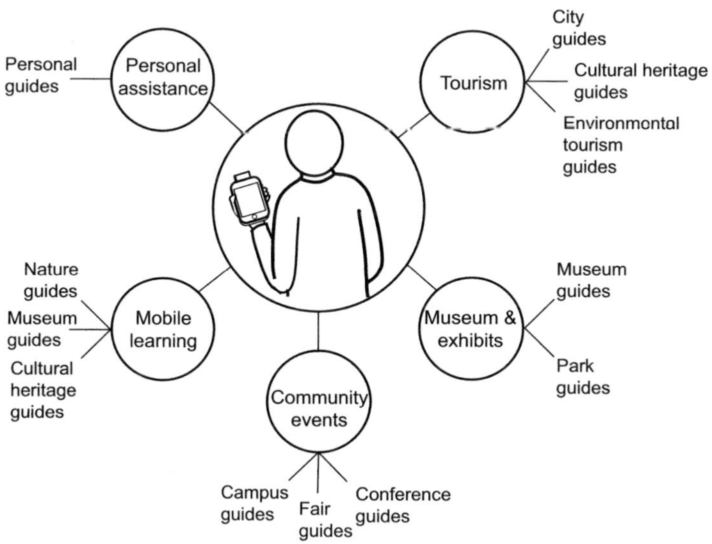

Figure 1.2: Taxonomy of Mobile Guides with the five major application domains and their related types of mobile guides.

The largest group of the reviewed mobile guides has been developed for the tourism domain. This includes the early mobile guide projects like Cyberguide [2] or GUIDE [62]. Mobile guides are considered as particularly suitable applications for the tourism domain [42, 50], since due to their context- and especially location-awareness, they have the potential to mimic or even augment the traditional "tools" of the domain (e.g. guidebook, map) [106, 260]. Bornträger and Cheverst [42] further argue that a lot of mobile tour guides are dealing with tourism, because it is a very interesting topic with quite good market chances and there seems to be a high acceptance rate among tourists. According to this taxonomy, systems for the tourism domain can further be differentiated into different types of mobile tourist guides (see Fig. 1.2 and Appendix A.2). Theses types have been defined as city guides, cultural heritage guides and environmental tourism guides.

Next to tourism another major application domain of mobile guide systems are museums and exhibits. Even though a museum visit can be part of a tourism experience, it can generally be considered an independent activity focused on the museum visit itself. Traditionally visitors read information labels, leaflets or brochures associated with the exhibits. Visitors also listen to guides, both human guides and recently also more and more audio guides and they use kiosk systems [245]. Mobile guides can provide means by which museums and exhibitors can communicate with their visitors in a more personal manner [106]. They can not only serve as reference information, but also provide expert guidance, dynamic advice, recommendations for further inquiry, and other learner supports that no paper-based guide can provide [106]. Depending on the area of deployment, mobile museum and exhibit guides have been further categories into museum guides and park guides, including natural exhibits.

In the process of the literature review a few guide systems were selected, which were designed to support groups of users as well as the individual participant in community events. Community members or participants have common needs, including scheduling, navigational assistance, context-sensitive information, and communication with peers [122, 140, 332]. Such community events can comprise conferences, fairs or campus life.

The growth of mobile devices and applications as well as their success on the market has not only inspired the business community and various experts from domains mentioned above. Mobile technologies have also been discovered by educators in schools, universities and other public institutions (e.g. museums) and companies as technologies that can have a great impact on learning [232, 334].

According to Naismith et al. [232] learning will move more and more outside of the classroom and into the learner's environments, both real and virtual, thus becoming more situated, personal, collaborative and lifelong. Mobile Learning was propagated as a new mobile extension of e-learning. The emerging field of mobile- or m-learning addresses a wide variety of issues related to the application of mobile devices for learning as discussed in Atwell and Savill-Smith [15] as well as in a review by [64]. All of the examined mobile guides support learners in informal learning settings (i.e. outside of the "classroom"). For this taxonomy five m-learning Guides have been reviewed and based on their educational purposes they have been attributed to cultural heritage guides, museum guides or nature guides.

An additional application domain for mobile guides is that of personal assistance. The type of mobile guide systems helping people to organize their personal life based on context-aware services

are here referred to as personal guides. However, these systems have only been briefly addressed by this review and have not been included in the results of the analysis. Systems like for example BNP (BMW Personal Navigator) [189] and ComMotion [209] have been categorized as personal guides.
Finally, it needs to be pointed out, that further application domains for mobile guidance devices, beyond the ones listed in this taxonomy do exist. This classification focuses on mobile guide systems which primarily serve the presentation of context-sensitive information. Other application domains include shopping assistants, recommender systems, reminder systems and general personal organization assistants. A more extensive review is provided by Buyon and Cherverst [51] as well as Koch and Rahwan [176].

Only a coarse overview on the application domains and the respective types of mobile guides can be given within the scope of this thesis. The remainder of this section will focus on those categories of mobile guides, that can be attributed to the mobile environmental guide system and are thus of immediate relevance to this thesis. Specifically representatives of environmental tourism guides in the tourism domain, park guides in the museum and exhibits domain as well as nature guides in the mobile learning domain, will be described.

Environmental tourism guides The segment of environmental tourism or ecotourism accounts for a significant proportion of world travel [83]. Ecotourism, is responsible travel to natural areas which conserves the environment and sustains the well-being of local people while providing a quality experience that connects the visitor to nature, increasing awareness and contributing to conservation [102]. Natural places that ecotourists travel to include protected areas but also other attractions such as the national parks, wildlife and biological reserves, coastal and marine areas [83]. Diamantis [83] points out, that ecotourism incorporates nature-based as well as educational aspects and also sustainable management issues involving economic and/or sociocultural issues.

Environmental Tourism Guides are considered as a possible instrument to encourage eco-friendly behaviors as well as a safe and more sustainable use of the environment [84]. In accordance with the definition of ecotourism they are designed to provide information on the natural area and the tourism region, which enables the planning of activities on vacation, offers entertainment, and also allows business transactions. A ecotourism guide is supposed to provide context-aware information based on the visitors' location, the time and the visitors' profile in order to fulfill the given information needs. Furthermore, these mobile guide systems should also serve as an instrument for park administrators for linking local people, visitors, and protected area, raising the visitors' understanding and awareness for the local natural environment as well as tool for monitoring visitor behavior.
Three Environmental Tourism Guides or Ecotourism Guides were compared for this review:

PARAMOUNT: The PARAMOUNT project (Public Safety & Commercial Info Mobility Applications & Services in the Mountains) aims at improving user-friendly info-mobility services for tourists in the mountains. Mountaineering and other tourism activities in the mountains bear particular risk especially to non-experts. By providing information, communication, navigation and safety functions, the PARAMOUNT system intends to support the mountain tourists and help reduce the number of accidents and casualties among this target group. In addition to services for the tourist this Ecotourism Guide also offers a monitoring tool for Search and Rescue teams.

ReGeo: The National Park Information System is a synthesis of the Austrian project VITA and the EU-project ReGeo ("Multimedia Geoinformation for e-Communities in Rural Areas with Eco-Tourism") [8]. The project ReGeo focuses on the development of a comprehensive regional information system based on a decentralized (virtual) geo-multimedia database [108, 323]. It is the objective of this Ecotourism Guide to provide services to ecotourists applying a combination of 2D and 3D representations of geo-multimedia information in order to combine sematic description and map-based representation of POIs and events. It is part of its rational, that especially navigational services will facilitate the management of visitor streams in order to avoid negative effects on the natural assets of the region. In addition the ReGeo system intends to provide an organizational framework for park managers and administrators as well as a marketing platform for merchants to sell their local goods [8]. The ReGeo system is one of the few systems also addressing the issue of the contribution and maintenance of relevant data for such a tourism information system. Due to the rural nature of the tourist region it is also difficult to find a central institution that is willing to maintain a central database. Hence the ReGeo approach suggests a virtual decentralized database incorporating the data of individual organizations [108].

WebPark: The WebPark system is specifically designed to provide personalized value-added LBS to visitors of protected and natural areas. Initially the WebPark consortium conducted an extensive assessment of the information needs of tourists in these areas as well as of park managers. Based on the results of this survey the conceptual development of the WebPark system focused on the provision of geographically relevant personalized services and the creation of new mobile-commerce (m-commerce) value-chains for the area administrations and for data integrators such as additional information services like weather updates etc. [24]. This Ecotourism Guide enables users to request information from several databases using their mobile phone or PDA and filters the information based on location, time and user profile relevance. Next to providing LBS to tourists, the WebPark project also aims at developing a tool for park managers. The system should serve as an instrument that facilitates the monitoring of visitor and that can help to achieve the goal of improving the environmental awareness among visitors. As a consequence this Ecotourism Guide offers apart from navigational services and information on infrastructure features like hotels and restaurants, also flora and fauna descriptions linked to the habitat the tourist is visiting [84].

The surveyed ecotourism guides have generally been designed for use on PDAs, even though some can also be used on smartphones or via an internet-based PC. All of the systems rely on GPS for user positioning. All systems use maps as part of their navigation service and also use a map as their main interface-metaphor. Next to custom 2D maps, two of the three mobile guides also present 3D maps of the natural areas. Furthermore, all of the ecotourism guides offer context-sensitive information usually based on location but two of the systems also apply user-context as an additional filter to adapt the provided content. In addition to navigation and information services, two of the mobile guides also offer a communication service as well as a diary service to add value to the tourism experience. A type of service that is exclusively offered by Ecotourism Guides is the emergency service that is offered by PARAMOUNT and WebPark. In accordance with the described requirements all ecotourism guides also offer a service to park administrators that allows visitor monitoring. It should also be noted, that all of the reviewed guide applications include some kind of m-commerce feature, either as pay-per-use service to access additional information or in order to purchase tourism merchandize. Finally, all of the ecotourism guides have been implemented as client-server applications. However, due to the limitations of the network connectivity in remote areas, the WebPark system specifies that an online connection is not always required for using the basic features of the system.

Park guides In contrast to natural areas, a nature exhibit (i.e. natural phenomenon in a park) generally still resemble a comparatively managed and rather controlled environment, sharing a number of characteristics with the museum and cultural heritage domain.

The category of Park Guides classifies systems that have been designed for use in botanical gardens and zoos as well as other public gardens and parks displaying natural phenomena similar to exhibits in museums or cultural heritage sites. These sites may also be located in urban areas and the actual exhibits like for example certain plants or the terrain of a particular animal species may be labelled. Still these systems should be discussed separately here due to their focus on natural phenomena. Nature exhibits usually contain large amounts of specimens (e.g. 22.000 species of plants in the Berlin Botanical Garden) and administrators intend to offer visitors personalized tours to a selection of this variety [47].
Two systems that match this category have been identified during the literature review. The eTour system [47], which provides individual tours on plants of a botanical garden in Berlin. Visitors can use the eTour guide to get multi-media information on plant species from around the world, spread over greenhouses and an extensive garden area. The BugaButler [35] is a nature exhibits guide designed for the national garden exhibit in Munich, Germany. The system guides visitors through the garden exhibits along predefined tours and supplies them with location-based information on current events. Both of the systems are PDA based. For positioning they use either GPS or manual positioning. It should be noted here that the BugaButler is one of the few systems that offers a GPS-based game as an additional service. Both nature exhibit guides utilize the ge-

ographic context for their information services, while the BugaButler also applies environmental context like time to inform the user of current events. Finally it should still be pointed out, that both of these nature exhibit guides are commercial systems. As a consequence little or no research has been performed with these systems and hence also very little or no documentation on these projects is available in the scientific literature.

Nature guides In addition to m-learning guides for museums and cultural heritage sites there are also mobile guide applications which were designed to support learners in natural environments. These guide systems, which can be employed in the field of environmental education, have been classified as nature guides. In contrast to ecotourism guides, which also support recreational activities enhanced by nature, these nature guides generally focus on nature-based activities. This incorporates the exploration and understanding of natural phenomena. Four nature guides have been reviewed in more detail:

Forest Education Support System: This system acknowledges, that learning through experience is an important process in forest education, through which people are encouraged to find and observe interesting things in the natural environment [1]. The system was developed to support field education programs in the forest education domain, assuming that using mobile computers would increase the enjoyment of learning in the field. This nature guide provides the user with nature information at certain POIs. The m-learning application uses quiz-type and sketch-type educational materials.

Ambient Wood project: This project offers a playful learning experience that takes on the form of an augmented field trip. Pairs of children equipped with a number of devices explore and reflect upon a physical environment that has been outfitted with WLAN and RF location beacons. The aim of this m-learning guide is to stimulate exploring, consolidating, hypothesizing, experimenting, reflecting. Children are supposed to collaboratively discover a number of aspects about plants and animals living in the various habitats in the wood. Their experiences are later reflected upon in a 'den' area where the children share their findings with each other and the facilitators [264]. Each of the teams has a remote facilitator, who they can relay information to using a walkie-talkie. The facilitator in turn can send the children information (e.g. images of plants and wildlife; illustrations of natural processes such as photosynthesis), which is displayed on the PDA. Alternatively he can pose questions to stimulate the children. The facilitator was also able to monitor the progress of the children through the woods using a GPS tracking system. What is particularly remarkable, is that the m-learning system manages to encourage children to actually interact with the physical environment. By furnishing them with a pinging probe that enables them to collect data on moisture and light levels, which are graphically displaying on the PDA, the system manages to induce interaction with the physical as well as the digital world.

DigitalEE II project: The "Digitally Enhanced Experience" project is an online digital platform for networked collaborative environmental learning. Okada et al. [243] have identified the need and at the same time challenge for social groups like students and educators or land developers and pollution victims to gather in "real" nature and consider environmental issues through the exchange of nature experiences, opinions, values, learning activities, preservation activities, and expertise. As a consequence, the project facilitates collaborative exploration of a natural environment between a visitor and an online-buddy (i.e. the virtual visitor). The learner on site is equipped with a mobile device that, besides a map for orientation, provides him with video camera to collect visual impressions of the natural environment and share these via a network connection with a remote participant at a desktop computer. Both share the experience, augmented by forest ecology information, via a photo realistic combined 3D and 2D interface. They can further communicate through a headset and the remote participant can virtually point out specific objects to his companion. The nature guide aims at stimulating various interactive communication and voluntary environmental observation.

Environmental Detectives: This is a platform for augmented reality gaming, engaging students in a real world environmental consulting scenario. Students role-play environmental scientists investigating a rash of health concerns on campus linked to the release of toxins in the water supply. Working in teams, players attempt to identify the contaminant, chart its path through the environment and devise possible plans for remediation if necessary. As students physically move about campus, data is provided to the students via location-aware GPS-enabled Pocket PCs, allowing them to collect simulated field data from the water and soil, interview virtual characters and perform desktop research using mini-webs of data. At the end of the exercise, teams compile their data using peer-to-peer communication (i.e. via BT and IR) and synthesize their findings. Klopfer et. al [174] use "handheld simulations" to augment the real world with virtual data granting the students a more holistic learning experience. A nature guide with such an augmented reality game allows students to engage in problem-based scenarios that would ordinarily be to dangerous and expensive to realize.

These nature guides have been designed for PDAs or similar handheld devices, potentially also in combination with other instruments. With regard to positioning technologies GPS is mostly utilized. Each of the systems offers context-sensitive information mostly as a pull service. The majority of systems offers some form of service to communicate with other visitors. Most of these systems offer at least visual and acoustic interaction modalities and are based on a client-server architecture.

1.3 Objectives and thesis structure

As has been illustrated above, environmental educators are searching for new concepts and instruments that can contribute to overcoming the current challenges, which implies

- motivating citizens, in particular young people to take part in EE activities
- employing ICT but still facilitate the direct experience of the natural environment
- promoting environmental literacy among participants as a foundation for sustainable development
- offering participants media-based but still personalized assistance, which can enable EE institutions to serve more citizens in spite of declining budgets and limited personnel resources
- providing EE institutions with an inexpensive instrument that can be easily maintained and updated.

State of the art context-aware applications such as mobile guide systems can provide new opportunities for EE and environmental tourism, since they allow for the combination of computer-mediated learning and direct experience of the natural environment on site.
However, existing mobile guide solutions for environmental tourism and EE focus mainly on issues of technological feasibility and information presentation without truly considering EE issues. There is a lack of concepts in informal EE for making effective use of mobile technologies accompanied by a deficit in the mobile guides community regarding the explicit support for EE methods. The few nature guide systems that have been employed in a field setting for EE activities, have commonly not been evaluated with respect to their actual effectiveness in achieving EE goals.

The objective of this thesis is the development of a new concept for computer-mediated EE based on mobile guide system technology. A new mobile guide system constructed on the basis of this concept should explicitly provide support for the application of EE instruments and should help assist educators in meeting the discussed EE challenges. In order to achieve this objective, the following tasks have to be accomplished:

- Identification of EE instruments that can effectively influence components of environmental literacy based on a comprehensive environmental literacy model
- Construction of a new model for mobile environmental interpretation representing the employment of mobile guide technologies for the application of EE instruments

- Specification of a conceptual design for a novel mobile nature guide based on the mobile environmental interpretation model
- Development of an information technological (IT) concept for such a mobile nature guide and implementation of a prototype system
- Evaluation of the proposed concept by testing the prototype system and its effectiveness in the field.

Chapter 1 provides next to the motivation for this work an introduction to the field of environmental communication and education, including an overview of computer-mediated methods in environmental education. It further discusses the state of the art of context-aware applications, in particular mobile guide systems.

Chapter 2 describes a new model for mobile environmental education that serves as the theoretical foundation for the conceptual design of a mobile nature guide system. An environmental literacy model, compiled from the literature, is analyzed to come to an understanding of how environmental literacy components can be effectively influenced by EE methods. Under consideration of these findings as well as the principles of environmental interpretation a new model for media-based environmental education is devised, according to which a mobile guide system can serve as an on-site mediator between the participant and the natural environment.

Chapter 3 presents the conceptual design for a mobile nature guide application rooted in the proposed model. The requirements for a mobile computer based EE instrument are derived from the analysis of a front-end evaluation. Different usage modes for such an instrument are identified and a task model is generated representing typical activities that participants engage in during the EE activities. Eventually, in accordance with the requirements and the task model, the concepts for components are described that are needed by a mobile nature guide system to comply with the proposed model.

Chapter 4 presents the IT concept for a mobile nature guide. The description of the system architecture is followed by the specification of services including the respective user and programming interfaces. Furthermore, the implementation of the MobiNaG prototype system is documented and issues identified during development process are being discussed.

Chapter 5 is dedicated to the evaluation of the applicability of the proposed concept utilizing a test application of the mobile nature guide prototype in the field. A test design was chosen that evaluates if using the mobile nature guide has an impact on the environmental literacy components of different target groups. It was further investigated if the effect of the new computer-mediated instrument differs from that of traditional instruments. Next to the presentation of the results of the study their implications for the mobile nature guide system and the field of environmental education are discussed.

Chapter 6 summarizes the results of this thesis and provides an outlook on future work.

Chapter 2

A New Model for Mobile Environmental Education

Foundation for the design of a mobile nature guide that can promote responsible environmental behavior should be a model for EE, which builds on the potential of novel mobile technologies. In such a new model for EE, which will be described in this chapter, mobile technologies should reinforce instruments of EE for influencing environmental literacy. Therefore the initial sections of the chapter will analyze concepts of environmental literacy and identify the respective instruments for the manipulation of its components.

2.1 A model for environmental literacy

In the EC and EE literature there are two crucial terms commonly applied, "environmental literacy" and "environmental awareness", both characterizing a complex of characteristics that affect a person's environmental behavior [67, 184, 336]. "Environmental literacy" is frequently used in the Anglo-American EE community, while in the European EE domain "environmental awareness" (see [184, 315]) is commonly applied.

Disinger and Roth (cited in [76]) proclaimed environmental literacy essentially as the capacity to perceive and interpret the relative health of environmental systems and take appropriate action to maintain, restore, or improve the health of those systems. Similar to the concept of environmental awareness this ability is also based on cognitive (knowledge- and belief-related), affective (emotion-related) and conative (behavior-related) elements [280]. According to these definitions the crucial difference between the two concepts is that environmental awareness is seen as the combination of components, which form the prerequisite for environmentally friendly behavior,

whereas environmental literacy also encompasses responsible behavior.
From here on, this work will focus on the more holistic concept of environmental literacy, since responsible environmental behavior, as the ultimate goal of EE interventions, also needs to be taken into account.

For this thesis, a new comprehensive environmental literacy model (see Fig. 2.1) has been compiled from a variety of existing models for environmental knowledge, behavior and literacy, discussed in the EE and environmental psychology literature.
This new integrated environmental literacy model is a theoretical structure model presenting the components and factors, which are hypothesized to effect environmental literacy. It is primarily based on empirical models but has not yet been confirmed by empirical evaluations itself. Several of the analyzed models from the literature [5, 107, 109, 138, 160] are based on the "theory of planned behavior" (TPB). TPB is one of the, to date, most widely sited attitude-behavior models and was originally proposed by Ajzen and Fishbein [4, 6, 7].

2.1.1 Components of environmental literacy

The proposed environmental literacy model encompasses the major components of environmental literacy:

- Knowledge
- Attitude
- Values and concerns
- Norms
- Self-Efficacy and perceived behavioral control
- Intention
- Behavior

The following section will describe the individual components along with their interrelations, as presented in Fig. 2.1.

Environmental knowledge Traditionally, researchers in the field of EE have claimed they can change behavior by making humans knowledgeable about environmental issues. It was also suggested that if one has more knowledge about the environment, the awareness level as well

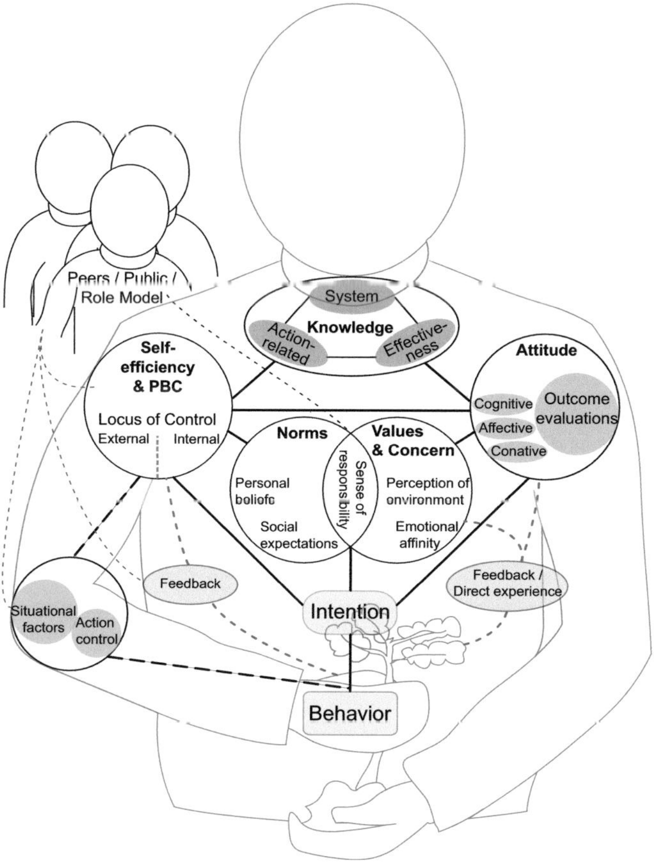

Figure 2.1: Components of environmental literacy (i.e. knowledge, attitude, values and concerns, norms, self-efficacy and perceived behavioral control (PBC), behavioral intention, behavior) and their interaction in influencing environmental behavior.

as the favorable attitude would be higher (Hungerford and Volk, 1990 cited in [148]). As a consequence traditional education programs are basically conducted by conveying certain information (i.e. knowledge) [109, 148, 259]. However, empirical studies showed, that knowledge explains only about 10% of the behavior variance [109]. Thus it has been widely acknowledged in recent years that knowledge has only an indirect influence on environmental behavior [46, 68, 109, 138, 148, 315].

Nonetheless, knowledge is still considered an essential prerequisite component for other environmental behavior related factors that lie closer to behavior itself, such as attitudes as well as perceived behavioral control (PBC) [109, 138, 148]. Frick et al. [110] suggest that environmen-

tal knowledge is not a one dimensional construct and propose a distinction into different forms of knowledge. Frick [109] determined three forms of relevant knowledge: system knowledge, action-related (i.e. behavior-related) knowledge and effectiveness knowledge. System knowledge means knowledge about ecological concepts, interdependencies in ecosystems and about the causes of environmental problems (e.g. the relationship between carbon dioxide (CO_2) and global climate change). Action-related knowledge is knowledge of behavioral options and possible courses of action to resolve environment problems (e.g. if I do not use my car, I produce less CO_2). Action-related knowledge is considered a better predictor of conservation behavior than system knowledge due to its relatively more behavior-proximal nature [110]. Effectiveness knowledge, finally, is knowledge of the relative ecological effectiveness of different behavioral options, i.e. the relative gain or benefit that is associated with a particular behavior (e.g. buying a new, fuel-efficient car, can be a better way to cut down CO_2 emissions than driving an old car less often). This sort of knowledge was repeatedly proposed as relevant for successful action [110].
Based on this concept of environmental knowledge, Frick [109] proposed a knowledge-structure model, which has been adapted and integrated as knowledge component into the proposed environmental literacy model (see Fig. 2.1). Accordingly system knowledge seems necessary to motivate a search for action-related knowledge as well as to generate effectiveness knowledge. Action-related knowledge, in turn, also co-determines effectiveness knowledge. Before you can seek to understand the relative conservational benefits of an action, you have to be knowledgeable about behavioral alternatives in the first place. It is also this knowledge about behavioral options that, according to Frick's findings [109], has an influence on the perception of the personal control with respect to a particular environmental issue. Thus as illustrated in Fig. 2.1 action-related knowledge provides the link to a person's perceived behavioral control. The displayed influence of knowledge on the attitude component via effectiveness knowledge is founded on the theory of planned behavior [6], which suggests, that a favorable attitude towards a behavior is established based on the expectation of a positive outcome of this behavior.

Environmental attitude Environmental attitude is considered one of the most promising concepts with respect to the question of how environmental behavior can be changed [160]. A variety of different approaches can be found in the literature concerning the formation and evaluation of attitudes [160]. While some studies argue for a single component approach, others promote that attitudes consist of multiple components [138, 160]. This approach is commonly based on the attitude model of Rosenberg and Hovland [278] that divides an attitude into three dimensions. As illustrated in Fig. 2.1 an environmental attitude comprises a cognitive, an affective, and a behavioral component. The cognitive component is related to knowledge as well as opinions and a person's beliefs about an object associated with this knowledge [138, 160]. The affective component is composed of the feelings or emotional responses to an object relating it to environmental concern, and the behavioral or conative component is represented by one's tendency to respond in some manner to an object or activity, linking it to behavioral intention. Thus like it is shown

in Fig. 2.1, attitude is postulated to be the first antecedent of behavioral intention [4, 148] and a mediating variable between effectiveness knowledge and intention [109]. Ajzen [4] suggests that environmental attitudes are determined by the individual's beliefs about the consequences of performing the behavior (i.e. behavioral beliefs), weighted by his evaluation of those consequences (i.e. outcome evaluations). The intention to perform a certain behavior is strengthened when he evaluates it positively [4].
They are, however, not only a mediating factor between knowledge and intention but they are further linked with self-efficacy or perceived behavioral control as well as with the complex of norms, values and environmental concern [4].

Environmental values and concern Schultz et al. [306] state that values and culture provide the lens through which our understanding of environmental problems is framed. As a consequence, values are viewed as underlying determinants of attitudes, behavioral intention, and beliefs (Olson and Zanna, 1993 cited in [306]).
Fransson and Gärling [107] define values as beliefs pertaining to desirable end states or modes of conduct that transcend specific situations and guide choices of actions. While attitudes are assumed to underly situational change, values are to a large extent considered to be stable across an individual's life, serving as blueprints or organizing system for expected consequences. This way they also influence knowledge acquisition and lead to an effect on attitudes and personal norms [107, 349], like presented in Fig. 2.1.

Closely related to the valuation of the environment is the notion of environmental sensitivity and environmental concern. According to Meyers [219], environmental sensitivity involves an increased emotional attachment and concern for the environment due to an appreciation for the aesthetic and use value of the environment. This is accompanied by an increased understanding for the necessity of environmental protection and an enhanced sense of indirect moral duty to protect the environment [219].
As proposed by Schultz et al. [307], the term environmental concern refers to the affect (i.e., worry) associated with beliefs about environmental problems, linking it to environmental attitudes as illustrated by the model in Fig. 2.1. Next to concerns about environmental issues, the Value-Belief-Norm theory (VBN), proposed by Stern and his colleagues (Stern, Dietz, and Guagnano, 1995 cited in [307]), suggests that values provide also the source of concern for proenvironmental behavior. In accordance with the VBN, Schultz [305] defines different types of environmental concerns. These include egoistic (i.e. value oneself above other people and organisms), alturistic (i.e. judge cost or benefit for other people/humanity), and biospheric concerns (weigh cost or benefit for all living organisms). Biospheric concern is likely to provide a broader motive for behavior beyond local and more abstract issues. Schultz [305] further proposes an extension of Batson's empathy-altruism theory to the study of environmental issues. Based on the work of Batson (Bat-

son et al. 1995 cited in [305]) on pro-social behavior, he defines empathy[1] as an other-oriented emotional response, congruent with the perceived welfare of another individual. Thus empathy for natural environment should lead to the activation of biospheric environmental concerns [305]. A related concept was proposed by Kals et al. [163] with the notion of "emotional affinity toward nature", which also stresses the importance of affect and emotions (i.e. positive feeling of inclination or 'oneness' with nature) in the generation of responsible environmental behavior. This hypothesis is based on the claim that building up emotional bonds toward nature can serve as a motivation to protect it (Fischerlehner, 1993 cited in [163]). A potential explanation for an inherent disposition for an "emotional affinity toward nature" may be given by the biophilia hypothesis by Wilson (see Wilson, E.O. cited in [158]). Kals et al. [163] make it clear that it is particularly the direct experiences of nature, such as observing animals, weather phenomena, or the change of the seasons, which help to build up emotional bonds toward and cognitive interest in nature. The research on "emotional affinity toward nature" further showed, that the sharing of experiences with significant others may function as an amplifier of the impact of stays in nature and thus increases the possibility that later in life the personal tradition to seek contact with nature is continued.

Next to an emotional bond toward nature and concern for the environment, many models of nature-protective or responsible environmental behavior assume a responsibility-related perspective [163]. This makes sense in regard to the common conflict between short-term interests of the individual in personal benefits and exploitation and long-term interests of the whole community in a healthy environment. This conflict should be overcome by taking over internal ecological responsibility and by accepting corresponding moral norms [163]. Based on Kaiser and Shimoda's [159] conception of environmental responsibility people can feel either morally or conventionally responsible. Moral responsibility feelings depend on a person's self-ascribed responsibility (i.e. a deliberate responsibility judgement) and guilt feelings. Moral responsibility feelings are related to moral concepts such as the welfare and rights of others, and fairness considerations [159]. Conventional responsibility feelings depend on the social expectations a person is aware of and his or her readiness to fulfill these expectations, thus closely linking the sense of responsibility to norms. Conventional responsibility feelings are grounded in social customs or traditions and appeals to authorities. Thus if people feel guilty for what they do or fail to do, they also feel morally responsible for the environment [159].
The norm-activation-model by Schwartz and Howard (cited in Schultz et al. [306]) links this sense of responsibility to behavioral intentions (compare Fig. 2.1). It predicts that an altruistic (i.e. responsible) behavior is more likely to occur when a person recognizes the possibility of harm to a valued other, and the person ascribes responsibility to himself or herself for these harmful consequences [159, 163, 306].

[1]The concept of empathy emerged in field of psychology in the early 20th century based on the concept of Einfühlung by Vischer 1883: Einfühlung = humans spontaneous projection of real psychic feeling into the people and objects they perceive. Tichtener 1909 was one of the first to define empathy as the process of humanizing objects, of reading or feeling ourselves into them [145].

Norms As previously mentioned, the "sense of responsibility" is closely linked to norms. Fransson and Gärling [107] define norms as an expectation held by an individual about how he ought to act in a particular social situation. They further differentiate between social norms and personal norms.

Social norms Social norms or expectations can prevent action if behavior is not socially acceptable (e.g. due to threat of punishment) or an individual will intend to perform a certain behavior when he perceives that important others (e.g. spouse, friends) think he should [4, 107]. These norms are sometimes also referred to as social knowledge, forming a common pool of information and directives on how to act [160].

Personal norms Personal norms or beliefs are frequently based on general norms or values, which have been internalized and do not need to be reinforced and thus form guiding principles on how to act [107]. Hunecke et al. [144] also speak of personal norm, as opposed to a social norm, as an inner moral conviction that is defended irrespective of the expectations of others. Personal norm manifests itself in a feeling of moral obligation and thus a sense of responsibility for the environment. Social norms on the other hand were found to intensify personal norms [144].

Based on Kaiser et al. [160] norms are also related to behavioral intention and thus as displayed in Fig. 2.1 have also an indirect impact on behavior.

Self-Efficacy and controllability Ajzen [4] points out, that people are not likely to form a strong intention to perform a behavior if they believe that they do not have any resources or opportunities to do so even if they hold positive attitudes toward the behavior and believe that important others would approve of the behavior. The same is true for people who belief that through their own behavior they do not have the ability to bring about change [148]. In accordance with Bandura's self-efficacy theory, Ajzen [5] differentiates between efficacy expectation (i.e., the perceived ability to perform a behavior) and outcome expectation (i.e., the perceived likelihood that performing the behavior will produce a given outcome).
Efficacy expectations, or perceived self-efficacy, are defined by Bandura [19] as beliefs in one's capabilities to organize and execute the courses of action required to produce given levels of attainments. Further Bandura [19] stresses that self-efficacy beliefs have a strong influence on how people feel (i.e. the affective component linking it to norms and concern), think (i.e. the cognitive component, linking it to knowledge and attitude), motivate themselves, and behave (i.e. the conative component, linking it to behavioral intention and behavior). Such expectations have been discussed in the literature in the context of the concepts of "self-efficacy", "perceived behavioral

control (PBC)" and "controllability" [2] [5,109]. Corresponding to self-efficacy the PBC of a person depends on beliefs that deal with the presence or absence of requisite resources and opportunities. These control beliefs may be based in part on direct or indirect experience with a certain behavior. Further control factors may include things as skills (i.e. conative), action-related knowledge (i.e. cognitive), emotions (i.e. affective). The more resources and opportunities individuals believe they possess, and the fewer obstacles or impediments they anticipate, the greater should be their perceived control over the behavior. The actual behavioral control does not necessarily correspond to the PBC of a person, since it is influenced by the factual situational factors [4].
While self-efficacy relates to the person's perception if he has the ability to perform the behavior, the notion of controllability [3] represents the beliefs if the person has the opportunity and control over performing the behavior. The individual may have total control when there are no constraints of any type to adopting a behavior [5]. In addition each person is assumed to have a specific locus of control, i.e. a belief on whether or not he has the ability to bring about change through his or her own behavior (Hines et al., 1986 as well as Hungerford and Volk, 1990 both cited in [148]). This locus of control can be internal or external. An internal locus of control characterizes an internal motive, which leads to the expectation that his activities are likely to bring about changes. Consequently, a person with stronger internal locus of control is more likely to participate in a specific activity. An external locus of control on the other hand relates to a belief that changes occur by chance or by the intention of powerful others such as God or government rather than by one's own behavior. One who has stronger external locus of control is, therefore, not likely to participate in activities to bring about changes (Hines et al, 1986 cited in [148]).
This complex of factors can influence behavior indirectly through behavioral intentions or directly in combination with situational factors. A direct path from PBC to behavior is expected to emerge when there is some agreement between perceptions of control and the person's actual control over the behavior and thus also takes the influence of situational factors into account [4]. Situational influences refer to constraints and facilitators on behavior beyond people's control [160]. A particular set of external factors are action control process such as incentives, reinforcements or reminders which may be applied by peers or members of the public [107,210].

Behavioral intention Behavioral intentions are considered the immediate antecedent to behavior [4]. A person is most likely to form a behavioral intention towards a responsible environmental behavior based on the mechanisms described above for the respective environmental literacy components.

[2]It should be noted, that there exist different views on the relationship and differences between self-efficacy, perceived behavioral control and locus of control in the literature, which lies beyond the scope of this thesis. For more information see [5].

[3]Controllability is employed here in the context of social psychology, based on the definition by Ajzen [5]. It should be differentiated from the concept of controllability in engineering, where in Kalman's general theory on control systems [162], controllability refers to the possibility of transforming the system to take on a desired state in finite time by the application of unconstrained control inputs (e.g. via an external controller).

- Knowledge on an environmental problem as well as a relevant action strategy and its effectiveness
- Attitude to belief that action strategy will lead to a positive outcome, given that it is conform with his or her value system
- Concern for respective natural phenomenon and driven by empathic feelings and personal norms has developed a sense of responsibility for the conservation of this phenomenon
- Belief that he has the skills, resources and opportunity to perform the action with an acceptable effort as well as the belief, that he can control the action and will achieve the positive outcome.

An intention to act can be understood as expressed willingness to act upon a certain behavior (Hungerford and Volk, 1990 cited in [148]). Yet, a number of alternative behavioral intentions may potentially be formed, which then compete for actual execution [210]. It is believed that the stronger a person's intention to perform a particular behavior, the more successful they are expected to be. It also needs to be taken into account, though, that intentions can also change over time. The longer the time period between intention and behavior, the greater the likelihood that unforeseen events will produce changes in intentions [4]. In general the realization of intentions into behavior still requires that situational factors permit it. As has been stated above, situational resources such as time, economic constraints, social pressures and environmental conditions [4] may constrain or facilitate this procedure along with the afore mentioned action control process also including verbal commitment, anticipation of positive consequences or reminders, cooperation by others [5, 107, 148, 210]. These can be used as prompts, providing cues to action for those people who have formed an intention but have not transferred it into action yet [210].

Responsible environmental behavior Environmental behavior or environment related behavior [107] as such can imply actions, which are detrimental or beneficial to the environment. The later case has been represented by a variety of terms in the literature ranging from pro-environmental [259], nature-protective [163], ecological [144, 159], environment-friendly [335], to responsible environmental behavior [194]. This type of behavior here generally referred to as "responsible environmental behavior" has been determined as the ultimate goal of EE and is seen as the essential virtue of environmental literacy. With reference to the definition of environmental literacy by Coppola [67], it is behavior that involves appropriate action to maintain, restore or improve the health of environmental systems. Such responsible environmental behavior may involve various types of personal or group involvement as they are propagated for environmental literacy frameworks in EE. This comprises for example Ecomanagement (e.g. more energy efficient transportation), Economic or Consumer Action (e.g. purchase environmentally friendly goods with respect to packaging), Persuasion (e.g. encourage others to stop environmentally unfriendly behavior), Conservation activities (e.g. activities to protect specific species or ecosystems), Political

Action (e.g. contact officials/politicians with regard to environmental issues), Legal Action (e.g. reporting violations of environmental laws) [14, 194, 336].

Feedback mechanisms Actual experiences with the behavior itself, be it firsthand or second-hand experience, may again have an influence on the discussed antecedents [4]. For instance the control beliefs of a person with respect to his perceived behavioral control can be affected by such feedback mechanisms [4] as represented in Fig. 2.1.
Direct experience of the natural environment also exerts an influence on the factors attitude and environmental concern, especially affinity towards nature [163, 219].

2.1.2 Instruments to influence environmental behavior

EE interventions that aspire to achieve a change in environmental behavior should take all aspects of environmental literacy, as proposed by the environmental literacy model (see Fig. 2.1), into account. For this thesis a variety of techniques were derived from the proposed environmental literacy model that can potentially be applied to influence a person's environmental behavior. These will be presented in the following section.

Increase knowledge Frick [109] stresses that the transfer of knowledge is still important since there appears to be generally insufficient knowledge about environmental issues. The results of Frick's [109] empirical study indicate that especially knowledge on ecological concepts (i.e. systems knowledge) including social and economical aspects should not be neglected. At the time system knowledge is primarily conveyed in formal settings like schools, making it less available to other fractions of the population albeit its significance for developing an understanding for environmental issues and for gaining an intrinsic motivation for responsible environmental behavior [109]. Nonetheless, the study also shows that in order to have even an indirect influence on environmental behavior, information-based educational interventions should also include information to increase action-related as well as effectiveness knowledge. Especially action-related knowledge is needed for daily activities related to the environment [109].
According to the constructivist pedagogical approach, knowledge cannot be taught but must be constructed actively by the learner [79, 220]. The acquisition of system knowledge frequently takes more effort and an intensive study of ecological concepts and environmental issues. Action-related knowledge on the other hand can be gained to some degree during daily life learning from role models (e.g. children learning from parents) or based on case studies. Effectiveness knowledge is more difficult to transfer since the contents are generally more technical and complicated. Feedback instruments concerning personal behavior are a key mechanism in this respect. It is, however, better to stress positive consequences of a person's behavior than only the nega-

tive ones [109]. Frick [109] as well as Hellbrück and Fischer [138] suggest that it is important to apply multimedia tools, including intuitive graphical representations, to communicate complex issues and concepts, which is of particular relevance to system knowledge contents.
Thus following these findings, knowledge-based education should focus on all three knowledge forms. In particular, education should foster valid expectations about the impact or effectiveness of one's own behavior as a necessary additional input to promote desired behavior [109].

Change attitudes Based on the proposed model (see Fig. 2.1) attitudes can be influenced by knowledge-based educational interventions. But the link between the two factors attitude and knowledge is a relationship which works in both directions. Thus a change in attitude can also induce the acquisition of knowledge [107].
One of the common instruments of EE, which is supposed to target peoples' attitudes is persuasion [336]. People should be exposed to persuasive arguments that will make them re-consider their attitudes and change their minds. Persuasion is probably most effective combined with social influences such as in interpersonal communication with significant others, nonetheless this instrument also needs to be considered in computer-mediated EE foremost with respect to the design of content and curricular.
To address the values and attitudes in developing environmentally conscious behavior, some authors like Pooley and O'Conner [259] also suggested that the key entry point for EE is via the affective domain. It is further the afore mentioned feedback mechanisms and direct experience of the natural environment, which may lead to a change in environmental attitude. This again underlines the importance of providing direct experiences of the natural environment during EE programs since it is assumed as the source affective-based attitudes [259]. It has already been discussed in the previous chapter that this aspect confronts computer-mediated EE with a deficit of current practises. Still the potential significance of feedback mechanism for the change of environmental attitude also offers a new possibility for the application of computer-based education instruments. Computer simulations could be applied to visualize potential outcomes of alternative behaviors and thus make even complex consequences of personal behavior more comprehensible. The same is true for the visualization of environmental monitoring data, measuring changes in environmental conditions linked to personal behavior. This again could influence a person's outcome evaluations.

Change values, concerns and norms In order to address the factor combination of norms, values and concerns, EE interventions should include a discourse on ecological responsibility and moral ethics [163]. According to Martens [210] it is not possible to change norms with short-term interventions, and attempts to influence normative beliefs frequently lead to a suspicion of indoctrination, restricting their potential effect. With reference to the environmental literacy model (see Fig. 2.1), it is especially the influence of peers and role models that affects a person's beliefs

of social expectations and with it also his norms. Hence it is especially the relationship between educator and the learner, which comes into play at this point. Role models and significant others are also of significance with respect to supporting direct experiences with nature with all five senses. Kals et al. [163] argue that this combination is crucial to promote emotional affinity toward and interest in nature, which should result in a change of values and environmental concern of a person. These contacts with nature should be generally integrated in EE interventions. The role models can fulfill multiple functions in this respect. The communication of feelings and the transference of positive social emotions to the natural environment and nature values all may contribute to the emergence of an emotional affinity. Security feelings mediated by significant others may prevent negative associations. Curiosity and cognitive interest may also be stimulated by the questions, hints, and information communicated by these role models.
This is an aspect where to a large extent interpersonal aspects of EE interventions are of relevance and thus computer-mediated EE currently can only have a limited influence.

Change self-efficacy and perceived behavioral control Next to the indirect effect through the transfer of action-related knowledge, self-efficacy and perceived behavioral control can be effected by EE interventions via the influence of peers or role models on control beliefs of a person [210]. Furthermore, feedback mechanisms tied to personal experiences of environmental behavior should be used by EE interventions to increase the feelings of control over the behavior and its positive outcome.

Finally, EE interventions can also have a crucial impact during the phase where intention is transferred into actual behavior. The above mentioned action control process can be used as an instrument to encourage environmental responsible behavior for those people who have formed an intention but have not transferred it into action yet [5, 107, 148, 210]. This, however, requires that the appropriate EE interventions can be provided just at the right time adapted to the participant's situation.

Change behavior based on instruments of behavioral psychology Next to instruments that attempt to influence individual antecedents of behavior, environmental psychologists have also identified a number of instruments from the field of behavioral psychology that are supposed to exert an influence on environmental behavior as a sum of its antecedents.
The intended modification of behavior can be achieved by reinforcers, i.e. consequences of behavior, which increase the likelihood of the reoccurrence of the particular behavior. Hellbrück and Fisher [138] differentiate between positive and negative reinforcers. Positive reinforcers are pleasant stimuli following a particular behavior, such as a reward or approval, leading to the repetition of the behavior (e.g. if you help collecting an invasive species, you are payed a reward). Negative reinforcers are consequences of a particular behavior that result in a negative stimulus and thus lead to the avoidance of the behavior (e.g. if you introduce a foreign species, you will be fined).

Another instrument is punishment, i.e. a negative stimulus following a certain behavior, without an option to avoid this stimulus. Immediate punishment can lead to an instant suppression of the respective behavior. However, punishment should only be considered as an adequate instrument if an alternative behavior is presented and learned employing different reinforcers. Otherwise it may result in aggression, especially if the person exerting it does not act as a role model himself. The intentional modification of environmental behavior is generally challenging, since environmental responsible behavior is frequently inconvenient in itself and the related consequences are often perceived as punishment. Benefits of responsible environmental behavior on the other hand commonly occur with a more or less large temporal delay and are thus not associated with the respective behavior (i.e. the time trap). This problem is further enhanced, since egoistic and frequently environmentally unfriendly behavior results in immediate benefits for the individual, reinforcing the behavior, while at the same time the consequences of the behavior, which effect other people as well, also occur with a delay and have to be dealt with by society. Attempts to punish behavior with the same temporal delay is, however, ineffective and only leads to confusion [138].

Another aspect, which needs to be considered is the cognitive dissonance theory. Festinger (Festinger, 1957 cited in [335]) suggests that holding inconsistent thoughts, perceptions or attitudes or to behave inconsistent with ones attitudes is emotionally disturbing. Since a person generally seeks internal harmony he will commonly be driven to adapt his behavior or other elements (e.g. attitude) to be consistent. According to a number of studies [259, 335] the induction of cognitive dissonance can be used as an instrument to produce environment-friendly adjustments in behavior. Hence, a change in attitude should lead to a change in behavior. Vice versa, getting a person to behave environmental responsible for a certain period of time may lead to an adjustment of attitude and thus a long-term change in behavioral practices. Still, this instrument only seems to be effective for those people who find the behavior morally important [335]. The instrument is further impaired by the fact that people tend to compensate cognitive dissonance with "small lies". They will use excuses and justification to avoid an adaptation. Since responsible environmental behavior is required in various segments of the daily behavioral practises people will frequently not attempt to act responsible in all these segments but, based on a low-cost hypothesis by Hellbrück and Fischer [138], will act in the area where their personal cost is lowest. Environmental activism organizations frequently have to deal with reactance, since their attempts to convince people to change their behavior are perceived as a restriction of their personal behavior and freedom of choice, which may lead to an avoidance or even a form of resistance or defensive behavior [138]. Still, EE interventions should attempt to motivate a change in the learner's environmental behavior. This could be achieved by inducing cognitive dissonance either by influencing a change in attitudes towards environmental friendly beliefs or by motivating them to engage in environmental responsible behavior. Despite the presented obstacles, environmental psychologists have identified a number of approaches to achieve this endeavor. One of the common strategies is the "foot-in-the-door-technique" asking people to take small steps, that require little effort, towards a promoted environmental behavior, followed by a public self-commitment to that behavior. According to the theory of cognitive dissonance, this should result in a change of attitude and eventually also a persistent change in behavior. Hellbrück and Fischer [138] also propose that this behavior should continuously be deepened and move from an externally motivated behavior to an voluntary intrinsically motivated experience eventually spreading to other segments of daily behavior [138]. As has been previously mentioned persuasion is also considered an important tool in promoting changes in environmental behavior. Hellbrück and Fischer [138] suggest the use of mechanisms of social learning such as the imitation of role models making the behavior perceivable and personal.

Potential EE instruments In summary the following instruments can be specified for EE to influence environmental literacy based on the proposed model and the review of EE methods:

Transfer of knowledge: The conveyance of knowledge is definitely the core instrument of traditional knowledge-based EE interventions. This should include all three types of knowledge (i.e. system, action-related, and efficiency knowledge). Computer-mediated EE approaches can play a crucial role in enhancing this instrument through the application of multimedia contents as well as intuitive graphical representations of complex issues. This should increasing knowledge, understanding and skills as the foundation for other factors of environmental literacy.

Visualize and simulate: In particular computer-mediated EE can be employed to provide the learner with visual impressions of behavioral consequences that can effect attitudes and outcome evaluations as well as perceived behavioral control.

Provide direct experience and encourage exploration: Based on the proposed model direct experience of the environment should be an essential part of EE programs, since it has an impact on the formation of values and concern, including an emotional affinity towards nature, as well as attitudes. New mobile ICT should be employed to effectively scaffold direct experience and exploration.

Provide role models: Be it parents, friends, educators or other significant characters, role models in EE can help to form norms and values as well as concerns and also contribute to establishing an emotional affinity towards nature. A role model can further influence attitudes, stimulating interest in the environment, and thus the acquisition of knowledge and offer a means to learn by imitation when displaying responsible action. Further a role model also has an effect on beliefs of self-efficacy and behavioral control. This calls for new interaction concepts in computer-mediated EE that allow the representation of role models.

Provide feedback and control behavior: Feedback mechanisms, like action control processes, should be employed by EE to increase feelings of control over the behavior and its positive outcome, as well as provide incentives to perform responsible behavior. Behavior modification techniques should be employed to promote environmental responsible behavior using reinforcers, the induction of cognitive dissonance, and persuasion.

2.2 A new model for mobile environmental interpretation

Based on the core concepts of interpretation introduced in Chapter 1, this thesis proposes a new model for a mobile environmental interpretation process. This model adapts the basic interpretation triangle of Lewis (Lewis, 1993 cited in [199]), which represents the process of personal-based interpretation (see Fig. 1.1) to model the process of media-based interpretation.

2.2.1 Model for media-based environmental interpretation

In the model for media-based environmental interpretation, illustrated in Fig. 2.2, the phenomenon remains at the center of attention as the subject of the interpretive presentation tailored to the participant, who is to interact with the phenomenon. This interaction, though, is not facilitated by a human guide but by an interpretive medium such as a brochure or sign. Still, the medium does not simply replace the human interpreter but serves as a mediator between him and the participant. Even though the human guide is not physically present during the interpretive activity, it is still him who creates the content and thus indirectly guides the participant based on his knowledge of and experience with the natural phenomena.

This model exemplifies the shortcomings of current media-based approaches pointed out by Ludwig [199]. He advocates, that personal-based interpretation has some key advantages over media-based approaches. For one the human interpreter can engage into a direct dialogue with the participant, which allows him to learn more about the person and as a result he is able to adapt the interpretation to participant's needs. On top of this, the human guide has the flexibility to take advantage of the uniqueness of a specific situation, by incorporating unexpected situational factors (i.e. phenomena) or by adapting the activity according to the feedback of the participants. Media-based interpretation seems to reduce the procedure to a one-way communication process. Although a certain amount of interaction between the participant and the medium is possible, like for instance a quiz in a brochure or an interactive display, the current media generally lack the flexibility and potential to react to feedback from the participant or to adapt to certain situational factors, such as seasonal changes. Generally, the prospects of a true personalization of the presentation to the needs of the participants are limited and frequently not a feasible option since they tend to be very resource-intensive. Thus current media-based interpretation approaches can only make limited use of the proposed EE instruments to enhance environmental literacy.

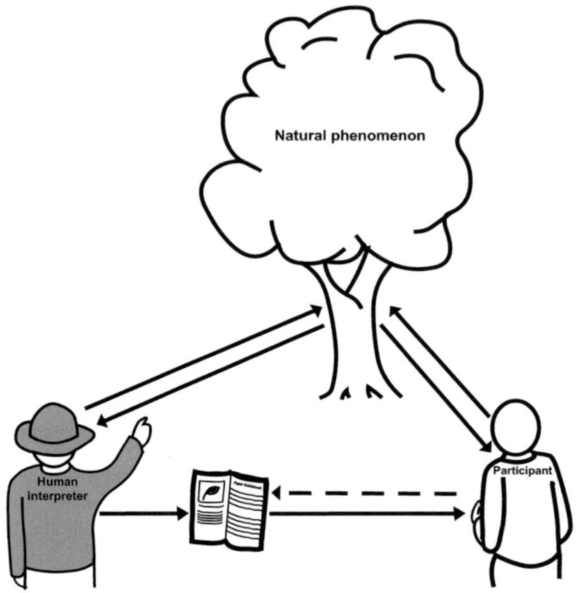

Figure 2.2: Extended interpretation triangle representing the interpretation process based on interpretive media (adapted from Ludwig [199]).

2.2.2 A model for mobile environmental interpretation

Fig. 2.3 presents a new concept for a media-based approach to environmental interpretation. The proposed model of mobile[4] environmental interpretation shows how the application of a mobile nature guide can potentially help to overcome the limitations of present media-based practices.

Among the core features of mobile guide systems that have been listed in Chapter 1, it is above all context-awareness along with the capability for adaptation, which distinguishes a mobile guide from traditional interpretive media. In contrast to a brochure or a sign, the mobile guide system can collect information from its environment, including the user, and adapt the services provided to the user based on this information. As presented in Fig. 2.3, this context information can comprise sensor based readings of the environmental and other situational factors. Next to context information on the natural phenomena a mobile nature guide can also collect information about the user. This can occur either by recording a history of his interaction with the device or based on direct input from the user and can be compiled to a user profile. The mobile nature guide can employ this user context, including user feedback, similar to a human interpreter, to tailor the presentation to his needs (e.g. asking specific questions to determine his prior knowledge on

[4]"Mobile" in this case refers specifically to mobile electronic media as opposed to all forms of media which the user can carry around including paper-based guidebooks.

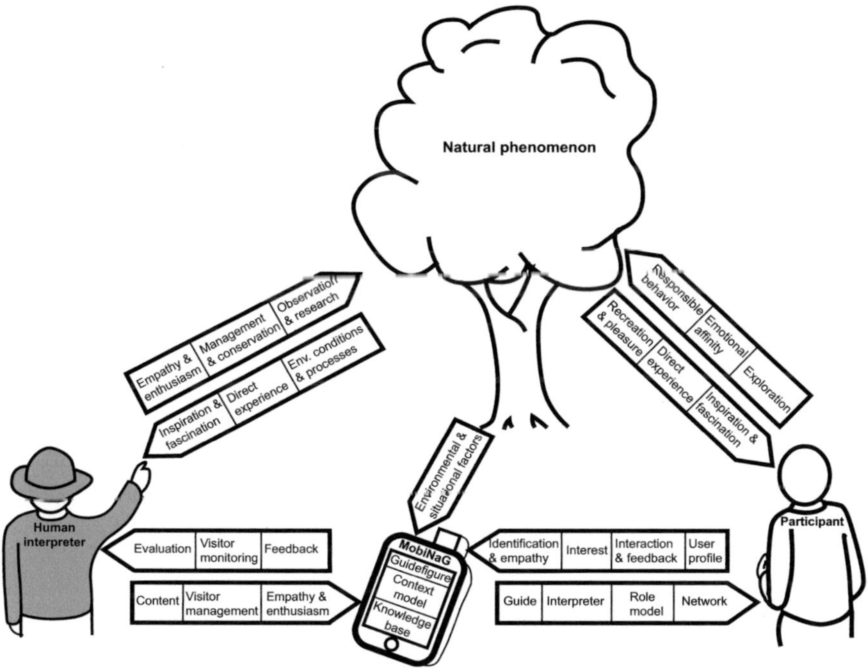

Figure 2.3: Extended interpretation triangle representing the interpretation process mediated by a mobile nature guide.

a theme). In order to provide an interpretive presentation adapted to the situation and the user, a mobile nature guide should next to a knowledge base contain a context model. This context model will be discussed in detail in Chapter 3.

Even though the proposed mobile nature guide can provide an interactive and context-aware interpretive presentation, it is still the human interpreter who has to contribute the content that forms the knowledge base of the system. Comparable to the production of traditional interpretive media, the educator or environmental expert needs to rely on his relationship (i.e. direct experience and interaction) with the natural environment to fulfill this authoring and editorial task. The aspects of this relationship, which are of relevance for this model, are also represented in Fig. 2.3. As a prerequisite for his profession and the dedication to the environment, the interpreter should have a high degree of environmental literacy, including enthusiasm for natural phenomena and also the capacity to empathize with nature. It is this empathy, along with enthusiasm and interest, gained through direct experience with nature, that the environmental interpreter wants to convey to the participant [129, 199]. Frequently the environmental interpretation activities are associated with the overall management and conservation efforts for a natural area.

Beyond extensive environmental knowledge the interpretation of natural phenomena requires detailed observations of environmental conditions and processes in the respective natural area. In order to provide the content for an interpretive presentation the educator or guide has to make the observed phenomena (i.e. processes or features) comprehensible to the participant who will commonly lack the expertise to identify and understand these phenomena by himself. Thus the act of environmental interpretation could also be seen as a form of translation. Hence biotranslation[5] as proposed by Kull and Torop [192] could be the ultimate ability for an environmental interpreter.
The continuously rising numbers of visitors to protected areas are accompanied by increased pressure and potential overuse of sensitive areas. As a consequence the respective managers and administrators are looking for more effective instruments for visitor management that can help to redistribute visitors (i.e. guiding visitors along certain trails to avoid overcrowding and as well as pressure on fragile areas) and change their resources consumption patterns [85, 272, 340]. Next to the mediated administration of environmental interpretation presentations, a mobile nature guide system can offer additional benefits for an environmental interpreter. It can be applied to gather feedback from a large quantity of participants and can further serve as an instrument to monitor and manage the activities and preferences of the visitors. Consequently, it is potentially also a valuable instrument for the evaluation of EE activities.

Still, the question remains how a mobile nature guide can make use of the proposed instruments of EE to improve environmental literacy, besides the adaptation of the presentation? The mobile environmental interpretation model (see Fig. 2.3) displays a number of functions that a mobile nature guide could fulfill for the visitor, based on the services offered by the system.
In order to provide direct experience and scaffold exploration, it should indeed serve as a guide for the participant. This implies providing guidance and advice when needed, particularly with respect to navigation and orientation but also regarding safety issues and the availability of specific ecotourism services. While current interpretive media can provide similar services, a mobile guide system can increase the assistance to the user and comfort of use by adapting the service to the momentary needs of the participant.
The core function of any interpretive medium is of course the provision of information on the respective environmental phenomena. Similar to other media, a mobile nature guide can provide the basis for knowledge construction by presenting the content that has been prepared by the author. With that it also remains to a large extent the author's responsibility to convey enthusiasm and encourage active involvement through the multimedia content he creates. A mobile nature guide as an electronic interpreter has, however, the capacity to go beyond the plain presentation of the content. As mentioned above, it has the capacity to adapt the content to the participant's needs, presenting the relevant information at the right place and the right time, in a form that is tailored to the preferences and understanding of the participant. In combination with interaction and feed-

[5] Kull and Torop [192] define translation as a means that puts signs in one language into a correspondence with signs in another language (cf. Barnstone 1994) given that the worlds of the two users should be functionally similar. Based on their work in biosemiotics they suggest, that these signs are not restricted to human languages but that such signs also exist in nature at least in other organisms [192].

back mechanisms, a mobile nature guide can provide a more flexible media-based presentation and apply strategies of motivating active involvement such as questioning the user about his prior knowledge or asking him to search for a particular feature in the environment [199]. Furthermore, based on the reviewed literature it can be assumed that a mobile environmental interpreter should have a strong motivational effect due to the fascination for technology [13, 147].

By making use of the full technological potential of present day mobile technologies, a mobile nature guide can especially grant new possibilities with respect to the visualization and simulation that can even exceed those of the personal-based interpretation. The utilization for example of augmented reality techniques could allow for the visualization of environmental conditions and processes, which are normally not detectable by human senses or at least require extensive training or experience. Simulations lend themselves particularly to visualize complex environmental processes and thus make them comprehensible also for non-experts. As has been pointed out before these technologies can not only be employed to transfer knowledge but they can also be used to visualize potential consequences to a certain environmental behavior, which in the real environment would only be observable with a considerable time lag. By these means a mobile nature guide can turn into an instrument for the participant to explore the environment and practise certain behavioral alternatives, which in practice would not be feasible or cause harm to the natural environment [174, 321].

Another important function that a human interpreter fulfills, which traditional interpretive media frequently lack, is that of a role model. An individual that the participant trusts and looks up to and that can encourage the direct experience of the natural environment, assists during exploration of the natural phenomena creating a positive atmosphere but who also can use persuasion or various action control strategies. Many of these aspects are currently restricted to personal-based interpretation. The mobile environmental interpretation model suggests as a potential solution the utilization of a guide figure as representation for the interpreter. The participant needs to be able to directly interact with this guide figure throughout the interpretive activity. It should be a figure that the participant can identify and empathize with. Thus the role model that conveys enthusiasm, generates interest and that can also facilitate the development of an emotional affinity towards nature.

A further beneficial function of a mobile nature guide is based on the network capabilities of mobile devices. Access to a local network or to the internet make it possible to connect to other participants exploring the same natural area or other natural environments around the globe, making it for example easier to understand and explore global environmental processes and issues. This way, the interpretive medium can become an instrument for sharing experiences with other participants or significant others. Considering that environmental interpretation is also a process of social interchange this is an important functionality. Furthermore, this functionality introduces that possibility for collaboration and collaborative learning or co-exploration on site or with experts around the world, another important skill of environmental literate citizens. Connecting to the internet would also provide access to web-based environmental information systems and consequently a seemingly unrestricted knowledge base.

2.2.3 A new model for mobile environmental citizen science

The model for mobile environmental citizen science presented in Fig. 2.4 evolved from the mobile environmental interpretation model (see Fig. 2.3) and represents a further step in the integration of ICT in the environmental interpretation process.

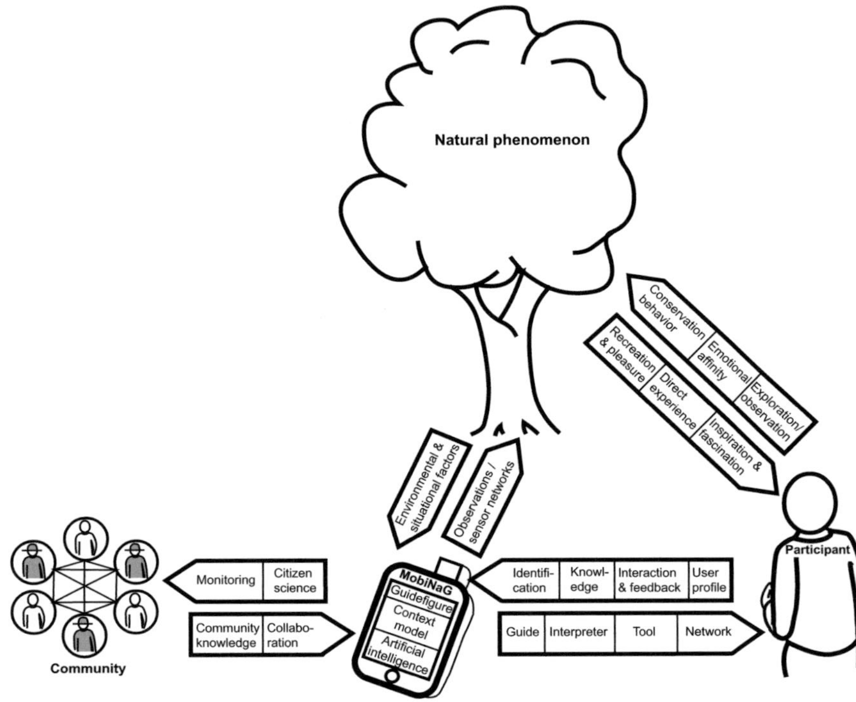

Figure 2.4: Extended interpretation triangle representing the exploration process assisted by a mobile nature guide connected to the environmental conservation community.

A more advanced version of a mobile guide for the exploration of the natural environment could supposedly be equipped with a more extensive array of sensors (e.g. optical, acoustic, chemical) or be connected to a local sensor network (e.g. see [202, 333, 351]). This means that, in accordance with the "mobile environmental information gathering" approach proposed by Antikainen et al. [10], the mobile guide should be able to make itself observations and collect data on the natural environment. On a visionary note, these data on environmental conditions and processes could be employed to build or extend its knowledge base, adapt the context model and subsequently the interpretive presentation. Based on artificial intelligence approaches such a mobile guide system should continuously learn more about its natural environment and built its own environmental model independent of a human interpreter as an author. Ideally, the mobile nature guide should have the capacity to generate and administer an interpretive presentation autonomously. The system could be initially trained in an area and would then continue to learn from activities with participants. The mobile nature guide would in fact become the interpreter itself, translating the natural phenomena for the participant. This of course raises further issues such as the question if such a system can or should in the long run replace the human interpreter. This and similar issues will be further addressed at a later point in this thesis.

This model further expands the application domain of a mobile nature guide. While the core functions of the mobile nature guide for the participant (i.e. guide and interpreter) persist, the interpretive medium could be used more extensively for the self-determined exploration of natural areas. This application may be of particular interest to spare-time naturalists, who already have considerable knowledge and a sound understanding of the respective environment. Since this group of participants will generally have a relatively high level of environmental literacy, the system needs to function less as a role model for responsible environmental behavior but rather as a tool to assist their exploration and especially allows them to document their field observations. Accordingly, the mobile guide can scaffold the naturalists' conservation behavior and provides them with an opportunity to supply their knowledge and thus make a contribution to the conservation of the area for instance by the collection of data on endangered species or monitoring the status of the natural phenomena in the area. Through their feedback and interaction with the mobile guide the participants could contribute to the knowledge base and "skills" of the system. Like this, the mobile nature guide can turn into an instrument for the environmental conservation community and the option of linking to a network with the mobile device gains additional significance. In accordance with the "citizen science" paradigm referred to for instance by Stevenson et al. [328], Trumbull et al. [339] as well as Brossard et al. [49], each citizen who is enthusiastic about nature and willing to engage in the exploration of the natural environment can supply data and share it with other members of the community (e.g. spare-time naturalists, educators and scientists). Through their participation in various monitoring initiatives they can at the same time help to conserve the natural environment. Equally, the community can also add to broaden the knowledge base of the mobile nature guide system with the collective knowledge of the community. Furthermore, this model enables the extension of the citizen science paradigm proposed by Stevenson et al. [328] as well as Brossard et al. [49] through facilitating remote interaction and collaboration between the public and professional scientists.

Chapter 3

A New Instrument for Media-based Environmental Interpretation

Based on the new mobile environmental interpretation model proposed in Chapter 2, this chapter will focus on the conceptual design of a mobile interpretive medium that can serve as a new instrument for environmental education. This entails the presentation of the requirements assessed for such an instrument and the discussion of concepts for suitable components and functionalities of a mobile nature guide system.

3.1 Mobile nature guide requirements

It is generally acknowledged, that prior to the conceptual design of an information system [157] as well as a traditional interpretive medium [57] it is necessary to assess the expectations and needs the users as well as administrators may have with respect to the activity and the usage of the medium, resulting in the requirements for the design [261].
For the conceptual design of a mobile nature guide as well as for the prototypic implementation of such a system, the Mobile Nature Guide (MobiNaG) project was conducted in cooperation with the Naturschutzzentrum Karlsruhe-Rappenwört (NAZKA), an environmental education institution and visitor center to a floodplain conservation area along the River Rhine, which also served as a test site for the project [94]. The assessment of the requirements for a mobile nature guide was one of the initial steps of the MobiNaG project.

3.1.1 Front-end evaluation

As a foundation for the requirements analysis a front-end evaluation was performed as part of this thesis. This is a procedure commonly used in the field of visitor studies and interpretation, in order to initially appraise the respective expectations and needs of visitors [57, 291]. It is the aim of the front-end evaluation to get to know and understand the potential users from different target groups, collecting information about their interests, needs, prior knowledge and ideas regarding the interpretive activity as well as the medium [57, 167, 179].

Target-groups Prior to conducting the front-end evaluation, the target groups for a mobile nature guide system were determined based on the experience of NAZKA administrators regarding the visitor population of similar environmental education institutions. Corresponding to the main visitor groups, the following target groups were selected:

- Students: School classes that take a field trip to a natural area accompanied by their teacher
- Families: Parents and their children who visit the natural area and visitor center on weekends for educational as well as recreational family activities
- Adults/nature lovers: Adults of various age groups who frequently have an inclination for recreational activities in nature, either as tourists or spare time naturalists, and thus visit the natural area on weekends, commonly accompanied by a partner or friends.

Materials and methods A variety of methodological approaches exist for visitor studies including quantitative measures such as standardized questionnaires as well as qualitative methods like general observations of visitor behavior. A comprehensive review of methods in the field of visitor studies can be found in Savage and James [291].

For the MobiNaG project the front-end evaluation was conducted based on a semi-standardized face-to-face interview. At first questionnaires were worked out in accordance with the recommendations by Savage and James [291] as well as Neubert [233] regarding questionnaire design. Different questionnaires were designed for students, teachers and adults each including open-ended as well as hybrid (i.e. semi-open-ended) questions (see Appendix B.1). The questionnaires were further pilot-tested and revised before starting the evaluation. A total of 42 adults (19 females and 23 males) and 48 students (25 females and 23 males) participated in the MobiNaG front-end evaluation. The interview methods are further described in Appendix B.1.

3.1.2 User profiles and requirements

The data from the front-end interviews were analyzed using a content analysis approach as proposed by Mayring [214] as well as descriptive statistics. The analysis led to a revision of the target groups for the system as well as the compilation of a detailed user profile for each group.
The results of the evaluation, especially the answers related to ICT skills and requirements regarding a future mobile nature guide, show that there tends to be a divergence between age groups, which suggests an additional differentiation of target groups. The student group can further be divided into primary and secondary school children. There also tends to be a difference between secondary school children of the grade levels 5-7 and grades 8-9. Subgroups can be identified among the adults (i.e. nature lovers) as well, where those younger than 50 years tend to differ from those that are 50 and above.
The generated profiles subsume demographic characteristics as well as habits, skills and expectations of the respective target groups (see Appendix B.1, Table B.1 and Table B.2). The most important aspects for the requirement analysis will be summarized in the following section.

Visit characteristics The recorded circumstances of the visits show that the majority of participants, regardless of the target group, do not come to the natural area alone but together with a group of people that they socially interact with (e.g. family members, friends or classmates). This stresses the fact, that the visit to a natural area or visitor center is commonly also a social activity. Thus it can be assumed that users of a mobile nature guide will not solely interact with the mobile device but also engage in social interactions, which, according to the findings of Kals et al. [163], are of significance for the establishment of an emotional affinity towards nature. Consequently, the system should actually offer opportunities to facilitate and enhance these social interactions.
With respect to the motivation for and the purpose of the visit certain differences exist between the target groups. School classes usually visit the natural area during a field trip as part of some form of science class. Hence the visit is frequently mandatory for students, who as a consequence need to be considered a captive audience, extrinsically motivated by their teacher. However, the majority of teachers, surveyed during the front-end analysis, ranked the motivation of their students during the visit as middle to high. Even though the purpose of the visit is provided by the teacher, who commonly intends to reinforce a curricular unit, it is generally the intention of the educator to provide the students with a playful and hands-on learning experience, during which the first-hand experience of natural phenomena is of particular importance.
Families and adults interested in nature (i.e. nature lovers) generally choose for themselves to visit a natural area. With the exception of some of the children, who may also involuntarily be brought along by their parents, members of these target groups can be considered a true non-captive audience, driven by intrinsic motivation. The purpose of their visit is accordingly related to the desire for recreation and restoration in nature as well as entertainment and shared activities but also the wish to learn more about the natural environment through first hand experience.

Another aspect of the circumstances of the visit, which is of relevance for the developers of a mobile nature guide as well as the authors of its content is the duration of the visit. Most of the surveyed school classes as well as some groups of adults spend 3-4 hours on site, while the majority of families and adults stay only for 1-2 hours. Therefore the device should be designed for at least four hours of continuous operation, which is particularly relevant with respect to the battery life of the hardware and the energy consumption of the mobile guide application. This also implies that the system needs to offer interpretive presentations of different length or recommended duration. It can also be assumed, that the visitors' willingness to spend extra time to familiarize with a guide medium will be low.

These new findings stress the need for the adaptation of an interpretive presentation to different target groups. As a result a mobile nature guide should be able to adapt to the user's context, including age or grade level, curriculum or type of educational activity, social context and interests. Furthermore, a mobile guide system needs to motivate primarily students and children, to engage in the direct experience of the natural phenomena.

User Knowledge, experience and interests In order to make the content of a mobile nature guide relevant to the visitor it is, based on the principles of environmental interpretation [129,199], essential for the author to be aware of the participants' previous knowledge, experiences and interests.

In this respect, the profiles that were compiled for this thesis (see Appendix B.1, Table B.1 and Table B.2) also provide important cues that can help to make the presentation relevant for the target groups. Cultural history aspects tend to be generally a good aspect to build on. For children and students it appears particularly important to relate to their knowledge about as well as experience with animals and to tailor the content to this interest. Combined with information on plants and ecosystems this could be a good basis for family presentations. Interpretive presentations for nature lovers should likely draw on somewhat more complex aspects of ecosystems and should also point out aesthetic as well as spiritual facets of the natural area. However, it needs to be mentioned, that the content of mobile interpretive presentations should not necessarily be restricted to these topics. They should rather be employed as anchors that help to communicate further environmental topics to the visitors. Furthermore, these findings underline the need for different types of content for different target groups along with personalization and adaptation functionalities.

ICT skills Other aspects of importance to the conceptual design include especially the visitors' prior experience with ICT in general and mobile devices in particular. The users' ICT skills are of significance for the usability of the device and thus also for the interaction design of the mobile guide. The results of this front-end evaluation show that most of the children are familiar with computers and also use them to access the internet. While secondary school children generally also own a mobile phone and sometimes even have had some experience with a PDA, primary school children frequently have only limited computer experience and some have no experience at all with mobile devices. The vast majority of adults stated that they have at least basic computer skills. Family members and adults younger than 50 years have mostly good to expert computer skills and frequently use the internet on a daily basis. The adults beyond 50 years rather tend to report basic computer skills and use the internet less frequently. The majority of adults also have had prior experience with mobile devices, mostly mobile phones. Almost 20% stated that they own or have already used a PDA. Consequently, it can be assumed that the visitors are generally familiar with mobile devices and willing to use them. This suggests that visitors should be able to familiarize themselves with a mobile guide. But the results also suggest that the different target groups, especially young children and adults beyond 50, should be provided with a specific interaction design. Since the majority of participants has had prior experience with the internet, interaction paradigms from web applications may provide a useful approach for the interface design.

Expectations - interpretive activities Corresponding to the results of the front-end evaluation, a mobile nature guide needs to offer interpretive services that provide personalized information related to the visitors location and on demand also background information. Since for students and children environmental interpretation is mostly about self-determined exploration of nature and adults also favor hands-on activities, such interpretive services need to provide instruments for self-determined exploration and thus promote opportunities for the direct experience of nature.

Expectations - mobile nature guide Finally, the participants of the evaluation were also asked if they were willing to use a mobile nature guide and which services they would expect from such a guide. All of the surveyed children and students stated that they were interested in using a mobile nature guide. With a mobile nature guide they generally anticipate to learn more about nature by interactive exploration and through multimedia presentations. The top three services selected by the students included explorative nature tours, games, and an instrument to document their experience such as digital souvenirs. Of the adult participants 91% stated that they were willing to use a mobile nature guide. Nevertheless, a common concern to all adults was that the device must be easy to use. Some of the participants of the group older than 50 stated that they would only feel up to using such a system if an initial tutorial or some form of assistance would be provided. Furthermore, the parents also suggested to offer environmental games as well as an instrument to identify animals or plants with the mobile guide. The top three services selected by all adults include navigational assistance, explorative nature tours and also the digital souvenirs. The analysis further suggests, that the target groups would like to use a mobile nature guide in different modes. While children and students as well as younger adults seek support for self-determined exploration, parents and adults beyond 50 would at least initially prefer to be guided along a pre-defined tour, with the option of switching between the modes as they gain more experience.

Several of the presented findings correspond to the requirements and information needs proposed for a mobile eco-tourism guide (i.e. WebPark) by Burghardt et al. [52] and Dias et al. [85, 86]. They also state that visitors to natural areas want a navigational assistance and seek personalized and location-based information on wildlife and tourism infrastructure. WebPark users also expressed an interest in digital souvenirs. Some of the other requirements, assessed for the WebPark system, may also be of relevance to a mobile nature guide. Its users are interested in safety related information such as warnings about changes in weather or the conditions of trails. Surveyed users preferred a silent presentation of information instead of an audio-based solution. Next to finding themselves on the map, users would also like a possibility to search as well as tag geographical features. Furthermore, based on the review of other mobile guides in Chapter 1, it can be assumed that groups of students or adults who may explore the natural area together using several mobile nature guide units, will anticipate a functionality that will allow them to communicate not only with administrators in case of an emergency but also with their peers.

3.1.3 Additional stakeholders and organizational requirements

According to the model for mobile environmental interpretation proposed in this thesis and the literature review of mobile eco-tourism guides [8, 83, 85, 86], a number of additional stakeholders can be identified that have to be considered in the conceptual design of a mobile nature guide. These additional stakeholders include:

- Administrators of natural areas and EE institutions
- Educators
- Communities.

Administrators The needs of administrators have been identified based on the collaboration with the model EE institution NAZKA and have further been completed by results from a study by Dias et al. [85].
Since EE is generally the main institutional goal, administrators are interested in interpretive media that can help them to offer interpretive services to an increasing number of visitors during times of limited budgets and thus also scares human resources. Next to increasing the environmental literacy of visitors and their appreciation for the respective natural area, interpretive media should also provide the participant with a pleasurable and recreational experience. Administrators are further worried about the distribution of visitors and their impact on the natural environment. Thus they want a tool to monitor the whereabouts of visitors and to manage their distribution [85] as well as control problem behaviors of visitors such as littering [205]. In addition, such an application for visitor management should ideally allow for the transmission of safety warnings to the

visitor to prevent accidents, and it should provide data to assist search and rescue missions [198]. Apart from aspects of visitor management, administrators would also appreciate a tool that would help them to evaluate the effectiveness of their EE activities along with visitor satisfaction. With the limited budgets of most EE institutions in mind, administrators are obviously also concerned about the cost of mobile nature guide systems, including the initial investment as well as cost of operation and maintenance. From the perspective of administrators the visitors should ideally be able to use mobile interpretive services on their personal devices and cover the cost for network connectivity through their private phone bill. Furthermore, managers are rather reluctant to install an additional technical infrastructure that may be required to establish a mobile nature guide system, especially because of its impacts on the natural environment and its potential cost.

Educators Obviously a mobile nature guide should be able to offer a successful environmental interpretation activity to the visitor based on the described principles of environmental education and interpretation in particular. Since with the current state of the technology it is still the educators responsibility to generate the content elements for the interpretive presentations, a mobile nature guide system also needs to equip them with a tool to easily edit and maintain the content of the application. Next to editorial aspects educators are interested in receiving the participant's feedback on the interpretive activities they designed. The mobile nature guide should ideally also include an opportunity to monitor the performance of the participants and evaluate the effectiveness of the interpretive medium with respect to the change in environmental literacy of the visitors. Teachers in this case can basically be considered a subgroup of environmental educators. Still they should be addressed separately, since according to the basic model of mobile environmental interpretation, they are not the ones that will design the content for the mobile nature guide. Nonetheless, next to giving students the opportunity for a direct experience of their natural environment, teachers are also interested in the content of the interpretive presentation, which should preferably match the curriculum of their course. A mobile nature guide should support situational learning approaches as a hands-on addition to the lessons in the formal classroom setting. The results of the front-end interviews with teachers further show, that a mobile nature guide must provide and interactive experience in order to hold the attention of the students. The teachers further recommend that the mobile guide needs to provide immediate feedback to tasks that the students have to fulfill during the exploration with the device. According to the comments of the interviewed teachers it is, however, also necessary to recapitulate the mobile interpretive activity with their students after returning to the classroom. For this purpose they are interested in a tool that will allow the students to document their experience during the visit and can serve as a basis for further discussion. Finally, similar to the interpreters and administrators, teachers are interested in monitoring the performance as well as the whereabouts of their students.

Communities Next to detailed location-based information on the natural environment, conservationists primarily seek tools to explore nature, identify natural phenomena and document their findings. For the purpose of connecting to the community of conservationists, it will be crucial to offer network and communication capabilities. Regarding local communities interested in offering eco-tourism related services the ReGeo project [108] can provide more details.

3.1.4 Environmental requirements

Next to requirements of the organizational environment Preece et al. [261] also recommend to assess requirements of the physical environment, which are commonly related to the environmental condition the system is deployed in. As has previously been pointed out, the application of mobile guides in natural areas brings about particular issues. In the course of this thesis the following particularities have been identified as environmental requirements which need to be taken into account for the conceptual design of a mobile nature guide:

Remoteness Natural heritage features are mostly situated in rural and remote locations such as wilderness areas, which are intentionally protected from anthropogenic influences and thus frequently lack the technological infrastructure that may exist at other tourism sites. This includes for instance a network infrastructure [229] or power supply. Moreover the remoteness of a natural area may even result in limitations of the organizational infrastructure. Thus tourism services including EE institutions and emergency services may be scares.
As a result a mobile guide for remote areas needs to rely on a powerful and energy efficient hardware. Network connections are on the one hand essential in order to request emergency services but on the other hand communications and positioning have to utilize technologies independent of a local infrastructure.

Unfamiliarity As the alienation from nature increases [45], a growing number of people will be less familiar with their natural environment. Given that natural phenomena are in many cases not as prominent as cultural heritage features, many visitors will have difficulties to identify them, since in most cases they will not be labelled or pointed out by signs. Some natural features such as wild animals may due to their mobility and shiness or simply their small size be difficult to detect even for experts. Because of their unfamiliarity with the natural environment visitors may also face particular challenges regarding orientation in the area. While in urban settings street signs and buildings can help their orientation, natural environments frequently lack such cues and there may not even be someone around to ask for directions [251]. This stresses the need for assistance with navigation and orientation in natural areas and that a mobile nature guide should have the capacity to point out important natural phenomena.

Sensitivity Frequently natural heritage sites comprise protected natural areas and phenomena, which are particularly sensitive to the impact of tourism [344], making it necessary to somehow manage the movement and behavior of visitors [205].

Natural dynamics Natural environments are generally subject to frequent changes [16]. There are processes like weathering which will continuously change the appearance of natural phenomena over the years. But there are also short term dynamics, among them the seasonal changes in appearance of the vegetation or the movement or migration of animals [94]. Thus the services of a mobile nature guide need to be adaptable to changes in the natural environment.

Outdoor usage conditions When employing mobile devices in outdoor settings certain aspects of the operational environment such as lighting, noise and dust, have to be taken into account [261]. This generally applies to outdoor usage of mobile guides but is especially relevant for mobile nature guides. In natural environments users and thus also the device will be exposed to variable and occasionally also unfavorable weather conditions, such as rain or high humidity and extreme temperatures, likely without the option to seek the shelter of a building. Lighting can also be considered a crucial aspect since it is particularly variable in natural areas. Next to seasonal and diurnal changes, the intensity of the sunlight varies due to different forms of shading, for instance caused by the canopy of trees. Especially direct sunlight may lead to reflections on the display of mobile devices which can reduce the readability of the presented content. Of course since the participants in environmental interpretation activities are supposed to directly explore the environment and so may literally "dig in the dirt", the device will also have to withstand the contact with this medium, be it soil or water. During such physical activities, children for example may climb on a rock the mobile guide may also have to withstand at least minor physical impacts and the system further faces the challenge that the user needs both hands free for an explorative activity or game but at the same time still wants to interact with the mobile system in order to receive feedback or further instructions.

3.1.5 Requirements analysis

The analysis of the requirements for a mobile nature guide system is summarized in a "requirements sphere", represented in Fig. 3.1. This novel representation of requirements displays a synopsis of the demands and constraints that can be extracted from the new model for mobile environmental interpretation (MEI) as well as those that resulted from the front-end evaluation and the literature review.

The requirements sphere is constructed of three different layers or circles (compare Fig. 3.1). The outermost circle represents the actual requirements expressed by the visitors and stakeholders rep-

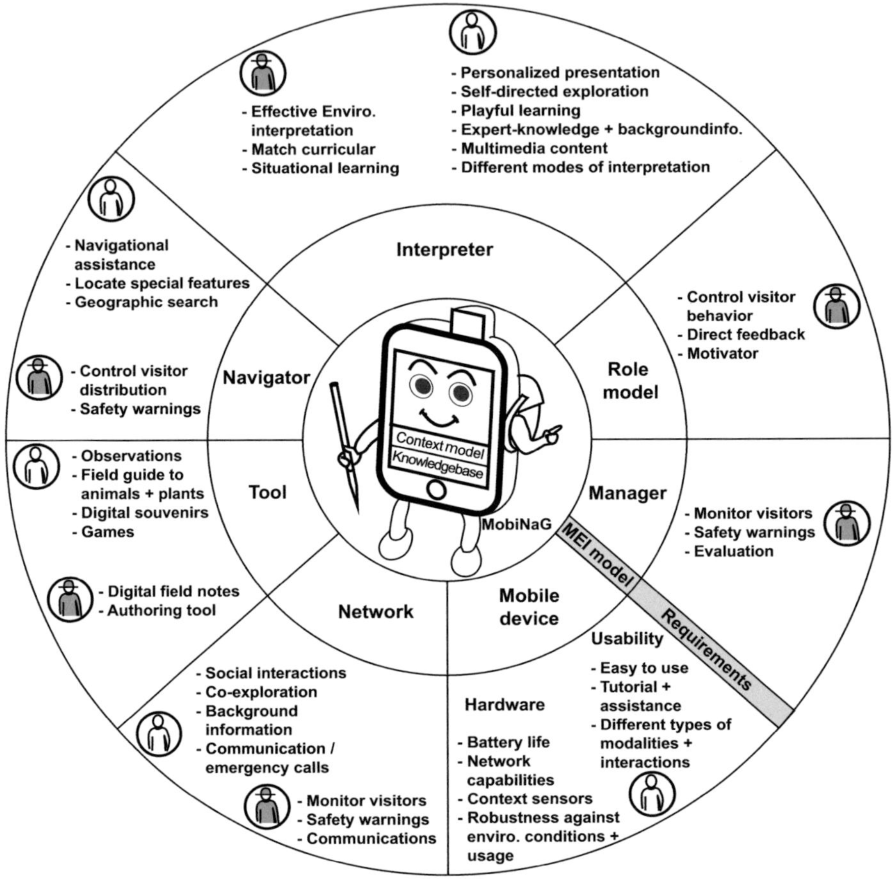

Figure 3.1: Synopsis of requirements for a mobile nature guide system (inner circle - MobiNaG) based on the Mobile Environmental Interpretation (MEI) model (middle circle - MEI model) and the requirements analysis (outer circle - Requirements) including those requirements assessed from potential users (i.e. white figure) and those from other stakeholders (i.e. grey figure).

resenting the future users and providers. These requirements can be grouped and assigned to the core functions of a mobile nature guide (i.e. circle in the middle), that were derived from the MEI model in the previous chapter. The actual mobile nature guide system, which should serve to satisfy the collected requirements by offering the proposed functionalities, is located at the center of the sphere.

As illustrated in Fig. 3.1, the collected requirements match up well with the respective functionalities envisioned by the MEI model and can in many cases be understood as a further specification of these concepts. Like the requirement that the presentation given by the system as an interpreter,

should include multimedia content. However, the front-end evaluation has also yielded a number of demands that have not been explicitly proposed by the model. This includes for instance the significance of social interactions, the desire for playful exploration and games, a digital souvenir function to document the experiences during the visit as well as a field guide that enables the user to identify unknown animals and plants. A whole new category had to be added for the mobile device. It encompasses usability as well as hardware requirements, such as the need for network capabilities for the device or the demand for a tutorial for novice users.
An important result of the front-end evaluation process is further the realization that visitors want to use the mobile guide in different modes with respect to the type of interpretative experience as well as the interaction modalities of the device. These different usage modes will be discussed in more detail in the following section.

3.1.6 Usage modes

The different usage modes anticipated by the users correspond essentially to different types of environmental interpretation as proposed by Ludwig [199]. He differentiates between a number of different types of personal-based as well as media-based interpretation. The two types, which are of relevance to the conceptual design of mobile nature guides, i.e. the interpretation path and the interpretation space, were specifically adapted for this thesis and are visualized in Fig. 3.2.

Figure 3.2: The two common types of environmental interpretation according to Ludwig [199]: the interpretation path (left) and interpretation space (right). They are presented along with the two corresponding usage modes a mobile nature guide should offer: the "guided-tour" and "self-determined exploration".

Interpretation path In the case of the "interpretation path", the interpretation connects several natural phenomena (e.g. outstanding trees) based on a common theme. The interpretive presentation is telling a coherent story about the theme that builds on the phenomena as thematic stops. These stops commonly need to follow a chronological order so that the causal relationships between the phenomena or topics of the story can be maintained. Frequently this type of interpretation follows a logical as well as a geographical line, which is given by the linear structure of some ecosystems (e.g. a river or coast line) or is determined by the trail system. The first usage mode proposed for the mobile nature guide system, the "guided-tour" mode, builds on this type of environmental interpretation. The visitor who decides to use the mobile nature guide in this mode, will be guided along a predefined route to a thematic selection of phenomena. For an effective environmental interpretation the user should visit all stops in the suggested order.

Interpretation space A media-based interpretation activity can also take place in an "interpretation space". The phenomena may still be related to one overall theme but the thematic stops lack the causal relationship and thus also the chronological order. The participant can freely move around between the phenomena and it is up to him, which of the designated features he chooses to explore. This approach constitutes the foundation for the self-determined exploration usage mode, during which the user can employ the mobile nature guide more as an instrument for finding his individual way through a natural area and for discovering the natural features of his interest for himself. Nonetheless based on the principles of EE and the requirements of administrators this self-determined exploration should still be limited to areas that are not sensitive to tourism impacts. It should further be noted that despite of the increased flexibility and support for self-determined and direct experience, the interpretive success of this usage mode depends to a certain extent on the intrinsic motivation of the user. It further signifies an increased challenge for the author of the interpretive presentation, who has to prepare a more modular interpretive presentation that can still meet the EE goals. This is one of the crucial advantages of the guided-tour mode where the human interpreter or author can, based on the principles of thematic interpretation by Ham [129], design a coherent and compelling story to communicate a main theme or message that is memorable for the participant.

3.1.7 Task model

Prior to specifying a detailed conceptual model for the components of a mobile nature guide system, a task model for the activities of potential users will be described. Based on the concepts of thematic interpretation by Ham [129] and Ludwig [199] as well as the "Flow-learning" philosophy of Cornell [69] a new task model for mobile environmental interpretation was developed in this thesis (see Fig. 3.3).

As illustrated in Fig. 3.3, the proposed task model describes the specific tasks, that a participant or a group of participants is likely to engage in, as well as the overall procedure of the environmental interpretation activity. It needs to be mentioned however, that this is a general task model from

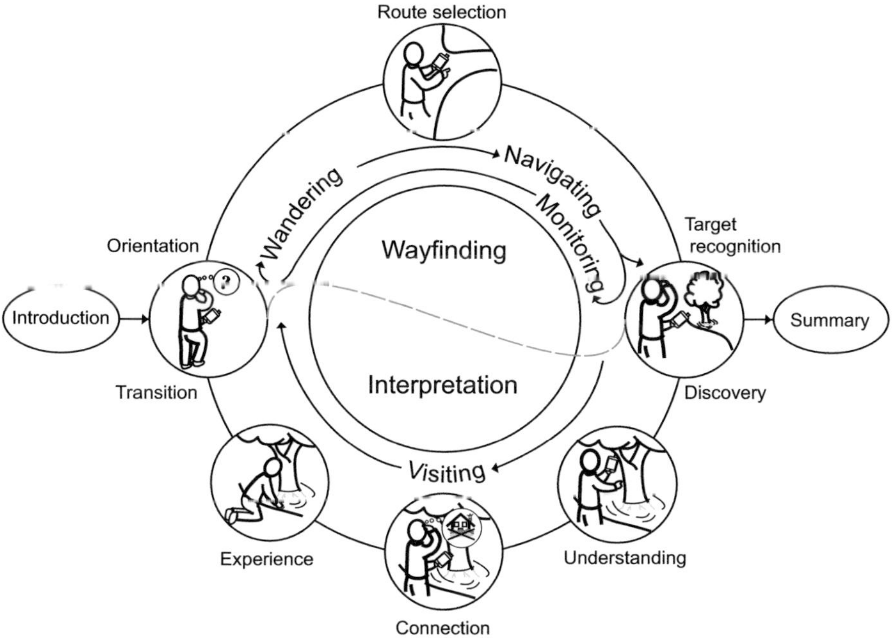

Figure 3.3: The task model describes an interpretive unit with the two phases "wayfinding" and "interpretation" including the tasks the participants are expected to perform during these phases.

the perspective of the user and author of the system that does not yet specify detailed interaction steps.

During his visit to a natural area a participant is expected to engage in at least one interpretive unit, which may be based on either the guided tour mode or the self-determined exploration mode. Fig. 3.3 shows that such an interpretative unit consists of different phases forming a cycle that can be repeated several times throughout the unit and the entire visit. The visitor is generally expected to enter this cycle via the introduction phase and complete the unit with the summary phase. Each cycle of an interpretive unit comprises two phases, the wayfinding and the interpretation phase. Throughout these phases, the user has to fulfill a number of different tasks.

With reference to theory of environmental interpretation [129] the visitor participating in an interpretive unit, is likely to ask the following questions during the introduction phase:

- What is the unit going to be about?
- Why should I participate in this unit? / What is there to explore, that is of interest to me?
- How much time is it going to take?

- What will be the physical requirements? / Are there any safety precautions that I should be aware of?

Wayfinding phase Following the introduction and familiarization phase, the user will in general first enter the wayfinding phase. Conroy [65] provides a comprehensive analysis of the wayfinding process. Wayfinding as defined by Conroy is the act of travelling to a destination by a continuous, recursive process of making route-choices whilst evaluating previous spatial decisions against constant cognition of the environment [65]. Based on the analysis by Conroy [65], the task model gives a representation of the wayfinding process. The first component of the wayfinding process and thus the initial task the visitor has to fulfill is the orientation task. Similar to the observations of Malaka and Zipf [203] for mobile guides in the tourism domain, the visitor to natural areas is likely to first ask a number of spatial questions:

- Where am I?
- Where is the natural feature that is of interest to me / that is the next station of the tour?
- How do I get to this target feature?

In order to answer these questions the visitor will generally need geographic information of some kind, traditionally provided in form of a map. While finding the answers to these questions based on the geographic information, the participant may still remain at the starting point where he went through the introduction phase. Throughout the front-end evaluation it has however been observed that during an orientation task the participants frequently start to move around. Yet, this type of movement is commonly undirected without a specific goal and is thus referred to as "wandering" [197, 350]. As he wanders around, the participant rather attempts to get an idea of the spatial dimensions of the area around him in order to align it with the geographic information he has been provided with. Once the visitor has answered the spatial questions and thus completed the orientation task, he has to select a route to the target feature. With the "route selection", the wandering movement passes over into another type of mobility, described in the task model as "navigation". It implies a directed movement along the selected route, towards the target location. According to the definition of Conroy [65] this process also involves constant cognition of the environment while moving along with an evaluation of the spatial decisions. This comparison of self-location with the selected route is represented in the model as "monitoring" and will continuously be performed while navigating along the selected route until the target feature has been recognized. In case of wrong spatial decisions, the participant may get lost, which implies that he has to return to the orientation task and the wandering process again. But if the user navigates the selected route properly, he can complete the wayfinding phase with the "target-recognition" task. With the completion of this task the user automatically passes over into the "interpretation phase".

Interpretation phase The task model for the interpretation phase is based on the concepts of thematic interpretation by Ham [129] and Ludwig [199] as well as the Flow-learning philosophy of Cornell [69]. Since during the interpretation phase, the participant is expected to remain fairly static with movement being mostly constrained to the vicinity of the selected natural phenomenon, the task model also illustrates a change in the type of movement from "navigating" to "visiting". Together with the spatial recognition of the target the participant also needs to discover the natural phenomenon, which will be interpreted by the mobile nature guide (e.g. a white willow tree standing at the bank of a bayou with its roots in the water). This "discovery" is considered the first task of the interpretation phase that should result in the visitor focusing his attention on the desired phenomenon. At the same time the participant should feel compelled to explore the surrounding natural environment and the phenomenon in particular. This is also the time for detailed observations that may lead to questions regarding the phenomenon, like for example:

- What kind of tree is this?
- Why does it grow here right at the bank of the bayou?
- How can it survive with its roots under water and thus deprived of oxygen?

The next task that the visitor is expected to engage in should provide answers to such questions and should lead to an "understanding" of the discovered phenomenon (e.g. "This is a white willow tree that likes wet 'feet' and can thus grow on the bank of the bayou where few other plants can settle. When its roots are flooded it uses a trick to supply its roots with oxygen. It grows very fine roots at the stem and branches which extract oxygen from surface water, which is rich in oxygen." [120]). The information for understanding the phenomenon can be offered either as a plain text or in a more interactive form, such as a quiz which promotes direct involvement and more of a constructivist approach to the "understanding" task.
After understanding the presented natural phenomenon it is still important for the visitor to comprehend the significance of this phenomenon for the overall theme of the interpretive unit (e.g. "Trees are wooden helpers that fulfill important functions in their ecosystems and also provide crucial services to mankind."). Making this connection between the phenomenon and the theme is crucial to the success of the interpretation even though it may make up a relatively small part of the entire unit. This is the case since the overall theme of the unit represents the key message of the interpretive presentation, that the participants are most likely to remember. This "connection" task is further of importance since it also helps the visitor to relate the observed phenomenon to his own life, making it more relevant and meaningful to him (e.g. "Due to its abilities this white willow can contribute to the stabilization of the banks of the river so that eventually other plants can grow there. Thus trees like this white willow fulfill an important function in their ecosystem as pioneer helping to establish the characteristic flood plain community of plants and animals. As part of this community the white willow plays for example an important role in slowing down

flood waters. Thus as a natural flood control measure it helps to protect us humans from flood disasters" [193]).
In addition the MEI model and requirement analysis have made it clear, that the direct experience of the natural phenomenon is of vital importance to the visitor and is also crucial for the promotion of environmental literacy. As a consequence the connection task should be followed by an "experience" task providing the participant with the chance to directly explore the selected phenomenon, assisted by the mobile nature guide. This is an opportunity for a playful and hands on learning experience to further comprehend presented information. Moreover, the experience task can be used to rehearse a certain behavior recommended during previous parts of the interpretation phase or it can also serve to practice perspective taking in order to achieve an empathy with the selected natural phenomenon (e.g. "Try to please carefully approach the river next to the willow tree and put your hand into the water. Can you feel the current? Imagine how much effort it would take to withstand its pressure during high water. How long do you think you could hold out? This white willow can survive in the flood water for up to 300 days.").
Following the experience task, the interpretation phase is concluded with the "transition" task. During the transition the interpretive phase should be briefly summarized and the participant should get prepared to move on to the next natural phenomenon. The transition also implies that the participant passes over into the orientation task, completing one cycle of the interpretive unit and entering a new one through the wayfinding phase. If the visitor uses the mobile guide in a guided tour mode, the proceeding target is already given and the transition/orientation task should include a short preview that can already spur the interest of the participant. In case of the self-determined exploration mode the participant enters the orientation task having to select a new target.
If, however, the participant has completed the entire interpretive unit or decides to terminate his visit at some point, the next target should be the end point, where the unit will be closed with a "conclusion". The conclusion should briefly summarize what has been seen and experienced and should above all reinforce the overall theme of the interpretive unit as a kind of take-home-message.

3.2 Components of a mobile nature guide

The proposed task model and the presented collection of requirements along with the results of the literature review in Chapter 1, will serve as a foundation to determine the components needed for a mobile nature guide system.
The findings of the preceding sections confirmed that a mobile nature guide should be designed as a context-aware mobile application. A system that should utilize given context information, such as location, to provide its user with certain services like an interpretive presentation. Accordingly the proposed mobile nature guide complies with the general definition of mobile guidance systems

provided by Baus et al. [22]. Consequently the components that will constitute a mobile nature guide correspond to a large extent to those found in a majority of other LBS and specifically mobile guide systems.

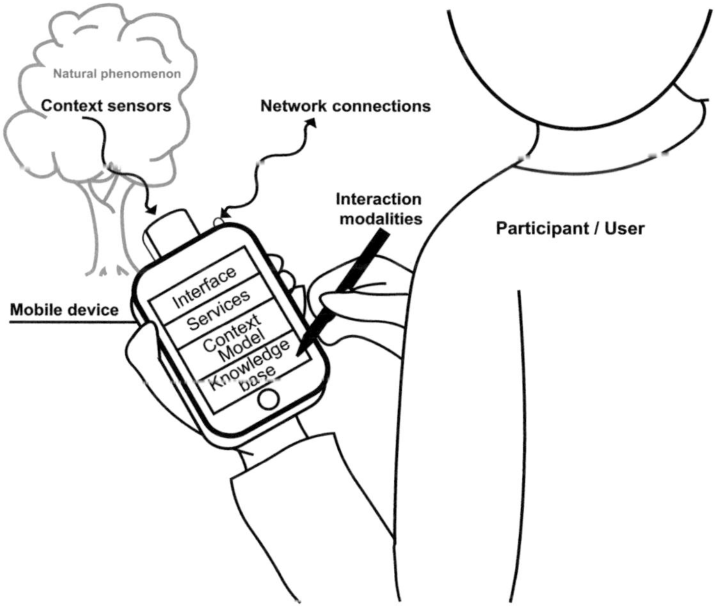

Figure 3.4: Basic scenario of user employing a mobile nature guide, represented by a PDA like mobile device labelled with required components for such a system.

Fig. 3.4 visualizes a basic mobile nature guide scenario specified for this thesis that offers an insight into the components proposed for such a system, which include:

- Context-based services
- Context model
- Knowledge base
- Interface and interaction paradigm
- Control center
- Technologies (Mobile device, network technology, context sensors especially positioning technologies).

Here a user participates in some form of mobile environmental interpretation mediated by a mobile nature guide system. Via the interface the user interacts through different modalities with the context-aware services, which should correspond to the required functionalities of the system. The content offered by theses services again is extracted from the knowledge base and adapted to the users needs according to the context model. A number of different technologies constitute the foundation for a mobile environmental interpretation and the respective services of a mobile nature guide. These technologies include the mobile device as core hardware component as well as various sensors to detect changes in context as well as the hardware components and an infrastructure that offers network connectivity. In the following sections these components will be discussed.

3.3 Context-based services

From the user's perspective the provided services are, apart from the interface, the essential feature of a mobile guide and should be supported by all other components. Which services a mobile nature guide system should offer, can be inferred from the synopsis of requirements (see Fig. 3.1) as well as the task model, presented above. Consequently the derived services include:

- Navigator
- Interpreter
- Encyclopedia
- Exploration tools
- Manager.

For each of these services a new conceptual model was specified as part of this thesis. The following sections will provide a detailed description for each of these conceptual models.

3.3.1 Navigator service

The underlying concepts of wayfinding

In order to provide adequate assistance for the wayfinding task described in the task model (see Fig. 3.3), an understanding of the cognitive foundation for the wayfinding process is needed, and

a brief summary of core concepts of the field of spatial cognition will be given. Research on spatial cognition has shown that people rely on cognitive maps throughout the wayfinding process [80, 170]. Kitchin [170] defines a cognitive map as a mental representation of spatial and environmental knowledge. Such a cognitive map is constructed from a person's knowledge concerning places, spatial relations between places as well as environmental attributive data. During the wayfinding procedure this information is retrieved from the cognitive map for spatial decision making (e.g. Where do I go to?, Which route do I take?) and transferred into spatial behavior. This allows the person to operate within an environment and to process environmental and geographical data [80, 170]. Still while monitoring the navigational progress, such behavior is of course constantly modified due to the interaction with the spatial environment, and the map is assumed to change as well based on the space-time context it resides in [170]. The knowledge needed to construct this internal model of the spatial environment is formed either based on experience from navigating through an environment or based on geographic information about the space [80]. Consequently, for newcomers to an area this geographic information can be seen as the foundation for providing them with navigational assistance [65].
As a result geographic information needed for assisting the wayfinding tasks should include a description of:

- Locations
- Routes
- Landmarks and choice points
- Environmental attributes
- Recommendations on spatial decisions.

Such geographic information can be presented in different forms. These include primarily linguistic forms like verbal survey or route descriptions and pictorial forms such as maps and photographs [80]. The wayfinding and geography literature suggest, that different spatial tasks are supported best by task-specific forms of information provision. Molnar et al. [225] state that it appears that some spatial tasks such as route following (i.e. getting from point A to point B) may be best accomplished through linear, verbal or text-based means rather than pictorial means [225]. For non-navigational spatial tasks (i.e. determining where one is in relation to the overall geography of an area), the pictorial provision of information appears to be best [80, 225]. According to findings of Devlin and Berstein [80] people performed best with regard to a given wayfinding task when they used a combination of forms, such as text supplemented with pictures or maps supplemented with landmarks. At the same time, differences in individual preferences suggest that any information designed to support a specific task probably should be available in different formats [225]. Devlin and Berstein [80] for instance also found, that despite of their level of

performance, people had high confidence in maps, generally assuming that maps will cause less difficulties in wayfinding.
It is widely acknowledged that maps remain the most popular communication language of spatial information [80, 218]. Meng and Reichenbacher [218] stress, that this is also true for mobile guidance applications. Maps are also recommended as a valuable component of media-based environmental interpretation, that helps to orient and guide visitors [57, 129, 241]. Consequently maps should play an important role in mobile environmental interpretation as well as core component of a navigator service, that can potentially be combined with other forms of presentations. In order to devise an adequate navigator service for a mobile nature guide, it is thus further necessary to take a closer look at the design of maps in general and of maps for mobile guide systems in particular.

Map design for mobile guides

Maps are a means of communicating spatial knowledge via an external representation of geographic information [208]. In order to be useful, maps must offer accurate information [88, 208] and at the same time provide an understandable message as well as aesthetical pleasure [208]. This can obviously not be achieved by a one-to-one representation of a given environment, but useful maps commonly represent the real world in an overview, reduced in size and containing only the essential information [208, 341]. Like this maps should focus the attention of the user on the information that is relevant for the user [88] and by that reduce his cognitive load and facilitate the completion of a spatial task [341]. According to Tversky [341] maps can offer this support by schematizing the world based on the following mechanisms, they: give an overview, omit information, use a mix of scales, use a mix of perspectives, exaggerate, distort, and visualize (for detailed description see Appendix B.2). Next to reduction and generalization, the readability and realism of maps can further be influenced by stylistic factors, like for instance the use of color and style of labels etc., which are discussed further in the cartography literature (see [80, 88, 269]).

The described tradeoffs between simplification and realism, readability and aesthetics are a traditional challenge of cartography [367]. However, additional challenges are introduced by the mobile use of digital maps on small portable devices. These are addressed by the field of mobile cartography, which according to Reichenbacher [267] deals with theories and technologies of dynamic cartographic visualization of spatial data and its interactive use on portable devices anywhere and anytime under special consideration of the actual context and user characteristics. This dynamic cartographic visualization is further challenged due to the resource limitations of mobile devices including screen size, colors, resolution, central processing unit (CPU) power, memory, and power supply. The effects on the visualization capabilities, which are of particular relevance for mobile cartography, are to date still limited compared to analogue representations (i.e. paper) and stationary digital devices [88, 268]. Next to the technical aspects related to the mobile devices,

further issues arise from the mobile usage situation. While the user is moving, he experiences a constantly changing environment and is frequently engaged in other tasks besides map reading like looking at sights along a scenic walk [218, 268]. At the same time the user may have to operate a technical device that he is not very familiar with. All this will consume cognitive as well as physical resources of the user (e.g. availability of hands for interaction). Thus this stresses the fact that for maps used in mobile devices it is ever more important that they include only the information that is instantly needed and effortlessly comprehensible, along with intuitive means of interaction [218].
Despite the extra challenges, it is particularly the application of mobile technologies, which can improve the communication of geographic information in mobile usage situations. Personalization of geographic information and its mobility can, likewise, be achieved by digital maps on a desktop computer and traditional paper maps respectively, but the key advantage of a mobile map is the capability of taking the user's context into account and adapting itself accordingly [218]. A specific context information, which is of course vital to the presentation of geographic information, is the user's current location (i.e. the system's location) based for instance on data from positioning technologies. Knowledge of the current location allows for an immediate adaptation of the map to the space-time context on site. This includes adaptation of information as well as its visualization [269, 270]. Thus mobile maps can potentially supply geographic information and services "just in place", "just in time" and "just needed and liked by the user", taking over an essential part of the user's mental effort [218, 267]. Mobile maps and their characteristics, in particular the crucial concepts of adaptation and egocentric geovisualization are discussed extensively by Reichenbacher [269] and Meng [217]. Further Reichenbacher [267], Reichenbacher et al. [270], Meng and Reichenbacher [218] as well as Patalaviciute et al. [251, 252] address issues related to mobile maps and "map-based mobile services" (see Appendix B.2 for more details).

Concepts for mobile nature guide maps

Yet, most of the research on mobile maps has been conducted on mobile guide systems providing assistance and spatial information on urban environments. The following section will discuss novel aspects of designing mobile maps, specific to the proposed navigator service for a mobile nature guide assisting users in natural environments. Table 3.1 lists the respective issues along with concepts that should be realized in the navigator service.

Remoteness Maps in mobile nature guide systems have to assist the visitor during wayfinding tasks in remote environments. As has been mentioned afore, this remoteness implies limited network coverage and potentially also restricted provision of positioning data (e.g. the GPS positioning signal may be degraded by tree canopy) [229, 251]. As a consequence the essential geographic information, needed to explore the natural area, should reside on the mobile device.

Table 3.1: Navigator service requirements and concepts for application in nature.

Natural area scenario	 Remoteness	 Unfamiliarity	 Sensitivity	 Natural features
Issues	- Limited network and positioning coverage - Limited guidance infrastructure (orientation & navigation cues)	- Difficulty with orientation - Difficulty with navigation (identifying cues, judging distances)	- Restricted access to fragile zones - Control visitor distribution + behavior	- Single phenomena (e.g. tree, rock, bird) - Phenomena extending over area (e.g. senic landscapes, habitats, lakes)
Con-cepts	- Essential geographic info on device - Alternative positioning mechanism - Visualize navigational cues	- Realistic representation of environment (match color scheme) + targets - Visualize natural landmarks + route descriptions	- Visualize fragile zones - Monitor navigation history - Present warnings + recommen-dations - Alternative routing	- Geocode and visualize phenomena as point objects + zone objects
Natural area scenario	 Natural dynamics	 Outdoor usage	 Visitors	 Direct experience
Issues	- dirurnal and seasonal changes of geoobjects - moving phenomena / targets	- Reduced readability due to lighting conditions - Restricted interaction capebilities during task	- Age group - Spatial skills - Familiarity with area - Preferences - Usage situation - Usage mode	- Exploration of environment with all senses - Meet expectations - Deepen emotional experience
Con-cepts	- Adaptation of style to changes in appearence - Geocode and visualize moving objects	- Adaptation of style and contrast to lighting conditions - Support hands-free usage (audio route directions) - Visualize forcast of environmental conditions	- Personalized maps - User adaptive + adaptable presentation + map interface - Guidance mode - Exploration mode - Adaptable routing	- Visualize phenomena for senory exploration - Emotionally appealing style

Furthermore the navigator service has to be able to resort to alternative positioning mechanisms, like the ones reviewed in Chapter 1 including manual positioning (i.e. self-positioning or visual selection) as well as strategies suggested by Kray et al. [183] to deal with the lack of positioning information (i.e. inference, interaction and compensation). Yet, the application of these strategies requires the recording of the user's navigation history and he has to be able to interact directly with the navigator service in order to select his position or to make navigational decisions based on cues in the environment. As has been discussed above, such orientation and navigation cues are commonly used in wayfinding employing traditional maps. They can comprise landmarks or also explicit marks [265], including spatial indicators such as street names or direction signs. Ross et al. [279] proposed that specifically the utilization of landmarks can also improve the navigational assistance of guide systems.

Using landmarks in mobile nature guides Extensive research on landmarks, landmark selection and landmark-based navigation was conducted in the fields of pedestrian navigation (e.g. Ross et al. [279], Hampe and Elias [132]) and especially in the field of robot navigation in natural environments (e.g. Betgé-Brezetz et al. [26] and Werner et al. [358]). This thesis will apply a selection of aspects from this research as basis for the concept of using natural landmarks for navigational assistance in mobile nature guides. According to a general definition by Werner et al. [358] landmarks are unique objects at fixed locations. The literature on landmarks further differentiates between individual landmarks based on a single remarkable feature (e.g. a statue, a building) and configurations of features (e.g. alignments of trees, panoramic views) [26,161,326]. For the selection of landmarks to support wayfinding tasks, a number of characteristics have been identified in the literature:

Salience The salience of a landmark is of fundamental importance in order to distinguish it from other objects in its surrounding [132, 141, 173, 326]. This salience of a feature does not solely depend on its individual attributes but on the distinction to attributes of features close by [265]. The unique properties can be either visual (e.g. visual contrast due to intensive color, large size, irregular shape) [141, 265, 350], acoustic (e.g. loud sound or specific melody) and even olfactory (e.g. characteristic odor) or haptic (e.g. specific ground surface cover) [358], but also structural (e.g. prominent location) [265]. Next to this perceptual salience, cognitive attributes can be of importance (e.g. biographical memories) [141,265].

Persistence A salient object can only be used effectively as a landmark, if it is persistent over time. This implies, that its overall appearance is stable and that it is fixed at its location, when the same person returns after a certain time or if another person follows the directions of for instance a guide system [141,358].

Informativeness A landmark should provide a navigator with a cue about his current location within the given environment as well as to what action he should take next to reach his target [132,322].

Still, even if an object qualifies as a landmark based on these characteristics, its selection by a person for a wayfinding task is further influenced by other variables related to the usage situation.

Hampe and Elias [132] stress, that the mode of travel of a person affects several aspects such as speed of travel, visual field and attention towards the environment, which have an impact on landmark selection. The car driver can frequently only observe landmarks that are located close to the road (e.g. gas stations, traffic lights) while pedestrians can look around for landmarks and may select buildings, shops etc. [132]. Further the direction of approach determines the characteristic view a person will have on a landmark and thus also its distinctiveness [132,358]. The familiarity of a person with a certain environment and his frequency of visit to a certain location is expected to influence his selection of landmarks as well [132,322]. The research by Steck and Mallot [326] further showed that different people may choose different landmarks based on different strategies of using them. Related to their usage the literature generally differentiates between two major types of landmarks [141,326,358]:

Global landmarks These landmarks are located in a distance from a route [141] and are hence also referred to as "distance landmarks" [358]. They are characterized by a good visibility from a large area (e.g. a tower or a mountain peak). They can determine a reference frame that does not change its position, when the observer moves a small distance. A global landmark can be used by the observer similar to a compass by keeping a constant bearing on the object [326].

Local landmarks This type of landmark, which is near the route is also described as "route mark" [358] or "way mark" [141]. Local landmarks are only visible from a relatively short distance (e.g. a phone booth or statue) [326]. Such route marks are considered to be a part of a route and are thus closely linked to route navigation. A local landmark can either serve as a decision point (i.e. the person has to choose among different directions), marking the end and the beginning of a route segment, or as a reference within a route segment [141,326].

Steck and Mallot [326] stress, that both types of landmarks are used in wayfinding. All landmarks may potentially function either as global or local landmark, depending on the phase of the wayfinding task.
The above description of landmarks also makes it clear that the selection of landmarks for navigational assistance is closely tied to the selection of possible routes and the creation of respective route directions. Since routes and the associated landmarks as well as directions will obviously vary considerably between environments, this may constitute an extensive task for the designer and author of a navigator service. As a potential solution Hampe and Elias [132] propose an automatical extraction of landmarks based on topographic datasets from geo-databases. An alternative solution could be developed building on findings from the field of robot navigation (e.g. [26,358]). Real-time Landmark-based navigation based on image recognition and a detailed environmental model of the environment could provide a further automated solution for navigational assistance in future mobile guide applications. These aspects however lie beyond the scope of this thesis.
Hence, landmarks can be considered to constitute a crucial component of alternative positioning

mechanisms also in mobile guides. That this is also the case for mobile nature guides, was documented as part of the MobiNaG project [251, 252]. Visitors provided with positioning data but with no further navigational assistance, still had difficulties navigating in-between targets [286] (see Chapter 5). But which landmarks should be selected to facilitate orientation and navigation for mobile nature guide users?

Darken et al. [75] confirm that there is a difference between navigation in urban and in natural environments. In remote natural areas the orientation and guidance infrastructure, established in urban areas (e.g. street signs and direction indicators), is frequently sparsely distributed or entirely missing. Even the tourist's last hope for directions, the local resident, is less likely to pass-by. Furthermore, if visitors are unfamiliar with the area, identifying locations in wilderness or forested areas might be especially challenging since one tree or trail may look almost like the other. In addition, there is an increasing possibility that someone visiting such an area has no prior training in navigational techniques [317] along with the general alienation from nature.

Consequently, the key question is still, which landmarks should be selected in natural areas, to facilitate orientation and navigation for mobile nature guide users. As part of the conceptual design for the navigator service, this thesis proposes a new classification of landmarks in natural areas (see Table 3.2).

The categories of landmarks presented in Table 3.2 should serve as a foundation for the selection of landmarks to support the wayfinding task of mobile nature guide users. The landmarks are grouped in reference to a concept of environmental cues by Soh [317] describing natural or man-made features in parks and wilderness areas that can help the visitor during wayfinding. Which specific features should eventually be included in a mobile map for a particular mobile nature guide, will obviously depend on the composition and structure of the specific natural area.

Natural objects that can serve as landmarks in natural environments include, landscape forms, water bodies and wildlife comprising especially vegetation but to some extent also animals. The proposed category of landscape forms (also referred to as "land contours" by Vinson [350]) can subsume features like a mountain peak forming a unique object based on its visual attributes like height but also smaller geological features such as a bluff or an erratic rock, distinct by their structural characteristics, like being in a prominent location or forming a barrier. The mountain and many other geological formations will, owing to their large dimensions, serve as a reference frame for the visitor helping him to determine his location from anywhere in the area and thus will function mostly as a global landmark. Other geological features such as the erratic rock, may only be seen upon approaching it and can thus serve as a local landmark. In contrast to many other natural objects landscape forms commonly change fairly slowly over many years and are considered to have a high persistence for the sake of this categorization.

Water bodies are, due their limited occurrence, almost always a visual contrast to their surrounding. But thinking of a waterfall for example, there are also acoustic attributes that can make them outstanding. Rivers frequently form a boundary or barrier that can only be crossed at certain points. Based on this structural attribute they can form an alignment that guides the visitor to a

Table 3.2: Classification of landmarks for navigational assistance in natural areas.

Category of objects	Type of environmental cue	Sample features	Salience	Persistence	Informativeness	Function
Natural objects	Landscape forms	- Mountain peak - Canyon - Rock formation - Panoramic scenery, disturbance	- Visual (height, form) - structural (prominent location, boundary) - cognitive (natural heritage)	- High (some seasonal variations) - Fixed	- Reference frame	- Mostly global - Sometimes local
	Animals (Fauna)	- Deer herd - Bird colony - Ant hill	- Visual (color, shape, movement) - Acoustic (animal calls, movement) - Olfactory	- Low to medium (constant variations) - Mobile	- Reference point - Decision point	- Mostly local
	Vegetation (Flora)	- Plant community (e.g. forest, brush, grassland) - Distinct individual (e.g. tree, flowering herb) - Structure (e.g. clearing, pile of dead wood)	- Visual - Olfactory - Structural - Haptic	- Medium to high (seasonal variations) - Fixed	- Reference point - Decision point	- Mostly local - Sometimes global
	Water bodies	- Lake, pond - River, stream - Waterfall, spring	- Visual - Acoustic - Structural - Haptic	- Medium to high (seasonal variations) - Fixed	- Reference frame - Reference point - Decision point	- Global and local
Man made objects	Trail system	- Trail - Junction, fork - Bridge, crossing	- Visual - Structural - Haptic	- High - Fixed	- Reference frame - Reference point - Decision point	- Mostly local
	Boundaries	- Railway tracks - Fence	- Visual - Structural	- High - Fixed	- Reference frame - Reference point	- Mostly local - Sometimes global
	Signage	- Info panel - Location indicator - Direction indicator	- Visual - Cognitive	- High - Fixed	- Reference point - Decision point - Direction indicator	- Mostly local
	Facilities	- Visitor center, hut - Picnic/Rest area - Campground - Ruin, Sculpture - Parking lot	- Visual - Acoustic - Structural - Cognitive	- High - Fixed	- Reference frame - Decision point	- Global and local

certain point. Depending on the characteristics of the water body as well as the regional environmental conditions a water body can either be highly persistent or may also vanish or change its course during certain periods. Water bodies such as rivers can serve as reference frame but other water bodies may also be reference or decision points along a route and thus constitute global as well as local landmarks.

To select animals as landmarks is rather unusual since their persistence is generally low, owing to their mobility. However, some animals also build distinctive housing or nesting structures like for example large ant hills. Beyond their visual and sometimes olfactory uniqueness, some species like certain birds may form seasonal colonies which can clearly be heard in the surrounding environment. According to theses aspects also animals or animal-made features can serve as mostly local landmarks.

Yet, a more regular category of living landmarks is supplied by the vegetation of a natural area. Still, it is literally possible "not to see the wood for the trees". Hence the ubiquity of some plants makes it also challenging to find features that stand out. These features can be on a larger scale entire communities of plants, like a forest or grassland, but also disturbances of these units, such

as a clearing. On the other hand there can also be single plants such as prominent trees or a flowering bush that may serve as a local reference or decision point. Plants can be perceived as unique based on the whole range of attributes visual (e.g. color of leaves or flowers, shape), structural (e.g. prominent location) but especially also olfactory (e.g. scent of flowers or entire plant) and haptic (e.g. walking through a wet meadow).
Due to the natural dynamics most natural features will vary in their attributes within relatively short time periods. In many regions of the world the major part of these features will undergo at least seasonal changes. This is especially true for the vegetation (e.g. flowering, defoliation), which may constrain the persistence of such features severely [350]. As a consequence the proposed navigator service needs to adapt either the representation of landmarks to the changes in appearance or use different sets of landmarks to support wayfinding during different seasons.

Despite their remoteness there are also a number of man-made objects in most natural areas; especially in those, which have been opened for tourism and recreation activities. The most basic infrastructure is the trail system, consisting of a network of trails and junctions but also bridges and crossings. Based on their visual, structural (e.g. width and design of trail) and haptic (e.g. surface cover) attributes, these features can of course themselves provide cues for wayfinding. Trail sections may serve as reference points while junctions are obviously decision points, both functioning mostly as local landmarks. The persistence of trail components can generally be assumed to be high.
With the increasing fragmentation of the landscape [146], more and more man-made linear structures, such as railway tracks or fences, cut through remote areas. They are likely to come to the attention of visitors, since they form boundaries that are frequently a barrier to their movement. For the purpose of wayfinding they can serve as reference points or also frame, as the visitor potentially has to follow the boundary to the next decision point.
In areas dedicated to environmental tourism some signage frequently does exist. It may consist of info panels as well as explicit marks for way finding like location and direction indicators. Signs are commonly visually as well as cognitively salient since their usage is based on common knowledge. Based on their explicit service for wayfinding and their usually high persistence they can be considered particularly useful as mostly local landmarks. A disadvantage of signage though is that they are subject to weathering and occasionally vandalism, thus requiring a potentially extensive amount of maintenance. The infrastructure established for tourism in many natural areas includes a variety of facilities, starting with parking lots and visitor centers but also including picnic/rest areas and cultural heritage features such as ruins or sculptures. To fulfill their purpose these have frequently been located at crucial nodes of the trail system, standing out visually and acoustically. Vinson [350] points out that orienteering experts rely most on man-made objects in natural environments. Just as a tree becomes a clear salient object in an urban setting, man-made objects stand out in natural areas and are highly persistent. However, the disadvantages of man-made objects include their impact on the natural environment (e.g. affecting the aesthetic experience visitors) and the cost for the installation and maintenance of explicit indicators. The representation of nat-

ural landmarks on the other hand can provide the visitor with important cues for the exploration of the natural environment. The mobile map can thus go beyond the support for wayfinding, helping the user to identify and comprehend natural features that he can observe in the real world.
In summary, landmarks should be adequately represented on a mobile map for a mobile nature guide. But landmarks need to be embedded in route directions in order to adequately support the wayfinding tasks [265].

Routes in mobile nature guides A route is defined as a chainlike collection or sequence of intermediate targets or nodes connected by route segment or edges [265,326]. Local landmarks are by definition part of a route, either located at a node, marking the end and the beginning of route segments or along the route within a route segment [141, 173, 326]. Located at nodes, landmarks represent a decision point, where the traveller has to choose among different directions and a re-orientation may have to take place. While travelling along route segments no orientation needs to take place but the person should travel from decision point to decision point. So landmarks located within route segments only serve as a reference point for the visitor to confirm that he is still on the right track [265]. Linguistic route directions are generally composed of a reference to a start location, represented by a landmark, a re-orientation instruction in form of a direction and a target location, also based on a landmark. In addition, distance measures and reference points may be included. A number of concepts exist to formalize such route directions in order to be automatically generate [141,173,240,265]. An example for a formalized route direction could be:

$$
\begin{aligned}
&[\text{START AT landmark}] \\
&+ [\text{TURN LEFT — RIGHT — MOVE STRAIGHT}] \\
&+ \{\text{MOVE ON FACING celestial direction}\} + \{\text{FOR distance measure}\} \\
&+ \{(\text{PASSING — CROSSING})\ \text{landmark}\}_{0...n} \\
&+ [\text{UNTIL YOU ARRIVE AT landmark}]
\end{aligned}
$$

'[ ]' denoting required elements, '{ }' optional elements, 'UPPER CASE' language elements, and 'lowercase' variables - adapted from [265]

Visualize sensitivity Besides issues of remoteness and unfamiliarity, the conceptual design for the navigator service also needs to address issues related to the sensitivity of natural areas. This implies that the service visualizes fragile zones in a way that makes the visitor aware of the restricted access to this area. Since these restrictions may be confined to certain time periods (e.g. breeding season of certain ground nesting birds), the presentation of these zones should also be adapted to the environmental conditions. Next to ecologically fragile zones visitors should be kept away from potentially dangerous zones (e.g. unprotected cliffs, highways) as well. Apart from visualizing these zones, the navigator service should also monitor the visitors location in case positioning data is available. If visitors actually enter a fragile zone, despite its respective status communicated through the map, the service should display a warning and should provide him with an alternative route to his target feature or at least out of the critical area. This mechanism of monitoring the visitor's navigation history and presenting him with location-based warnings via the navigator service is also an important instrument for managers who seek to monitor and potentially control the distribution of the visitors. But also visitor safety and the conduction of potential search and rescue missions can benefit from such a mechanism [198].

Adaptation to natural dynamics A major issue when designing a navigator service for a mobile nature guide is the existence of natural dynamics. This aspect has to some extent already been taken into account by the concept for the adaptive presentation of natural landmarks and route directions. These dynamics are further of relevance for the overall presentation of geographic information by the navigator service. Kokkonen and Peltonen [177] stress that a map for outdoor tourism should resemble the image of the respective landscape in a way so that it can help to fulfill the visitor's expectations regarding the landscape as well as potential activities and thus grant a satisfying experience. With respect to designing maps for mobile nature guides, this entails the presentation of season specific maps, as proposed by Kokkonen and Peltonen [177], Patalavitiute et al. [251] and also by Nivala and Sarjakoski [239]. This can for instance be achieved based on the adaptation of stylistic elements of the map like matching the color scheme to the predominant colors that can currently be perceived in the natural environment. In correspondence to the concept of culture specific coloring proposed by Zipf [367], this concept can be referred to as "environment specific coloring". Next to the color scheme, the symbolization should be used to enhance the detection of relevant features [177]. Another approach could be to only display features that are accessible during this season (e.g. mountain trails in the summer). One can further envision the dynamic visualization of constantly changing environmental conditions overlayed on the map. For example based on remote sensing data the distribution of precipitation, floodwater, tides or air temperature values in the natural area could be displayed.

Next to cyclic changes, the issues of natural dynamics also address objects that are dynamic with regard to their location. This includes especially mobile environmental features, such as animals, and also managers and educators patrolling the area. As far as compatible with management and conservation goals, the navigator service should geocode (i.e. assigning geographic identifier to map feature) and visualize these moving objects. For example Edwardes et al. [96] proposed an approach to visualize the distribution of animal populations on a mobile map, based on census counts and probability distributions. Another possibility would be to use data of individual animals that are monitored with telemetric methods.

Adapt to outdoor usage conditions The adaptation of the map style can be necessary as well to enhance the readability of the map in order to compensate outdoor usage effects such as light-

ing conditions. The map style should be designed to provide sufficient contrast between map elements, either in direct sunlight or in the shade. A straightforward approach would be to offer user-adaptable contrast settings. A more complex solution would be to provide adaptive contrast based on context data, either from a light-sensor or from positioning technologies in combination with a complex environmental model of the area.
Other unfavorable outdoor usage conditions may include precipitation (e.g. rain or snow) or extreme temperatures. All will likely result in restricted interaction abilities with the system since users will either put the device or their hands in their pockets for protection. The interaction with the system via the display may also be restricted, as the user focuses on other tasks like observing or directly interacting with the environment. In all these cases a hands-free interaction mode is needed in order to continue to provide navigational assistance to the visitor. This may be achieved for example by audio route directions as well as notifications of events and warnings.
The dynamic visualization of constantly changing environmental conditions, which has been previously envisioned, could also be used to display a forecast of outdoor usage conditions.

Adaptation to visitor needs Users of a mobile nature guide will belong to different target groups and will thus differ in spatial skills and familiarity with the natural area as well as in their interests and preferences. Further the individual usage situation, including aspects such as location, orientation, time, should be taken into consideration. The aspect of user adaptive as well as user adaptable map-based services has been subject to extensive research in the field of mobile cartography and a variety of concepts have been introduced by Reichenbacher [269], Nivala and Sarjakoski [239] and Zipf [367]. The resulting concepts of adaptive visualization include information adaptation (e.g. selection, adjusting level of detail, classifying and grouping) as well as the adaptation of visualization (e.g. scale, generalization, style and form of the presentation of geographic information) [269,270]. Of further interest are concepts related to map alignment and orientation such as the idea of "egocentric maps" as presented by Schmidt-Belz and Hermann [300], where the map is continuously being aligned with the orientation of the user. Accordingly the navigator service should offer user-adaptive geographic information of the natural area. As part of the MobiNaf project Patalavitiute et al. [251] present a concept for personalized maps for mobile nature guides. This includes aspects of personalized map styles like for example target group specific symbolization of map objects.
A navigator service for a mobile nature guide also has to take the usage mode (i.e. guided-tour or self-determined exploration), selected by the user, into account. For a visitor who chooses to participate in a guided-tour the navigator needs to offer a guidance mode. This involves the visualization of a precalculated route belonging to the selected tour as well as the presentation of corresponding route directions. Furthermore the map can be automatically adapted to optimal settings for the tour (e.g. specific scale or thematic style). In case the visitor decides for the exploration mode, the routing service needs to calculate the appropriate (e.g. shortest or most interesting) route from the users current position to a selected natural phenomenon. In case the

user does not select a specific location but chooses to explore while wandering around the area, the navigator service should adapt the map to best support this task (i.e. adapt scale, increase level of detail in the surrounding of the visitor) and notify him of interesting objects in his vicinity. The navigator service should, however, also honor the principle of adaptability proclaimed by Reichenbacher [269], so that if desired the user should be able to adapt the visualization to the same amount as the adaptive behavior of the system. For example if the user decides to abort a tour, the service needs to be able to calculate an alternative route based on the users needs.

Representing natural features The previously introduced concepts of natural landmarks and fragile zones have already shown that the map needs to represent singular objects as well as configurations of objects and features that extend over a certain area. The same is true for natural features, which have been selected as phenomena for interpretation. These can be single point-like phenomena (e.g. a tree, an erratic rock or a squirrel's nest) or phenomena that extent over a certain area (e.g. a lake, a scenic landscape or a specific habitat). Consequently, the navigator service needs to geocode and visualize both, point objects as well as zone objects. Furthermore next to zone like natural features the overall landscape surface may play a significant role when travelling in an natural area. Since during the interpretive activities a visitor may engage in extensive hikes, it is particularly the relief of a landscape that has an impact on his physical endurance. Thus when suggesting routes, the navigator service should not only consider the distance and the available time to travel this route but also the differences in altitude as well as the physical fitness of the visitor. Therefore the relief of a landscape should in certain natural areas also be visualized.

Support direct experience Finally, the navigator service should also encourage and assist with the direct experience of the natural environment. On one hand this is sustained via the provision of navigational assistance, which helps the visitors to reach and identify natural phenomena and linking to an interpretive activity that involves direct experience. In accordance with Kokkonen's and Peltonen's [177] concept of "mediated seeing" (i.e. inducing realistic expectations and giving a preview of the available activities), the symbol for a natural phenomenon should give a realistic representation of the specific object and may also provide simple instructions for first hand experiences. Yet, direct experience can also be tied in with the wayfinding task. During wandering as well as navigation the visitor perceives the natural environment around him. So the map should also visualize opportunities to experience the natural environment with all senses. A guided tour for instance could be based on a "sensory map" that would indicate zones or points along route sections that have specific olfactory, haptic or acoustic qualities that the visitor may consciously explore. This may be conveyed for instance through visualizing specific symbols or animations. The result could be an interpretive map, which beyond providing navigational assistance, can serve as an interface between the user and the system but also between the user and the natural environment.

Kokkonen and Peltonen [177] further propose that, via its "look and feel", a map can also be used to form persuasive affections about a destination and to deepen the emotional experience. By contributing to a positive experience the navigator service should ideally strengthen an emotional affinity as proposed by the model for a mobile environmental interpretation. Further approaches in geography to communicate affective elements are addressed by the discipline of "emotional geographies" and a comprehensive survey is given in [41].

Components and functionalities of a navigator service

Based on the discussed issues and the described concept for a mobile nature guide navigator service, a number of components and functionalities for the navigator service were devised in the course of this thesis. Table 3.3 lists the different tasks associated with the wayfinding phase and presents a simplified adaptive visualization of the map layers with their respective components and functionalities. Further included are references to user input and context information that can be employed for the adaptation of the service.

Basemap Initially a user decides to visit a certain natural area and to engage in an interpretive activity. With this decision he selects an Area of Interest (AOI) (see Zipf and Häußler [368] for definition of AOI in OpenLS standards) that corresponds to a specific environmental model of the area. During the introduction phase and the early orientation phase of the wayfinding task (see Fig. 3.3) the visitor wants to get an overview of the selected environment. Accordingly the navigator service displays the AOI in form of a predesigned base map. Apart from the extent of the AOI, the base map usually visualizes other geometric references with a high persistence and fixed position [268]. This includes for instance basic tourism infrastructure such as parking lots as well as major buildings offering first aid and tourism information. With respect to reducing the impact of visitors it should include restrooms and trash cans as well [129]. Still, the base map may also include components, which should be adapted to the user's preferences as well as the environmental conditions, such as the basic trail system accessible during this season. Furthermore the map should be styled in accordance with the user preferences, the outdoor usage conditions and environmental dynamics as discussed above. The basic map style can in addition visualize overall aspects of the landscape structure for instance by hill-shading.

Positioning and tracking layer Next to the base map the the navigator should also visualize the more dynamic components, which may be changed more often. In accordance with the concept commonly used in geographic information systems (GIS), these separate types of dynamic spatial information are grouped and superimposed as layers onto the base map [90].
The first of these components to be added, should help the user to get oriented by showing him

Table 3.3: Components and functionalities of navigator service for a mobile nature guide.

User tasks	User context + adaptation	Map layers / visualization	Map components + functions	Context information
1) Orientation [Wandering] - I want to visit this area! - What does this area look like?	User selection User preferences	***Basemap*** 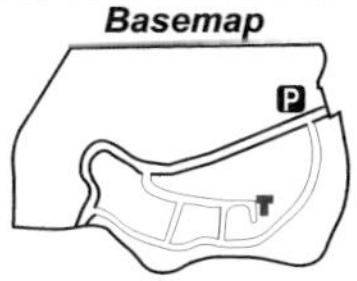	Area of Interest (AOI) - Trail system - Basic infrastructure - Style - Relief	Environmental model Environmental conditions
- Where am I? - Where am I headed?		***Positioning + tracking layer*** 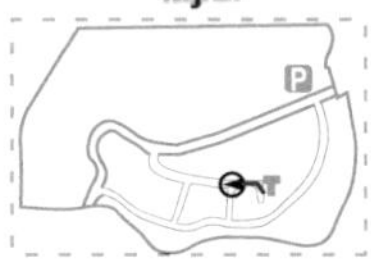	- Position + orientation + tracking indicator - *Map alignment*	Position + orientation data Navigation history
- What can I see here? - Where are natural features of interest to me?	User preferences	***Objects of Interest layer*** 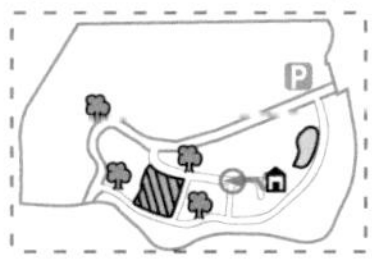	- Natural + man-made objects = - Points of Interest (POI) - Zones of Interest (ZOI)	Environmental conditions
- How do I get to the feature?	User selection User characteristics	***Routes layer*** 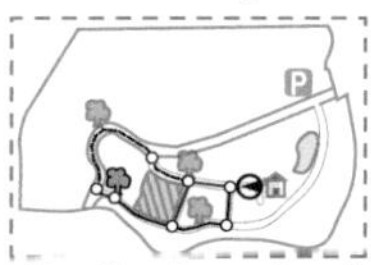	- Starting point + end point - Routes = route section$_{1+..+n}$ + decision points	Environmental conditions Visitor management
2) Route selection [Navigation] - What features are part of this tour?	Usage mode User selection User preferences / characteristics	***Routing layer (tour)*** 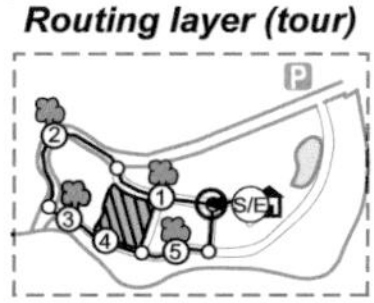	- Starting point + end point - Stations = OOI$_{1..n}$ - Calculated Route = route section$_{1+..+n}$ + decision points	Environmental conditions Tour description
- Where do I have to go next? - How far is it to the target?	User preferences / characteristics	***Routing layer (explore)*** 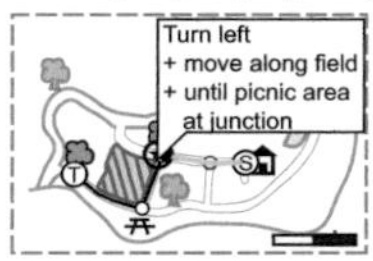	- Adapt Scale - Landmarks - Directions - Scale / distance indicator	Position data
3) Target recognition [Monitoring] - Am I still on track? - Have I reached my target feature?	User behavior User query	***Event layer*** 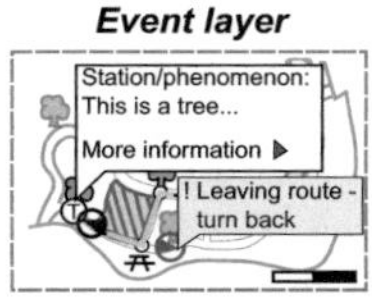	- Fragile zones - Warnings - Target/event info - Link to services	Position data Environmental conditions Visitor management

his current position and the direction that he is headed. Next to the position indicator, which can also display the heading, a tracking indicator should be included in a positioning and tracking layer. The tracking indicator as a trace of the user's position can be visualized through continuous plotting of the user's current position or from the recorded navigation history. A tracking indicator can also help the user to familiarize with the navigator service and with issues of map alignment. The orientation data can further be employed to provide an adaptive map alignment functionality as proposed by the "egocentric maps" concept [300].

Objects of interest layer During the orientation phase, the visitor also wants to get an idea of what he can see in the selected AOI and specifically where he can find the natural features and related interpretive activities that are of interest to him (compare task model in Fig. 3.3). In order to suite this demand the navigator needs to add a layer to the map that visualizes Objects of Interest (OOI). Current mobile map-based applications commonly represent geographic features that are of interest to the user as Points of Interest (POI). POI are defined as geo-referenced point objects, used for the representation of location-based information [125]. This POI standard (see POIX standard by W3C [164] and OpenLS standards in Zipf and Häußler [368]) does, however, not adequately represent features that can not be categorized as point objects due to their larger spatial extent and potentially the lack of a single defined access point [125]. According to Haid et al. [125], such features are regularly found in rural tourism areas and they correspond to the zone objects proposed in the previous section. Haid et al. [125] therefore propose the introduction of a new standard in the form of OOI, that is more suitable for the description of POI as well as Zones of Interest (ZOI). The categories of OOI, which are to be displayed by the navigator service, need to be filtered out based on user preferences and environmental conditions. For nature lovers interested in plants, who visit the area in the spring, the categories may for example include flowering trees, meadows with spring-flowering herbaceous plants and information facilities.

Routes- and routing-layer As a prerequisite for the route selection task, the user needs to know how he can get to a selected OOI. In a route layer the navigator should visualize all routes that connect the starting point (i.e. current user position) with the end point (i.e. location of the OOI) by means of a series of route sections connected by decision points. Which route sections are actually presented also depends on the user characteristics (e.g. physical ability of the user, desired duration of visit etc.) as well as the accessibility of routes based on current environmental conditions and management objectives.
With the selection of a specific route/section the user enters a new phase of the wayfinding task, during which he wants to navigate to the selected OOI. The information needed for this activity needs to be provided via adding a routing layer. Here, the usage mode plays an important role for the functionalities of the navigator service. The map components presented in a routing layer will differ to some extent if a visitor decides to take a guided tour or go on a self-determined

exploration. Defined by the description for the selected guided tour, the service needs to display all stations (i.e. predefined collection of OOI) that are part of the tour and connect them with a predefined route. The choice of tours offered to the visitor once more depends on the user profile as well as the environmental conditions. Once the user has chosen a particular tour and/or route, he can now focus on the part of the map that contains the selected features. Therefore the scale may automatically be adapted to optimally present this section of the map.
In order to offer optimal navigational assistance, apart from the self-location and -orientation indicator, the routing layer should also include routing directions at each decision point, and associated landmarks should be visualized. The form of directions and type of landmarks to be used should also be adaptable to the user's characteristics and preferences.
Additional support can be given by displaying the scale of the map or including explicit distance measures. This extra reassurance is needed as part of the monitoring process and target recognition task, during which the user may repeatedly wonder, if he is still on track and/or if he has already reached his target. In order to prevent the visitor from getting lost and to help him avoid fragile and dangerous zones, the navigator should visualize such zones, depending on environmental conditions as well as visitor management goals. Warning messages should be displayed upon leaving the calculated route.

Event layer Together with other location-based events these warnings should be displayed on a separate event layer. The key location-based information event is of course the arrival at a target feature, which should result in an OOI message being displayed (see Table 3.3). This target info box should include the name and a basic description of the feature as well as a link to more extensive information provided by the Interpreter service. In order to support hand-free operation such event messages should be provided either as audio message or accompanied by an acoustic alert.

Finally, a further functionality that should be integrated in a navigator interface is a status display for the availability and accuracy of positioning data to avoid confusion and improve confidence in the service [187]. If no positioning data is available, the user should be notified. In this case the routing service can either be continued based on alternative positioning mechanisms discussed above or the user can continue to navigate towards his target without route directions relying on the map and his own spatial skills.
Another functionality, not represented in Table 3.3, is the visualization of the landscape relief. The visualization of altitude poses a particular challenge. It is traditionally achieved by the display of contour lines, which are however difficult to interpret for unskilled map readers. On a traditional 2D overview map it can further be achieved by stylistic methods such as hill-shading [177]. A variety of mobile guide systems visualize the relief by a change in perspective (e.g. 3D panorama view or bird-perspective [8, 298]). The elevation information is primarily needed to represent the

slope of a trail, it can also be communicated on a 2D map using symbols indicating steepness or via an elevation profile [52].

3.3.2 Interpreter service

The new interpreter service is responsible for the presentation of location-based, non-geographical information. Since it delivers the actual interpretive content, it can be considered the core instrument for mobile environmental interpretation. More precisely it should be designed, to support the visitor during the interpretation phase of the interpretive unit (compare task model in Fig. 3.3).

Concepts for a mobile interpreter

The conceptual design for the interpreter service also needs deal with the additional issues, linked to the deployment in natural areas. Table 3.4 lists the respective issues and concepts for their resolution.

Remoteness Next to limitations in network coverage, the remoteness of natural areas also results in a limited information infrastructure. Hence in contrast to museums, only a small number of natural phenomena is likely to be explained by labels or even information panels. In order to cope with the potential unreliability of the network infrastructure, the essential interpretive content should reside on the mobile device. This way the mobile nature guide can actually help to compensate the lack of an information infrastructure. Ideally the interpreter service should be able to answer questions that the visitor comes up with while exploring the natural environment and should give feedback to observations made. Due to the complexity of the natural environment, the diversity of features and the probable variability of individual questions such a highly flexible answering mechanism constitutes a major challenge that lies beyond the scope of this thesis.

Compensate unfamiliarity It is again the proposed alienation from nature [45], which results in a fair amount of visitors to natural areas, having difficulties to identify many of the natural features. This is accompanied by a lack of prior knowledge and with it a limited understanding for natural phenomena and processes. Obviously the employment of a mobile nature guide should help to meliorate this unfamiliarity. In accordance with the principles for environmental interpretation introduced in Chapter 1, it is necessary to adapt the presentation to the prior knowledge and level of understanding of these visitors, to grant them a positive experience. In order to make the interpretive presentation easier to comprehend it is generally recommended to illustrate the message with visuals (e.g. images, graphics). Especially for visitors unfamiliar with the area, a visualization and clear description of the respective phenomena is going to be important to facilitate the identification and avoid confusion and frustration.

Make aware of sensitivity One concern closely associated especially with media-based interpretation, is that of the visitors' impact on the respective natural environment, since each interpretive activity will lead to some new impacts [129]. Without doubt, the sensitivity of a natural area also has to be considered in the interpreter service. While the navigator service attempts to prevent the visitor from entering (i.e. having an impact) fragile zones, the interpreter should make this and other conservation measures understandable, enhancing the users concern for these zones. But the interpretive service should not only increase knowledge as well as general environmental literacy of the visitor but also promote responsible environmental behavior on site (e.g. avoid littering, discourage the collection of living plants or animals).

Interpret natural phenomena The service should have the capability of "interpreting" all types of natural features that qualify as natural phenomena suitable for interpretation. According to the proposed model of mobile environmental interpretation, the selection of natural phenomena for the interpretive presentations is still done by the author of the interpretive unit. Both Ham [129] and Ludwig [199] suggest that the planner of an interpretive unit should ideally select environmental features that naturally attract the attention of the visitor. Such outstanding features can include point objects as well as zone objects and can be representatives of the following categories: plants, animals, geological formations, water bodies or cultural history features [129]. A certain similarity between such features and natural landmarks is apparent. Different forms of salience can contribute to attracting attention. Persistence, especially with respect to a fixed location, is a definite benefit but not a must. Informativeness plays a very important, even though different, role. Instead of providing a spatial cue each phenomenon has its own story which it presents to the careful observer in form of a variety of messages [199] (e.g. "The white willow previously used as an example sends out the message:'I can stand with my roots in the water and have survived many floods'."). Which phenomenon is actually selected by the author, and which of its message is used for the interpretive presentation can depend on: its compatibility with an overall theme, its suitability for the location or likely relevance for the visitor as well as its suitability in supporting the overall EE goals.

Adapt to natural dynamics The occurrence of natural dynamics, such as diurnal or seasonal changes, is also of relevance to the interpreter service, at least if the changes affect characteristics of the described phenomenon. One option for the author is to keep the description very general so that it may still fit the phenomenon regardless of changes (e.g. a tree will generally remain in its prominent position on a cliff despite of a change in seasons). This approach, however, only makes partial use of the potential of interpretation and does not work for all phenomena. For an effective interpretation the interpreter service should adapt the presentation to the different dynamics. This, however, also results in a challenge for the author, who may need to prepare multiple content modules for each phenomenon and thematic framework. The application of concepts for potential technical solutions to this challenge, such as "interactive digital storytelling", are discussed in detail by Spierling et al. [320]

Table 3.4: Interpreter service requirements and concepts.

Natural area scenario	 Remoteness	 Unfamiliarity	 Sensitivity	 Natural features
Issues	- Limited network coverage - Limited information infrastructure (labels + panels)	- Difficulty to identifying natural phenomena - Limited understanding + prior knowledge	- Control visitor distribution + behavior	- Single phenomena (e.g. tree, rock, bird) - Phenomena extending over area (e.g. scenic landscapes, lakes)
Con-cepts	- Essential interpretive content on device - Capability to respond to questions about environment + give feedback	- Visualize phenomena + illustrate messages - Adapt presentation to prior knowledge	- Increase understanding for conservation issues - Promote responsible behavior on site	- Enable presentation of phenomena as point objects + zone objects - Selection similar to natural landmarks
Natural area scenario	 Natural dynamics	 Outdoor usage	 Visitors	 Direct experience
Issues	- diurnal and seasonal changes of phenomena - moving phenomena	- Reduced readability due to lighting conditions - Restricted interaction capabilities during task	- Target group - Level of enviro. literacy - Prior knowledge - Interests / preferences - Usage situation - Usage mode	- Exploration of environment with all senses - Playful learning - Deepen emotional experience
Con-cepts	- Generalize presentation or adapt it to dynamics - Incorporate dynamics as content feature (e.g. visualize as animation)	- Adaptation of style and contrast to lighting conditions - Support hands-free usage (e.g. audio presentations)	- User adaptive + adaptable presentation - Different types of interpretation for different usage modes	- Relate to all senses - Provide instructions for hands-on activities - Facilitate perspective taking through stylistic elements

Since these constant dynamics are an inherent part of natural environments, they should also be integrated into an interpretive presentation as content features. Similar to the navigator service the interpreter could also visualize such dynamics for instance by showing animations of these dynamics and their effect on a particular phenomenon, along with an explanation.

Adapt to outdoor usage conditions Constraints resulting from outdoor usage conditions obviously have also an effect on the interpreter service. Potential solutions are very similar to those discussed above for the navigator service. This encompasses an adaptation of style of the presentation (e.g. improve contrast, increase font size, change colors). The presentation of text-based content as audio files should be provided as an alternative with the additional benefit that the user could simultaneously observe the phenomenon and consume the information about it.

Adapt to visitor needs According to the proposed principles of environmental interpretation (see Chapter 1) the interpretive presentation should be relevant (i.e. meaningful and personal) to the participants. This can be achieved by providing a user-adaptive and also adaptable interpretive presentation. The level of detail and the complexity of the presentation should fit the user's state of environmental literacy. Ham [129] for instance suggests mechanisms such as self-referencing and labelling to make the presentation more personal. The collection of content modules (i.e. tours, phenomena), offered to the user should be tailored to his interest and preferences. Research conducted as part of the MobiNaG project has confirmed, that different target groups require different styles of content presentation [94]. For instance students and children favor a large diversity of visual media including photos and animations, while for senior adults, the font size should be increased to improve the readability. As has been shown by the requirements analysis the different users prefer different usage modes for a mobile nature guide. Therefore the interpreter service also needs to offer different types of interpretation such as presentations (e.g. thematic tour, games etc.).

Support direct experience Finally the requirement to support the direct experience of the natural environment must be met especially by the interpreter service. Hence the presentation of a phenomenon should relate to opportunities for experiencing the environment with different senses, afforded by the respective feature (e.g. the bark of an old tree can be observed but also touched and its wood may produce a particular sound). To make the presentation more engaging it should include actual instructions for hands-on activities that provide an opportunity for playful learning and also help to deepen the relationship with the phenomenon. This relationship and ideally the anticipated emotional affinity should be promoted by using stylistic elements that facilitate perspective taking, such as writing the text from the perspective of the phenomenon.

Types of interpretive presentations

In order to optimally support the user during the interpretation phase of his visit a mobile nature guide should adapt the content to the respective usage mode selected by the user. Hence for this thesis a collection of types of interpretive presentations has been compiled which need to be considered in the conceptual design of the interpreter service. Fig. 3.5 presents a variety of interpretive presentation types that are anticipated to match the interpretive activities to be offered by the system.

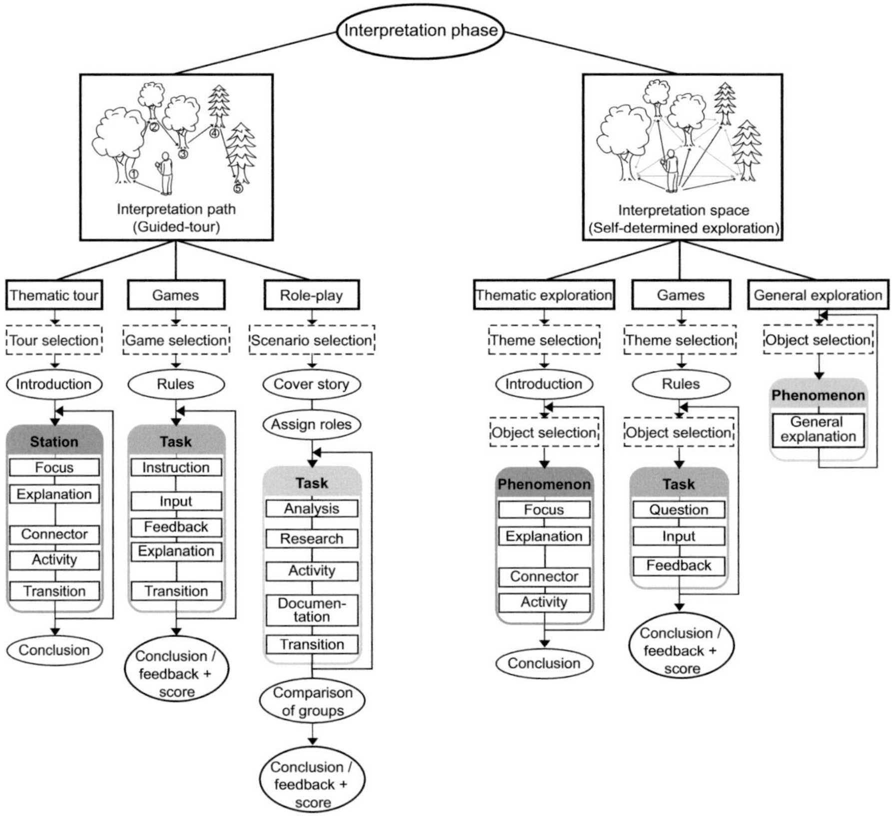

Figure 3.5: Corresponding to the proposed usage modes of a mobile nature guide, the interpreter service should offer different types of interpretive presentations. Content modules related directly to a phenomenon are enclosed in grey boxes. Larger distances between these boxes indicate that more content should be included in these modules.

Which type of interpretive presentation should be given by the mobile nature guide, first depends on the usage mode chosen by the user, as can be observed in Fig. 3.5. If the visitor chooses to engage in an interpretive activity based on a guided-tour (i.e. interpretation path), he will generally seek to follow a specific route to a predefined sequence of phenomena. In this case the thematic-tour, based on the guidelines for self-guided tours proposed by Ham [129], can generally be considered the appropriate type of presentation. Whereas, if the visitor is seeking a more playful and interactive but still sequential interpretive activity, he may prefer a game-based tour. An additional alternative for an interpretive presentation in the guided-tour mode is the "role-play". This is, however, a form of presentation that is envisioned as a kind of role play for larger groups of visitors, preferably school classes.
The visitor who prefers the more independent, self-determined exploration within an interpretation space, can decide between the exploration of phenomena that have a common theme (i.e. thematic exploration) and an exploration of all available phenomena (i.e. general exploration). As a variation of the thematic exploration the visitor can also engage with the phenomena based on a more game-like presentation. These types of presentations mainly differ in the combination and structure of content modules.

Thematic tour The thematic tour consists of an introduction to the selected tour as well as a sequence of stations and ends with a conclusion. Each station is represented by five content modules, which correspond to the interpretive tasks presented in the task model (see Fig. 3.3). The focus module should help the visitor to discover the phenomenon that will be presented. It is followed by a brief explanation of the phenomenon. The connector module should relate the phenomenon and its theme to the overall theme of the tour. This is followed by the activity module, which is supposed to support the visitor during the direct experience with the phenomenon. Finally, the transition module should prepare the visitor to move on and spur his interest in the next phenomenon. The conclusion is presented to the user after he has visited all stations that are part of the tour.

Game-based tour A game-based tour, such as for example a scavenger hunt, should start off with an introductory module that explains the rules and background to the game (e.g. "Many animals in this forest live on fruit from different plants. On this tour we will look for such plants and you will have to decide which animal eats these fruit."). Thus throughout the tour the user is confronted with a series of tasks associated with the different phenomena. At the beginning of each task an instruction module is presented, informing the user what he has to do or presenting a question he has to answer. The user can enter or select his solution via an input module. From an EE perspective, it is important that the user will receive an immediate feedback on his performance given in a consecutive feedback module. This again is followed by a more extensive explanation of the phenomenon and the correct solution to the task. The transition, once again, gives the user a preview of the next task. A game-based tour is ended with a conclusion module, which should give the user feedback about his overall performance (e.g. score).

Role-play In the field of game-based pedagogy a number of different situated-learning approaches have been proposed in the 1990s, among them Problem-based Learning (PBL) [20,292], Anchored Instruction [72] and Goal-based Scenarios [295]. It is their objective to provide students with an authentic problem to motivate organized learning in collaborative groups and an opportunity to practice and apply acquired knowledge as well as problem solving skills. The role-play concept is founded on these approaches. Similar to the "Environmental detectives" project [321] (see Chapter 1), groups of students should use mobile nature guides to explore the natural environment and collaboratively solve an environmental problem based on a realistic scenario. The student groups are supposed to take on the role of a stakeholder in the scenario and use the tools, provided by the mobile nature guide, to accomplish a certain task. Initially the students should be presented with a cover story describing the scenario. They are assigned to the different stakeholder roles afterwards and presented with a series of tasks. The first content module within the task should be attributed to assisting the group with the analysis of the problem. The next module should provide access to background information as a basis for researching problem solving strategy. The activity module should offer tools and information to complete the task. Participants should also be encouraged to document their results for later discussion, this requires an additional interactive module that allows for the input of information. Individual tasks are again bridged by a transition module. The comparison of the results should grant a kind of synthesis of the experiences of different groups and allow an analysis of the overall performance of the groups, plus the presentation of a solution. Ideally, the results should be compiled and stored on a server so that the comparison can also be conducted in the classroom later on. The role-play presentation has been assigned to the "guide-tour" group in Fig. 3.5 assuming that most of the tasks will be tied to different phenomena and will be addressed in a defined order, potentially also building on each other. Still this predefined order is not necessarily mandatory thus a role-play presentation could also be envisioned in the self-determine exploration mode.

Thematic exploration In the group of self-determined exploration presentations, the thematic exploration mirrors the thematic tour in the guided-tour group. Yet, in this case the visitor initially selects a theme and, following the introduction, picks an OOI from a list of phenomena that are connected to the selected theme. The presentation of the phenomenon is otherwise similar to that of a station except that the transition module is missing. Since the phenomena can be visited at random, there can be no reference to a specific consecutive phenomenon before the user has made a new selection.

Game-based exploration The same is true for the game-based, self-determined exploration. The user can select phenomena without being restricted to a logical order. Fig. 3.5 presents a simplified version of an exploration game such as a quiz, with the task consisting of only three modules. Upon arrival at a selected phenomenon the user is presented with a question, which he can then answer via the input module. As an additional option, the user could be granted access to background information that may be of help in answering the question. However, with the requirement for direct experience in mind, it seams reasonable to devise questions that can be answered based on careful observation and experience. The unit is completed with a feedback module including the correct answer along with a brief explanation and an encouraging comment on the performance of the user.

General exploration The most basic form of location-based information presentation is represented in form of the "general exploration". In this case the visitor can simply roam the AOI and select phenomena that are of interest to him, independent of a theme and without chronological order. This implies, that the user receives neither a content related introduction nor a conclusion. The presentation of the phenomenon cannot be based on the principles of environmental interpretation but consists of a general description and explanation of the phenomenon.
It should further be noted that Fig. 3.5 does not represent an exclusive list of presentation types. Other combinations and structures are principally conceivable.

Interpreter service components and functionalities

In accordance with the task model (see Fig. 3.3), this thesis presents a scenario for an interpretive unit (see Table 3.5). Corresponding to the structure of a thematic tour, introduced in Fig. 3.5, the table illustrates the presentation modules that the service should offer to assist the user with the respective tasks of the interpretation phase. It further demonstrates which components are needed for each module, and which data is necessary for its adaptation and the presentation.

In an initial step the visitor has to select a tour. The interpreter service should present the visitor with a list of tours, adapted to his interests and preferences as well as to the environmental conditions. For example the hobby botanist, visiting in spring time, should be presented with a list of tours on plants that are especially interesting during this season. Next to the list of available tours, the visitor needs some additional information on the tour, to make his choice. Using information from the tour profile, the service should provide the visitor with a preview of the tour, including a sketch of the route layout as well as an abstract of the tour (i.e. brief summary, length and difficulty). Following the selection of the tour the visitor is presented with the introduction to the interpretive unit. Using the identifier for the specific tour (i.e. Tour_ID) the interpreter is expected to load the tour description and display the introduction module. This should contain the overall theme for the tour as well as various media files with the content of the module. The user should be provided with a choice of text and/or audio files accompanied by different forms of visuals. The presentation should be adapted to the user's characteristics and preferences. After providing the visitor with an overview of the tour and spurring his interest in the tour, the interpreter service should link the user to the navigator service. Employing the tour identifier, this service should help the user to find his way to the first station of the tour. Upon arrival at the station, the navigator should handover the OOI identifier (OOI_ID) of this station to the interpreter service. Based on the OOI_ID the service can extract the station information from the tour description.

Table 3.5: Interpreter service components based on the example of a thematic tour.

User tasks	User context + adaptation	Presentation module	Interpreter components	Context + content sources
A) Tour selection - What is the tour about? - How long is the tour going to be? - How difficult is the tour going to be?	User interests + preferences User selection	***Selection + preview*** Tours: Wooden Helpers; Animal Apartments; Mountains are not for.. Wooden Helpers Summary: ... Length: 1,5... Level: med...	- Tour list - Themes - Logos - Tour previews - Map image - Tour abstract	Environmental conditions Categories of tours Tour profile
B) Familiarize - What is there to explore? - Why should I participate?	User preferences / characteristics	***Introduction*** **Wooden Helpers** On this tour we will explore how trees fulfill important roles in their habitats and also for ... Forest ...	- Tour_ID - Overall theme - Introduction module - Text / audio - Visuals	Tour description Media files
		Tour_ID **Navigator service** OOI_ID / Station		
1) Discovery - What can I see here? - What is special about this phenomenon?	User preferences / characteristics	***Focus*** **Wet Feet** What kind of tree is this? This is a white willow tree, who likes wet "feet", that is why ... White w...	- OOI_ID - Title - Focus module - Text / audio - Visuals - image of phenomenon	Environmental conditions System resources Media files
2) Understanding - How does it work? - Why is it located here? - I want to know more	User preferences / characteristics User query	***Explanation*** **The Root Trick** When the white willow is flooded it uses a trick to supply its roots with oxygen. ... Flooded ...	- Theme - Explanation module - Text / audio - Visuals - Animation, close up image, graphs	Environmental conditions System resources Media files
		Term / OOI_ID **Encyclopedia service**		
3) Connection - What does it have to do with the tour theme? - Why is this phenomenon important for me?	User preferences / characteristics	***Connector*** **Living Flood Protection** Like other trees the white willow is a wooden helper. It stabilizes the river bank ... Floods ...	- Title - Connector module - Text / audio - Visuals	Environmental conditions System resources Media files
4) Activity - How can I explore/experience the phenomenon directly?	User preferences / characteristics	***Activity*** **Feel the Current** Experience the current yourself: 1) Carefully approach the water ... Experienc.	- Title - Activity module - Text / audio - Visuals - Animations/ video with instructions	Visitor management Environmental conditions System resources Media files
		Task / Tool Feedback **Exploration tools**		
5) Transition - What am I going to see next?	User preferences / characteristics	***Transition*** **First Aid in the Wild** We have seen how trees can help to ...	- Title - Activity module - Text / audio - Visuals	Tour description Environmental conditions Media files
		Tour_ID **Navigator service** OOI_ID ***Station*** $_{2..n}$ **Navigator service**		
C) Conclude - What is the message? -What can I do?	User preferences / characteristics	***Conclusion*** **Endangered Helpers** Trees are wooden helpers that fulfill ...	- Title - Conclusion module - Text / audio - Visuals	Environmental conditions Media files

The first module is the focus module that should assist the user during the discovery task. The module includes a text and/or audio component accompanied by a pictorial medium. This is basically true for all following presentation modules offered by the interpreter service. In oder to meet the proposed requirements for the interpreter service, the style of the presentation (i.e. font-size, color, contrast etc.) as well as the style of the content (i.e. length and complexity of the text, type of image based media etc.) need to be adapted to the preferences and characteristics of the user as well as the environmental conditions. For the selection of media being used the system resources available for the application (i.e. processing power, battery state etc.) also have to be taken into consideration.

A difference between the presentation modules exists with regard to the image-based media or visuals that should be used to best support the individual tasks. As a general rule the visual aid should always correspond to and illustrate the information given in the textual component of the module [129]. For the focus module it is recommended to use a photograph or other image of the OOI that helps the user to identify the specific phenomenon. In the explanation module a variety of pictorial media can be used to support the understanding of the phenomenon. Visual aids are significant since they generate associations and can also trigger certain emotions, which help to promote a positive experience and hence also deepen the emotional relationship. For the explanation of process-based phenomena, animations can be particularly helpful. The same applies for phenomena that cannot be directly perceived by human senses, such as features at a microscopic scale. Small scale phenomena can also be visualized via close-up pictures. The comparisons of features, may be an important instrument to make the interpretation more meaningful. It can be supplemented by graphics that show the elements of comparison next to each other. For the connection task a picture related to the overall theme or to some aspect of the user's daily life should generally suffice. For the activity task, the visual aid can be particularly significant to clarify the instructions to the activity. A video or animation could for instance be included as a kind of tutorial. The visual used in the transition module should of course relate to the succeeding phenomenon. Finally the image-based medium in the conclusion module should be an inspiring and memorable image, which should visualize the underlying message of the tour.

Next to the interaction with the navigator service, for navigational assistance in-between stations, the interpreter should have the capability to interact with other services, such as the encyclopedia and the exploration tools. Particularly while using the explanation module, the user may query the system for more detailed information on the phenomenon or the clarification of specific terms. This request is then passed on to the encyclopedia service that should provide a definition or more extensive information on the respective topic. The activity module should primarily provide instructions for a direct experience of the phenomenon. Besides these instructions, such an activity can further be assisted or entirely build on sensor instruments or virtual tools furnished by an exploration tools service. The exploration tool can, in return, give feedback to the interpreter service regarding the results of the performed activity. Of course all instructions for activities need to be adapted to the given visitor management goals. A further tool that should be accessible

from the modules of the interpreter service, is a documentation tool or diary service, allowing the user to record and annotate a personal log of their visit.
The interaction between these services can also be of particular relevance in case of the role-play and game-based presentation types. It has been shown by Fig. 3.5 that these presentation types further use additional modules, not represented in the thematic tour scenario. Especially the input module requires additional components, like a form element, which allows input as text or via checkbox or radio button. The feedback module may also call for additional components such as a score indicator, which calculates and displays the user's performance.
The connector and the transition module, which have not been pointed out explicitly, rely on the components already introduced for the other modules. This is also true for the conclusion module which ends the thematic tour after the cycle of station modules is completed for all stations.

The interpreter service can in general provide only relatively brief content parts for each presentation module. For one the display size of most mobile devices is relatively small, thus for instance reading long texts requires a lot of scrolling, which is frequently cumbersome with the limited interaction modalities provided. Yet, even if the hardware related limitations may be overcome e.g. by audio presentations or gesture input [207], there are also pedagogical guidelines, which recommend the confinement of the length of the presentation. Ham [129] for instance states that a self-guided walking trail taking 45 minutes is considered a long trail. He further suggests that the text to each unit should not exceed 60 words. Consequently visitors who desire more detailed and extensive information on a phenomenon or environmental issue have to be supported by the encyclopedia service.

3.3.3 Encyclopedia service

The encyclopedia service should help the visitor with two kinds of tasks. One of them is to assist the user in understanding the presentation by giving him an explanation for unknown terms. If the user has difficulties understanding a content related term like the word "oxygen", selected as an example in Fig. 3.6, he should be able to trigger the encyclopedia service. This pull service should present an explanation of the term to the visitor based on a glossary of environmental terms. The other task is to supply the visitor with more extensive information on a particular topic.

Sources of additional information One crucial issue is that of the source of additional information. It can either be generated specifically for the mobile nature guide by the content author or it can make use of other, already existing, information resources. The benefit of the former is that the style of the information can be consistent with the rest of the presentation and the information can also be tailored to the specific target groups. Further, the author has also control over the content in the sense that he can match the glossary entries and topic to those actually used in the

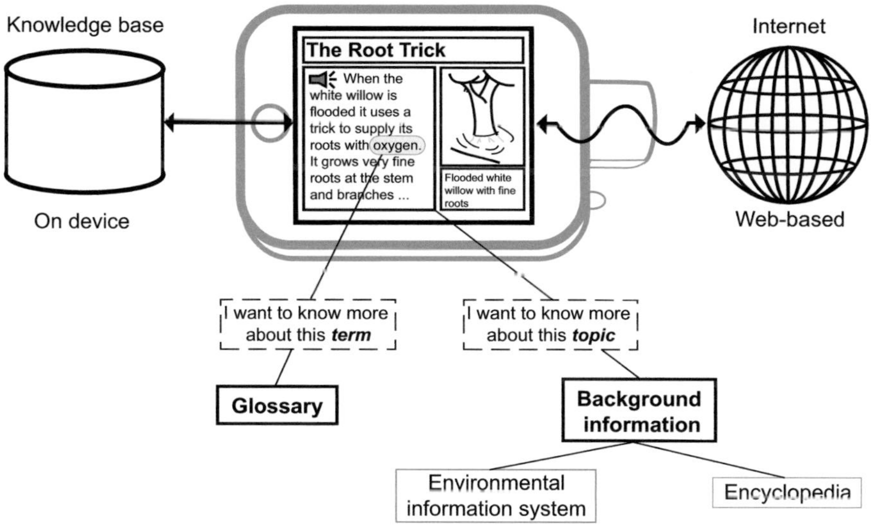

Figure 3.6: The encyclopedia service should provide the user with more detailed information on topics presented by the interpreter service and explanations of terms. This information can reside on the device or it can be accessed via a wireless network/internet connection.

system. In addition the update of the resources can be synchronized with changes to the presentation. Yet, particularly the maintenance and the initial creation of the comprehensive resources may cause a considerable effort for the author. External resources could include environmental information systems (EIS) of various kinds and digital encyclopedias on a commercial as well as a public domain basis. These sources may be less target group specific than desired, but they will likely include a wider spectrum of information. Furthermore, they can be included with less effort and are being maintained by the provider or community of these resources.

Access to additional information There are two approaches to supply this additional information. The content source can be maintained on the device, as part of the knowledge-base, or it can be accessed via a network connection on a remote server, for example on the internet. Since the storage capacity of mobile devices is generally still restricted, it seems reasonable to opt for a server-based or web-based solution. Still the limitations to network connectivity in remote areas could make such an entirely network-based service unreliable. As a consequence, a hybrid solution may be preferable. In this case an encyclopedia application or package could be downloaded, onto the personal device from a desktop PC, prior to the visit. Another option would be the download of an encyclopedia package adapted to the tour content during periods when network connectivity is available.

For an encyclopedia service of a mobile nature guide the following combination of the discussed

options is considered most suitable. Information that is considered essential for the understanding of the interpretive presentation should reside on the mobile device. This criterion should mainly apply to glossary entries. The terms used in an interpretive presentation could be entered into the knowledge base by the author as part of the content generation process. As an alternative these terms could be filtered from an existing environmental glossary and then loaded onto the device as an exchangeable glossary content package. The provision of more detail information (i.e. background information), which also requires considerably more storage capacity, is considered an auxiliary service, which does not necessarily have to be accessible at all times. Thus for making background information available, a remote access solution is considered feasible. It has already been mentioned that the collaboration with existing web-based information resources may even open up further options. For instance the access to web-based EIS would make a wide variety of environmental information accessible to the visitors. This could include resources of various levels of detail that are also tailored to different target groups. Lay visitors could for example use information from public EIS like the "Theme Park Environment" system of the State of Baden-Württemberg [282]. Naturalists may even be interested in expert information offered by expert EIS such as the "XfaWeb" of the State of Baden-Württemberg [115]. Next to EIS, the utilization of web-based encyclopedias such as "Wikipedia" of the Wikimedia projects [361], should be considered as well. This or similar public domain platforms have the benefit that they can be integrated into the host system on a collaborative basis. Such an integration approach has for example been realized in public EIS like the Theme Park Environment system [92, 93, 115, 116]. Furthermore the open source approach grants the author at least a certain amount of control over the content of the encyclopedia and he can contribute actively to updates and the maintenance of the resources. The fact that there is no additional cost tied to the usage of public domain systems, makes them even more appealing for EE institutions with tight budgets.
An additional issue associated with the integration of external web-based information resources is related to the style of the presentation. Web-based information systems are generally designed for desktop PC, which leads to restricted readability on mobile devices with small displays. This problem may be accompanied by long download times depending on the network connection. This layout or stylistic problem may, however, be solved by using presentation templates specifically designed for viewing the information on mobile devices, such as for example the "Wapedia" for Wikipedia. Another solution could be the application of a migration environment as proposed by Bandelloni et al. [18], which dynamically converts the presentation of content and interface elements of a web-based system to best suit different target platforms.

Finally the ultimate but more complex version of an encyclopedia service should offer the user a wizard interface that can provide answers to any content related question the user may have. The implementation of such a functionality, however, poses a number of challenges, among others related to issues of input modalities and the processing of natural language questions.

3.3.4 Exploration tool service

As has been mentioned afore, the capabilities of a mobile nature guide should be used to offer the visitor additional tools to promote and enhance the direct experience of the natural environment. Thus an exploration tool service should support the following tasks:

Perception The tool service is expected to open up new dimensions and perspectives on the selected features. This implies making phenomena perceivable that usually can at least not be readily sensed by the visitor [275].

Identification Even if a phenomenon is perceived by a visitor, he still needs to be able to identify it in order for it to become meaningful and comprehensible. This is for example the case for unknown species of plants, fungi and animals but also for land forms and geological formations.

Documentation Tourists commonly seek to memorize their experience by taking some form of record based on photographs and/or videos or by compiling a travel diary [2]. This is also true for visitors to natural areas as indicated by the requirements analysis. So the tools service should also offer an opportunity for the collection of souvenirs.

Also the experienced human interpreter will usually pack some additional gear that will help to make his interpretive presentation more interactive and engaging [129]. Spare time naturalists frequently carry similar equipment that will be of assistance during exploration. Fig. 3.7 presents a collection of such utensils that should serve as a model for the new mobile nature guide tool service.

Perception tools Several utensils can be applied to make phenomena perceivable. The visual experience can for example be enhanced by a pocket-lens, which can help to visualize small scale objects such as insects or particular details of a flower. Objects in a distance may only be vaguely perceived. But by using binoculars such objects, like a bird and its nest high up in a tree, may be detected clearly. A digital camera equipped with a macro- and a tele-lens can potentially serve the same purpose. Some acoustic phenomena may not be perceivable as well, either because they are emitted in a frequency spectrum that lies outside the one hearable by humans (e.g. ultrasound calls of a bat) or they only occur irregularly or during certain times of the day (e.g. chirping of a cricket or the rustling of leaves in the wind). Such phenomena can for instance be made audible by replaying a sample from an audio recorder. Other tools can be used to measure environmental conditions in order to make them perceivable or easier to comprehend. A tape-measure for instance can be used to measure the dimensions of certain features or distances between certain objects, making them easier to compare to similar objects or familiar features. The readings of a

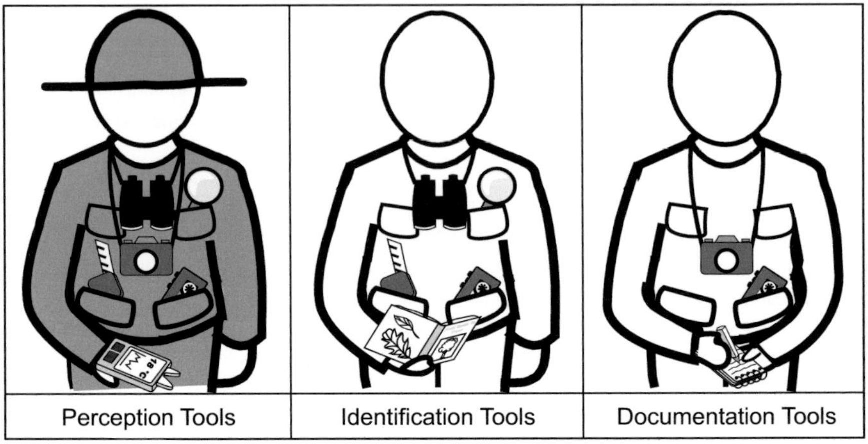

Figure 3.7: Collection of utensils that traditionally support interpreters (left) and spare-time naturalists (middle and right) during the perception (perception tools ↓: pocket-lens, binoculars, camera, cassette player, tape-measure, temperature probe), identification (identification tools ↓: pocket-lens, binoculars, audio recorder, tape-measure, field guide) and documentation tasks (documentation tools ↓: camera, audio recorder, scrapbook), while visiting a natural area.

thermometer or a pH-probe make it possible to visualize these conditions and thus making them easier to comprehend, which is the basis for the understanding of certain processes.

There are different ways to integrate such perception tools into a mobile nature guide. On the one hand the mobile device can be equipped with or connected to additional, physical sensors that collect environmental data from the real environment. On the other hand the service can be entirely based on virtual instruments that present pre-recorded data or information from a virtual environment based on the internal environmental model.

With respect to the visualization of such a service, a form of Mixed Reality (MR) display seems a reasonable choice for a tool that is supposed to enhance the user's perception of his environment. MR refers to a subclass of Virtual Reality (VR) related technologies that involve the merging of real and virtual worlds [224]. According to the conventionally held view, a VR environment is one, in which the participant-observer is totally immersed in and able to interact with a completely synthetic world. The real environment, as the opposite extreme, consists solely of real objects (i.e. objects that have an actual and objective existence), and includes for example what is perceived in direct observation or what can be viewed of a real-world scene via a conventional video display [224]. The virtuality continuum in-between the two extremes is covered by different facets of MR (see Appendix B.3). Both facets of MR, Augmented Reality (AR) and Augmented Virtuality (AV), can potentially be applied in a mobile guide system. However, as Mountain et al. [228] point out the particularity of the mobile guide system is that the user is always present in the real environment and should either directly or indirectly see the real world. Even in a partially immersive AV environment, the user could see both the real world scene around him, and a computer generated scene on the device. Research investigating the application of MR displays on mobile guide systems was conducted for example by Chittaro and Burigat [63] and Mountain et al. [228]. Mobile augmented reality systems, especially in the field of environmental education and tourism, were investigated by Fisher et al. as part of the "Wearable Environmental Media project (WEM)" [105] as well as by Rogers et al. in the "Ambient Wood" project [275] and Okada et al. in the "Digital Environmental Education" (DigitalEE) project [243]. For more information on these systems see Appendix B.3.

It is also the goal for a perception tool of a mobile nature guide, to augment the user's perception of the environment with computer generated or real data. Taking the mentioned projects into consideration, it seems reasonable to also apply MR technologies for the realization of the perception tool.
Since it is the primary goal of this service to promote the direct exploration of the natural (i.e. real) environment, it should be based on an AR display that still allows direct viewing of the natural phenomena. For many of the presented tools, the most sophisticated solution would be granted by a head-mounted display (HMD) with see-through capabilities, where the additional information can be directly superimposed onto the real-world scene. But according to the documentation of the WEM project [105], the necessary set up still requires a considerable amount of high-priced equipment. Thus for the time being this is not likely to be a feasible solution, neither for EE institutions nor for the average spare time naturalist. A less complex and more affordable solution could be the selection of a non-immersive hand-held AR display similar to the PDA solution applied in the Ambient Wood project [275]. The user would remain immersed in the real environment but could use the mobile device as a window to an augmented environment (as displayed in Fig. 3.8).
Similar to the modes proposed by Rogers et al. [275], a tool service of a mobile nature guide should offer both user-initiated and environmentally-initiated digital augmentation. Which mode will actually be available, will likely depend on the type of interpretive presentation that the user participates in. As part of a thematic tour it will mostly depend on the station (i.e. environment), which tool can be used for a proposed activity. During a self-determined general exploration the user may theoretically choose freely when and where to use a perception tool. Some digital augmentation tools may, however, be location-based thus constraining the users free choice.
With respect to the applied media or types of augmentation, the perception tool should make use of real data but potentially also of virtual data. The information can either be recorded live by physical sensor (e.g. probe, camera) or it may be pre-recorded and displayed on demand or following an event. While live data is assumed to be always a real documentation of environmental conditions (e.g. temperature reading or close-up digital photo of a flower), pre-recorded information can either be real or also based on computer generated virtual data (e.g. simulations of a flooding event, an ice-age scenery at the location). Regarding the usage of media, acoustic or visual, it has already been stressed by Rogers et al. [275] that digital augmentation should not

overload the user with information and consequently distract him from the exploration of the real environment. It should rather provoke and trigger him to think of how it relates to what is around him and help him to comprehend this experience.
The pocket-lens as well as the binocular tool could be based on pre-recorded images of particular features. Ideally, to provide a realistic presentation, these images are bound to the specific location of the recorded feature. If the mobile device is equipped with a digital camera, the system could be turned into a "video see-through" display, and these two optical tools can also be based on live images using the zooming capabilities of the camera. A tool to make acoustic phenomena audible requires an auditory display that presents pre-recorded, real sounds or phenomena, which have to be transformed to be audible (e.g. ultrasound) as well as entirely synthetic sounds natural processes such as "root uptake". An audio player integrated in the device could be used for this tool. The auditory display could potentially also rely on live recordings. This would, however, only make sense, if non-audible phenomena could be recorded and transformed near simultaneously, which would likely be constrained by the processing power of the device as well as the sensitivity of the microphone. A tape-measure tool to determine distances and dimensions could be offered in form of a pre-recorded parameter set of particular objects. It would of course be more sophisticated to use an on-site calculation of distances and dimensions, based on image processing algorithms. The user should be able to mark a distance on the "video see-through" display and the system should be able to calculate its actual length based on the camera zoom factor and the pitch of the device. Other measurement tools for perceiving environmental conditions (e.g. humidity, temperature) can either be simulated as virtual instruments or they can be based on live data from physical probes. Virtual instruments could for instance rely on average values from prerecorded long-term data sets. If possible the use of actual sensors, providing live readings of environmental conditions, would of course be more satisfying and compelling for the user. This data could be received either from probes directly connected or integrated in the mobile device or from a sensor-network distributed in the natural area [105, 318].

The capability to superimpose information onto the real environment could reveal new possibilities for the interpretive potential of a mobile guide. It would make it possible to offer tools that go beyond mimicking existing tools and could potentially even exceed the abilities of the human guide with respect to the visualization of "invisible" natural processes. One possibility would be the addition of "augmented labels" that would help the detection of hidden phenomena and also facilitate the discovery task. With respect to hidden natural processes the "X-ray vision," envisaged by Abowed et al. [2], could be realized and grant the user a virtual "insight" into natural phenomena, such as a tree, to observe the transport mechanisms and facilitate the comprehension of the transpiration process like displayed in Fig. 3.8.
Furthermore, for a game-based interpretive presentation an augmented reality display can also offer additional possibilities for playful interaction with the natural environment and at the same time avoiding detrimental effects on the natural features. Such MR games are already being employed as entertainment feature on mobile phones like for example "Kick-Real" by c-lab/Siemens [274].

Figure 3.8: Based on a "video see-through" hand-held display the visitor's view of the environment (here a tree) can be superimposed with "augmented labels" and a "x-ray-vision" tool can augment the visual perception to reveal the vascular tissue inside the maple tree, which transports water from the roots to the canopy (i.e. xylem) and the tissue that transports soluble nutrients from the leaves down to other plant organs (i.e. phloem).

It is assumed that they can also be utilized on mobile nature guides, helping to build a bridge between the real natural environment and the virtual environment.

Identification tools Next to making natural phenomena perceivable and granting additional playful interaction, the proposed AR perception tool should ideally also help the user to identify specific phenomena. This could for instance be achieved by adding "augmented labels" to the phenomena. Yet, augmented labels are at the current state likely to be limited to OOIs, documented in the environmental model of the system. In order to adequately support the visitor in the truly self-determined exploration of the environment an additional identification tool should be provided. This corresponds to the results of the requirements analysis, which document the visitors' desire to identify animals and plants.
Especially the identification of biological species is a significant element of wildlife watching, which (in particular birding) has become a popular spare time activity. The ability to identify an observed organism and, based on its name, learn more about it, encourages people to reconnect with nature and therefore to value it [328]. The naming and classification of species has tradi-

tionally been the undertaking of the biological discipline of taxonomy [56]. Several methods are available for aiding this process of identification (i.e. finding the correct taxonomic unit [73]) (compare review by Pankhurst [247]). The most important methods are conventional identification keys (e.g. traditional dichotomous keys) and interactive keys. The usage and constraints of these conventional keys is discussed by Dallwitz et al. [73] and Stevenson et al. [328]. It is stressed that taxonomic keys are notoriously difficult to use, and primarily applied by experts. An entirely different approach was developed for the public: the field guide. This is basically a browseable picture guide, that translates taxonomic information into a form that the public can use [328].
Some of the constraints of the conventional approaches may be overcome by new forms of computer-based identification tools. These include interactive keys as taxonomic expert systems. An interactive key is an interactive computer program, in which the user enters attributes (i.e. character-state values) of the specimen, usually based on a more flexible synoptic key instead of the hierarchical dichotomous key. The program eliminates taxa whose attributes do not match those of the specimen. This process is continued until only one taxon remains [73]. The features and advantages of interactive keys as well as examples for different software packages are extensively discussed by Dallwitz et al. [73] as well as Wilson [363]. However, the basic difficulties for non-experts with using taxonomic tools persist. One of the issues relates to the utilization of technical language that facilitates the precise description or biological objects, but forms a barrier for non-experts. Next to the expert systems also computer-based versions of field guides have been developed recently. Stevenson et al. [328] provide a comprehensive review of "electronic field guides". Such electronic field guides are available in form of a variety of software types such as CD-ROM, stand-alone media applications for desktop computers as well as web-based systems that either use a specific client application or a web-browser. Some systems have been developed for mobile devices as well. Either as stand-alone application (e.g. Birds-Canada) or as web-based applications that can be accessed via mobile devices (e.g. the Electronic Field Guide System (EFG) [327]). According to the review, important advantages of these electronic field guides are, that they allow more flexibility in their production than paper-based field guides. In addition they allow an efficient updating of the information. Furthermore, they can offer the information based on a larger variety of media, including sounds and movies. Plus the integration of interactive training tools that help to increase familiarity while avoiding a cognitive overload and confusion. Moreover, they grant the ability to conveniently compare similar species dynamically, side by side. The fundamental benefit though of electronic field guides is that they can tailor their presentations to the preferences, expertise and current situation of the user, which includes:

- Adaptation of information display to skill, experience and age
- Pre-selection of specific list of taxa based on environmental conditions (geographical area, season etc.)
- Error tolerance and log of identification process

- Advice on further queries to make identification more reliable.

Accordingly, especially context-awareness can be assumed a major benefit of mobile electronic field guides and they can thus be considered a natural addition to a mobile nature guide. For the design of electronic field guides Stevenson et al. [328] generally suggest, that a user friendly field guide should be primarily based on visuals and should use step-by-step point-and-click choices. The number of identification decisions should be limited to as few as possible (approx. 1-6 levels). Appropriate characters to be used should include size, color, and unusual features. A challenge, especially with regard to interaction design, is of course the presentation of the field guide information on a small hand-held display. A good example for an intuitive solution is given by the Cyber Tracker project [34].

The options for the integration of an electronic field guide into the mobile nature guide correspond by and large to those proposed for the encyclopedia service. An electronic field guide or identification service could reside entirely on the mobile device (i.e. "stand-alone") and be thus always available, independent of variabilities in network coverage. Keeping in mind though, that the electronic field guide content is to a large extent based on images and other multi-media information for a potentially vast amount of species (e.g. Peterson Field Guide to Eastern Trees of the United States comprises more than 400 species alone [255]), it can be assumed that the limited storage capacity of the device will rapidly be exceeded. This limitation could for instance be compensated by providing packages of identification data specific to the AOI, which can be downloaded, as a plug-in for the knowledge base, prior to the visit. The alternative solution for the integration of an electronic field guide would be to furnish an access to web-based field guide resources via a wireless network. This could either be realized as a hybrid solution, with a specific electronic field guide client on the device requesting content on demand, or by providing access to various field guide resources on the internet using a web-browser. A stand-alone as well as a hybrid solution could mean a better integration into the overall design of a mobile guide as well as a high compatibility with the other services. This implies the preparation of a species list that is specific to the AOI and complies with the interpretive presentations. The compatibility with other components, especially the perception tools could provide a crucial benefit, since some tools like the pocket-lens, binoculars, tape-measure and audio player can be helpful instruments for the identification process. But a challenge of this approach is the generation as well as maintenance of electronic field guide content. This content commonly consists of at least a simple, broad and shallow key and various species accounts based on illustrations as well as text and audio descriptions [328]. The design of this content usually requires a fair amount of experience and considerable expertise about the specific taxonomic group. In this respect an external, web-based solution would be preferable. Similar to the concept for the encyclopedia service, this approach could include the integration of community platforms and thus would allow to benefit from the contributions of the community of professional taxonomists and naturalists. Stevenson et al. [328] take the community approach one step further and suggest the adoption of the "citizen science" paradigm for electronic field guides. This corresponds to the mobile environmental citizen science model proposed in this thesis as an extension of the mobile environmental interpretation model. The prerequisite for this approach is, however, that the identification tool service also allows the user, next to the identification of a feature, to record his findings and submit them to the community platform. The CyberTracker project should be named as a remarkable example for involving citizens in the mobile collection of biodiversity data [34]. In any case the implementation of the identification tool needs to take emerging standards in the field of biodiversity informatics into account. These include for example the electronic interchange standards, which are being established by the Taxonomic Database Working Group (TDWG) and Integrated Taxonomic Information System (ITIS) [328]. On a futuristic note, one can envision the merging of the MR perception tool and the identification tool, including the automated identification of specimens. Research on a mobile augmented reality system for botanical species identification conducted by White et al. [360] as well as a review on automated species identification by Gaston and O'Neil [114] show that initial work to realize this vision is on its way.

Documentation tool Taking records is not only relevant for collecting biodiversity data, but visitors commonly also want to document what they have observed and experienced. The preceding review of existing mobile guide systems has shown, that a considerable number of mobile guides offers some form of documentation service in form of a personal tour diary, public notes or annotations as well as digital souvenirs or a personal brochure of the presented content. A documentation tool for a mobile nature guide should serve the following purposes:

- Collection of memories and sharing with others
- Post-visit exploration, elaboration and reflection
- Bookmarking and recommendation.

It is a common desire of tourists to collect or buy souvenirs that will serve as a reminder of their visit after their return. In addition to souvenirs, they also take photographs or buy postcards that help them to memorize a place and assist them in sharing these experiences with others at home [50]. This "taking the visit back home" aspect is also an important element of ecotourism and EE, because it supports the establishment of an emotional binding with the environment and the lasting attribution of a positive value to it. This tourist habit can, however, also have its downsides. While the shooting of photographs and video is usually less intrusive, the urge to collect souvenirs can have a negative impact on the natural environment. The removal of plants, animals and rocks by herds of visitors can in total endanger local populations as well as ecosystems and is thus usually discouraged by park managers. Therefore, the documentation tool should encourage and enable the visitor to collect digital souvenirs. This could for instance be achieved by providing the visitor with a digital camera tool that allows him to take photographs or record videos.

In addition the user should be able to take written notes and/or make audio recordings either of personal comments or in order to document acoustic phenomena.
Besides souvenirs, the visitors may also be interested in taking home certain content parts from the interpretive presentation. Not all of the textual and visual information may be entirely consumed or comprehended during the visit, and some people may want to go through the information once more later on. Especially parents and teachers are frequently interested in a post-visit exploration and elaboration with their children or students. This is also an effect anticipated by educators since the repeated use of the information can lead to an increase in knowledge and provides the opportunity to reflect on the theme and the experience, which may induce a change in environmental literacy. Therefore, the user should be able to select elements of the interpretive presentation, which he wants to add to his diary of the visit. He should further be able to link personal annotations to these content parts.
In addition to taking personal notes, visitors may also be interested in public annotations of certain features. Especially first time visitors frequently appreciate recommendations of features or tours that people with similar interests may have enjoyed. The application of this kind of social navigation approach [143] may, however, be only of restricted use for a mobile nature guide. It may serve its purpose in particular for the self-determined usage mode, where the degree of guidance is minimized. In case of the guided tour mode, though, managers and educators prefer to be in control of the presentation. On the one hand they may want to preserve the interpretive presentation as a consistent unit and on the other hand they may want to preserve the "pureness" of the environment also as an information space and avoid an uncontrolled information overload. Yet, this public annotations instrument gains in importance in the light of the proposed mobile citizen science approach, that explicitly involves the visitor in the design of new content. Especially for regular visitors it may be of interest to leave also geographical bookmarks as suggested in the WebPark project [95], which would make it easier to resume a previously interrupted interpretive activity.

Digital souvenirs and content parts selected for post-visit exploration can be combined to a diary. Such a diary can basically be considered a temporal and spatial log supplemented by multi-media content, which is generated by the user. Consequently like it is shown in Fig. 3.9, the core elements of each diary entry are the current location based on positioning data and the current time based on the system clock. Digital souvenirs and notes can then be added to the time stamp and geographic coordinates. A content element of an interpretive presentation can further be linked to a diary entry and vice versa. A bookmark should essentially consist of a geographic coordinate and a text-based comment added by the user, while public annotations basically represent a public diary entry. Yet in order to completely realize these tools, it is also necessary to provide the visitor with an option to take the information home or to add public annotations and bookmarks to a central system were they are available to all users. The information can either be transferred to a central server via a wireless network connection or they can be "downloaded" from the mobile device via a cradle during the next synchronization procedure. If the central server is a web-server, the

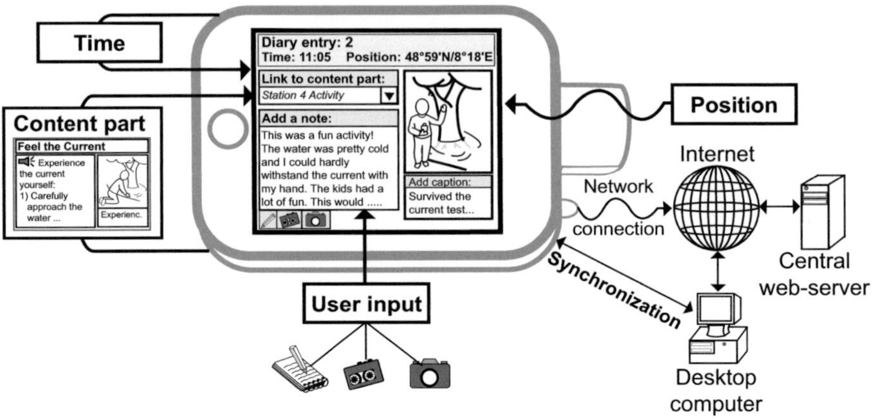

Figure 3.9: The diary tool provides the user with the opportunity to collect digital souvenirs, integrated into time- and location-based diary entries. The user should be able to make personal or public annotations to specific content parts using different documentation tools. These annotations should then be transferred either to a central web-server, the user's desktop computer or a memory card.

visitor can eventually download the personal documentation to his desktop computer at home or at school via the internet.

3.3.5 Manager service

Apart from the previously described services for visitors, the manager service should be designed to support managers and educators in the following tasks:

- Monitoring visitors
- Managing visitors
- Evaluating the mobile nature guide.

The components and functionalities of the manager service are presented in Fig. 3.10. The manager service differs in certain aspects from the services designed for the visitors. It can be considered a meta-service that collects information from all other services and can also supply additional information to these services.

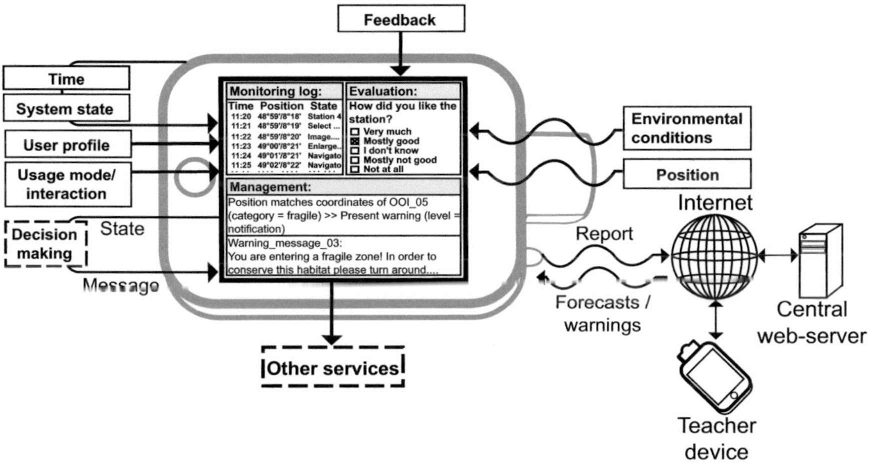

Figure 3.10: The manager service should function as an instrument that enables managers to monitor and manage the activities of the visitors. Furthermore, it can provide a tool for managers and educators to evaluate the performance of the system and the user, as well as gathering feedback to asses user satisfaction.

Monitoring The monitoring component should function as a master log. It should contain the core information available on the user, his usage of and interaction with the system as well as data about the environmental conditions and the status of the system itself. Basic elements of a monitoring log entry should be: the current position of the user and the current time. The current environmental conditions (e.g. weather, lighting, flooding, danger of avalanches etc.) can either be based on recordings from sensors directly connected to the device or reports from sensor networks as well as other external sources via a wireless network connection. Since the state of the environmental conditions in the AOI should be maintained in the environmental model, this model can also be used as a source for the log entry.

Data related to the user (e.g. age, physical fitness, interests etc.) can be extracted from a user profile. This profile has either been entered by the user himself or can be based on a user model resulting from the target group analysis. Entries on the usage of the system can be related to the use of the content (e.g. type of interpretive presentation, content parts consumed etc.) or to the use of system functionalities. This information can be gathered based on records of previous selections by the user or his interaction with interface elements. In order to monitor the user's satisfaction with the mobile nature guide, the manager service could prompt the user with specific evaluation dialogs that request feedback on content and usability. It further needs to be noted here that the collection of user data, especially logging the users position, is always associated with privacy issues [308], which may constrain the data that can be collected and/or used for the manager service. However, the discussion about privacy issues lies beyond the scope of this thesis

(For a detailed discussion on this issue see e.g. Mountain and Raper [229]).
The information collected by the monitoring component in turn forms the foundation for the management and the evaluation component.

Managing As has been previously described, remote management instruments that can be applied via a mobile nature guide are essentially "indirect" and "soft" tools, based on information and education [205]. The management should accordingly generate context-based messages like warnings or recommendations that are then displayed by the service currently used. The management service should for example compare the current position with coordinates of fragile zones, documented in the environmental model. In case these coordinates match up, the manager component should generate a warning message (e.g. "You are entering a fragile zone! Please turn around and leave this fragile area."), which is then to be displayed by the navigator service (see examples in Fig. 3.10 and Table 3.3). If this form of automated monitoring and management on the mobile device is reliable, it could bring a considerable benefit to managers of natural areas. It may either save time and manpower dedicated to visitor management or it may put managers into the position to actually monitor and manage visitors in the first place. As an alternative or addition to the automated on board solution, the monitoring log could be continuously transferred via a network to a mobile nature guide server at a visitor center. The decisions on generating management messages could then be processed server-based or managers could manually monitor the visitor and send out messages themselves. Still, this approach would neither save resources nor does it seem feasible for large amounts of visitors. Nonetheless it may be necessary to revert to this approach, if decisions on more extensive management steps have to be made or in case of an emergency. As part of the Momuna project Sauer and Osswald [290] proposed that such a remote, online monitoring and management component is also interesting for teachers and educators who need to supervise their students or group members.

Evaluating Managers as well as educators are also interested in evaluating the system as well as the user's performance (e.g. his spatial and temporal behavior) and his satisfaction with its content. Next to the monitoring log and a record of management activities, especially the feedback dialogs are important for this task. Since the actual evaluation has to be performed by trained personal on a collection of documented user experiences, the evaluation component should transfer a report with appropriately formatted data to the server. This can take place after the user finishes the application, either via a wireless network connection or as part of the synchronization process. Beyond this more summative evaluation the component can potentially also be used for a kind of formative evaluation throughout the usage of the mobile nature guide. This way, if the user's performance or feedback is not as expected, the evaluation component could initiate an adaption of the interface and presentation to the skills of the user.

3.4 Context model

In order to influence the functionality of the services as illustrated, the context information needs to be represented in the system in form of a context model.

3.4.1 Context

Dey and Abowd [81] emphasize, that designers of mobile applications need to understand both what context is and how it can be used. This should enable them to choose what context to use and what context-aware behaviors to support in their applications. The use of the word "context" tends to be vague and varies between disciplines, because everything in the world happens in a certain context [59, 266]. The reviews on context and context awareness by Dey and Abowd [81] as well as Chen and Kotz [59] illustrate, that a variety of definitions have emerged. A commonly used definition of context is an operational one established by Dey and Abowd [81]:

> Context is any information that can be used to characterize the situation of an entity. An entity is a person, place, or object that is considered relevant to the interaction between a user and an application, including the user and applications themselves.

In addition to this broad definition Schmidt et al. [299] specify, that a context is identified by a unique name. For each context a set of features is relevant and for each relevant feature a range of values is determined, implicitly or explicitly, by the context [299]. A widely used example for a context is "location". This may for instance be the absolute location of an entity such as the user, which consists of a set of geographic coordinates.
Several of the reviewed publications on context propose some form of hierarchy, typology or categorization [89, 98, 238, 269, 299] to group related features. Context categories represented in most of these typologies are: "user" (with features such as identity, preferences and knowledge), "physical environment" or "situation" (e.g. location, physical conditions, infrastructure), "time" and "system" (e.g. device used, network type and bandwidth). Dey and Abowd [81] propose the existence of two different types of context features. Accordingly location, identity, activity and time are "primary context" types that answer the "where, who, what, and when" questions. These primary contexts do not only constitute a context information in themselves, but can serve in the context model as indices to other "secondary context" types. The identity of a user for instance as primary context can link to the users preferences or skills documented by the user model.

The context categories and features that need to be considered in a context model for a mobile nature guide are those that were identified in the previous section as information that is required to adapt the services to the user's needs. Fig. 3.11 displays the context model incorporating the respective features.

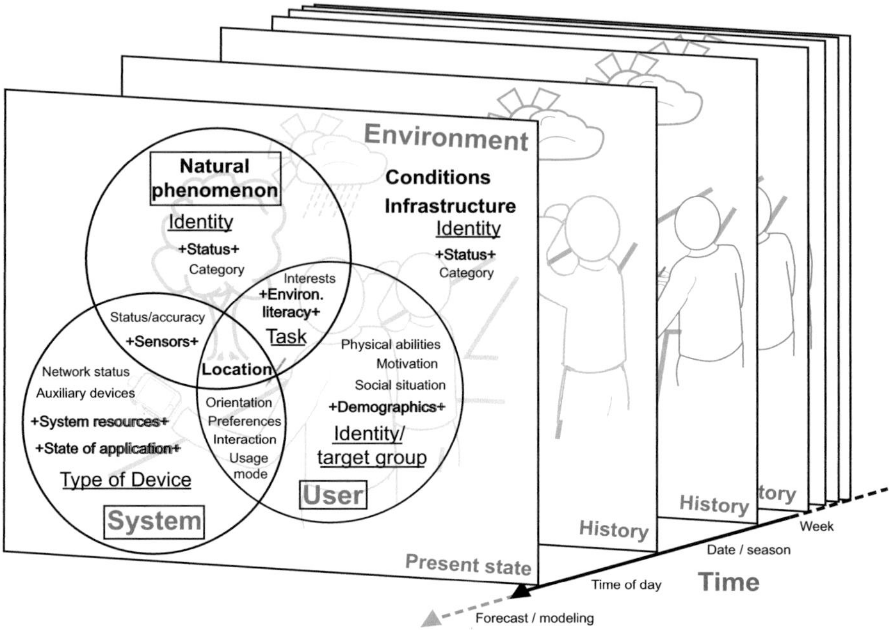

Figure 3.11: The MobiNaG context model consists of four major categories of context, the environment, setting the frame for application of the mobile nature guide, the user, the system and the time. **Primary context features** that can serve as indices to other context information are underlined and set in bold. **+Context feature groups+** are marked with "+" signs.

3.4.2 Mobile nature guide context model

This novel context model, which was specifically adapted for this thesis, presents a categorization of context information needed in a mobile nature guide application. Similar to the reviewed work on context, it differentiates between four major context categories: environment, user, system and time. Each major category may consist of further subcategories and a number of context features belonging to this group.

The environment context sets the frame in which the interpretive experience takes place and the application is used. The context information that is attributed to this category can further be grouped in environmental conditions, infrastructure and natural phenomena. The crucial role of environmental conditions (i.e. outdoor usage conditions) including features like weather, lighting or noise, has previously been discussed. The infrastructure, also part of the environment, plays a role for interpretive activities, for instance in the form of the trail system or parking lots. Next to the location and identity of such an infrastructure object, its status and category are also relevant. Its status includes features such as opening-times, accessibility or cost of usage. For a mobile

nature guide, however, the context features related to the natural phenomena can be considered the key aspects of the environment context.
The context model represents the immediate interaction between the three entities user, system and natural phenomenon, by the intersection of their context spheres. This also illustrates an overlap in context features. The core context feature common to all three entities is location. The location of the mobile guide system can be assumed to be the same as the location of the user. Obviously the basis for the location-based service is that the location of user and device matches up with that of the natural phenomenon. Other pieces of context information shared by the user and the system are the orientation, the preferences of the user regarding the usage of the system, the interaction with the system itself, including the user's performance and satisfaction, as well as the usage mode. The context feature that relates user and natural phenomenon is primarily the task or activity of the user which during an interpretive activity should mostly focus on the natural phenomenon. But also the user's interest in the phenomenon and his environmental literacy are of relevance to both entities. With respect to the intersection between the system and the phenomenon, it is the system's sensor equipment that determines, which characteristics (i.e. context features) of the natural phenomenon can be detected.

User context The user's situation can be described by a number of context features relating to his characteristics, present activities and status. If the mobile nature guide employs a user specific profile, the user's identity can act as a primary context type answering the "who" question and can serve as an index to other context features relating to the characteristics of the user. These include demographics, physical abilities as well as interests and environmental literacy. The generation of user specific profiles may require the user to go through an extensive profiling procedure, which might be cumbersome on a mobile device with limited input modalities, especially for a leisure time user. Furthermore, the maintenance of specific profiles may take up a certain amount of valuable memory on the device. Thus the user specific profiles may be inefficient for mobile nature guides rented out by an environmental interpretation institution, dealing with many first time, and frequently also onetime, visitors. As an alternative, target group specific profiles could be preconfigured based on front-end evaluation data. An obvious disadvantage of target group specific profiles is that they can only represent average values for the context features and thus will not allow a perfect personalization. Still, a user will only need to select a target group as part of the profiling process, and these profiles can also be maintained more easily.
Some elements of the user context are actually groups of characteristics that can be broken down further into individual context features. Demographics for instance includes features like age, gender, profession etc. The same is true for environmental literacy (e.g. knowledge, attitudes, values etc., see Chapter 2). Next to features related to characteristics, it is especially the current task and related activities, which should effect the services and functionalities offered by the system. The task can also serve as index to other context features such as interaction capabilities. The interaction with the system will further be influenced by the user's social situation. If the mobile nature

guide is used by a group of people like a family, the information should be presented for instance as an audio message, perceivable to all participants. From an environmental interpretation perspective, the motivation and performance of the user is also of importance and the system should have the capacity to adapt the presentation in a form that will maintain or increase the motivation.

System context The "type of device" can serve as a primary type of context for the characterization of the system. Which type of device is used, can for instance already indicate which system resources are available. The system resources further subsume context features like size of the display, battery lifespan, processing power and input modalities. In addition to the capabilities of the system itself, the availability and status of auxiliary devices such as loudspeakers or digital camera, has an impact on the functionalities the mobile guide can offer. According to the description of the context-aware services in the previous sections, a number of potential functionalities are based on a network connection to the internet or a central mobile nature guide server. Thus the status and bandwidth of such a network connection is also an important context feature. Comparable to the user's task, the state of the application is of relevance to many of the services. This includes for example the type of interpretive presentation that the user is engaged in and which part of presentation the user interacts with.

Natural phenomenon context Next to a location a natural phenomenon has also an identity (e.g. an OOI_ID), which can work as an index to other characteristics like the category of phenomena that it belongs to (e.g. plants). Further the status of the natural phenomenon is an important context information for the mobile nature guide services. The status group may include context features like the protection status, mobility or distribution data as well as special events or states (e.g. flowering in the spring).

Time context Of course the state of the context features is not static, especially in an outdoor environment, but changes with time. Fig. 3.11 shows different states of the context model lined up along the time line. Consequently time is represented as one of the major context categories. Individual context features of the time category are: the time of the day, day or week of the year and along with the date, the season of the year, which can be considered to be of particular importance for offering context-aware information about natural phenomena. Furthermore, the record of context states over time, here referred to as "history", can be an important context for a mobile nature guide application. As previously discussed the navigation history, based on the record of the users location overtime, can be used by services like the documentation tool. Other features like the usage or interaction history can be used to further tailor the presentation to the user's needs and increase the usability.

3.4.3 Context life cycle

The presented context model specifies which context features are needed in a mobile nature guide application. In addition it should be specified, how the context information should be captured, processed and applied by a mobile nature guide system.
Abowd et al. [2] propose a kind of context life-cycle in which the context information initially has to be collected before it is analyzed. This can take place either by treating it as individual variable or by combining it with other information collected in the past or present. The system should then perform some kind of action based on the analysis. This procedure should be repeated continuously. Following this and other work on context-aware applications by Salber et al. [287], Pascoe [249], Chen and Kotz [59] and Real et al. [266], this thesis proposes a context life-cycle model adapted for a mobile nature guide (see Fig. 3.12).

Prior to the usage of context in the system, the context has to be acquired. This context capture can occur explicitly, via input by the user. Alternatively it can also take place implicitly, using embedded sensors or based on an external infrastructure of sensors [81]. Raw sensor data enter a processing module as low-level context information, whose quality should be evaluated in an initial step. The application may not be interested in the low-level information like the lighting level in lux but may only want to know that the brightness level is "dull". Thus low-level context information commonly needs to be interpreted resulting in high-level context information [17,82], which can be stored in the context knowledge base in accordance with the context model. Still, also single high-level context elements may not be of relevance to the system by themselves but they need to be combined to an even higher abstraction level. For example the context elements time and location can be combined by the aggregation process to form the "length of stay". As part of the reasoning process, some context information such as "motivation" of the user, that cannot be sensed directly, can be inferred from other context like for instance the "length of stay" or the "utilized content parts" [17, 266]. Next to the inferring of deduced context, the reasoning module should also resolve context conflicts, maintain the consistency of the context knowledge base and make decisions on adequate reactions of the system, based on certain rules [17]. This finally results in the application of the context.
Varying categorizations of context applications were proposed in the literature by Schilit et al. [297], Pascoe [249] as well as Dey and Abowd [81]. For the conceptual design of the mobile nature guide, this thesis proposes the following application categories as a synthesis of the above approaches:

Context presentation The mobile nature guide detects context features like location or environmental conditions and simply presents this information to the user (e.g. position indicator on the mobile map).

Contextual adaptation and execution The context information captured by the mobile nature guide results in an automatic reaction of the system. This can either be the adjustment of

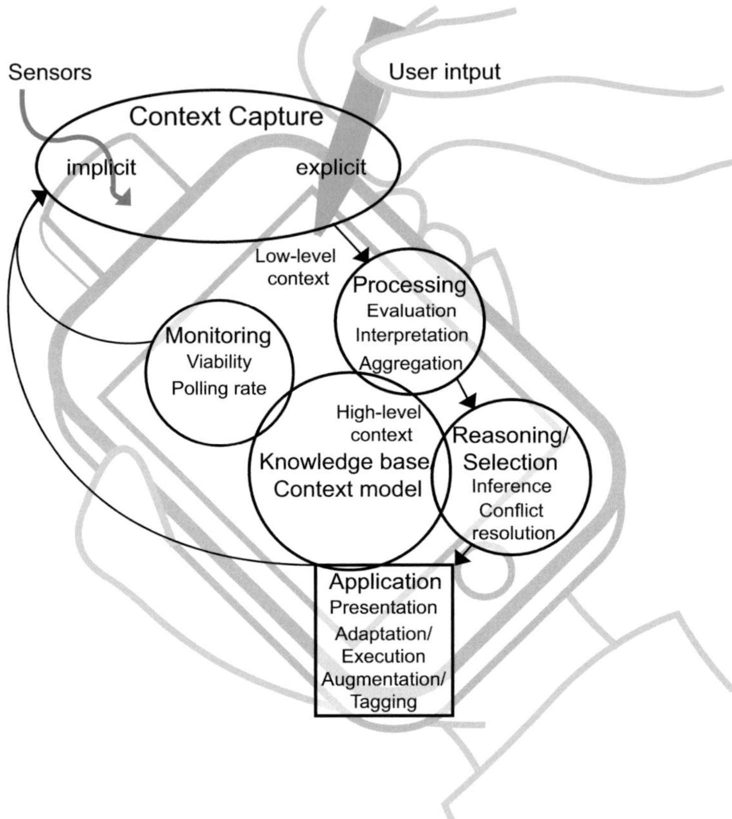

Figure 3.12: The context life cycle illustrates how context has to be captured and processed before the context information can be applied, following the reasoning and selection steps. Context information is maintained in the knowledge base in accordance with the context model, and a monitoring unit can initiate the update of context features in accordance with their viability.

the interface or services (e.g. low lighting level leads to increase in background light) or the execution of a service functionality (e.g. context-based information presentation upon arrival at an OOI).

Contextual augmentation and tagging The mobile nature guide augments the user's sensory capabilities by collecting information he can not perceive. Thus the provision of context information does also further the user's awareness of environmental phenomena that are part of the environment context. The system can also augment the user's physical environment, by superimposing it with digital information. Further information can be tagged with context for later usage.

Context features can be either static (e.g. the user's identity) or dynamic (e.g. user's location) [139]. In a natural area many of the captured context features can be expected to be dynamic, especially due to temporal dynamics of natural environments. A monitoring module should apply a polling mechanism that initiates repeated context capture, in correspondence to the respective context viability [287].

3.5 User interface and interaction paradigm

Baus et al. [22] stressed that the user interface along with the means of interaction generally play an important role in mobile guide systems, as they are the parts that are most apparent to its user and thus greatly influence a user's perception of the system. Since most of the visitors to natural areas are non-expert users, belonging to a non-captive audience, the interface for a mobile nature guide should grant an intuitive and enjoyable interaction without prior training or reading of the manual

3.5.1 Interface metaphors

According to Nadin's semiotic paradigm [231] of interface design, an interface is the meeting place between two different entities, which are to communicate. In the case of the MobiNaG system there are two dimensions to this interface concept. On the one hand there is the software interface as the meeting place between system and user. On the other hand, just as proposed by the environmental interpretation model, the MobiNaG system itself connects the user and his natural environment. Prior to designing the actual "look and feel" of the interface, it is important to develop a conceptual model of the system. This model describes a person's concept of what the system is: what its components are, what properties they have, and what interactions one can enter into [261]. As one way to design an appropriate conceptual model Preece et al. [261] suggest the selection of an interface metaphor. One of the most common interface metaphors applied, is that of the desktop. However, Marcus [206] stresses that for mobile devices new fundamental concepts are needed and it is obvious that the desktop metaphor does not apply to mobile guide applications. A suitable interface metaphor should combine familiar knowledge with new knowledge in a way that will help the user understand the conceptual model proposed by the designer of the system [261]. Generating an appropriate interface metaphor one should look in the user's description of the task for metaphors that are commonly used in the user's application domain [101].
Accordingly, the central task of the potential users of the mobile nature guide were analyzed in the course of this thesis. The exploration of a natural area is the main task of the users as part of the interpretive activity. Traditionally visitors to natural areas make use of a paper-based guidebook

or brochure as well as signs, in order to perform this task. As a more interactive alternative visitors frequently seek the assistance of a human guide, leading them through the area [129]. Since the guidebook and the human guide are concepts familiar to most users, they lend themselves very well as an interface metaphor to a mobile nature guide. A combination of the two metaphors should satisfy the evaluation criteria for an interface metaphor as proposed by Erickson [101], granting the users an intuitive interaction with the system. Depending on the usage mode, the user should be able to apply the system similar to a guidebook mostly self-determined. But he should also be able to rely on the system as a guide who leads and advises based on the recommendations of the authors of the system.

3.5.2 Iconic guidebook interface

The selected interface metaphor sets the framework for the actual screen design of the mobile guide system. Still in addition to the metaphor aspects, the handheld devices, used as hardware platform, offer unique challenges for the designers of mobile guides [206, 355]. It is generally acknowledged that a key limitation for the interaction design of mobile devices, is their limited display size, accompanied by restricted input capabilities.

Consequently as one solution for the interface of a mobile nature guide this thesis proposes a mostly icon-based guidebook interface. The ISO / IEC 11581 Standard defines Icons as a graphic displayed on the screen of a visual display that represents a function of the computer system [152]. They are a part of a graphical interface that can facilitate the user's ability to learn, understand and remember functional elements of the system and aid in the manipulation of these [235]. Icons have several advantages with regard to their deployment as interface controls in small display applications. A key advantage of icons with respect to the limited screen real estate is, that they take up less space than equivalent text. Further they should be intuitively recognizable and they are a language independent mean of communication [235]. Disadvantages, however, can be that the meaning of icons can be ambiguous and difficult to interpret by the user. In a variety of mobile guide systems, which make use of a direct manipulation interface, icons were used in interface design to some extent [113, 136]. For a mobile nature guide, interface icons should be used to the maximum possible extent. Other interface controls that can be used in the design of a nature guide interface include pop-up menus or lists as well as check boxes and radio buttons for user input.

3.5.3 E-Interpreter interface agent

With respect to the model for mobile environmental interpretation, it does not suffice to simply grant an intuitive interaction via an iconic interface and present factual information. Based on the novel mobile environmental interpretation model (see Fig. 2.3 and the performed requirements analysis (see Fig. 3.1), the system should also serve as a role model for user. This requires the employment of an affective interaction instrument that can provide immediate feedback and is able to facilitate a direct and meaningful experience, which will ultimately result in empathy with the natural environment. As a viable solution to this challenge, this thesis proposes the application of an animated interface agent, representing an electronic interpreter or "e-Interpreter".

There has been a diversity of research proposing that animated interface agents offer new opportunities to provide non-expert users with a more intuitive interface paradigm [30, 70, 150]. As embodied conversational agents (ECAs) these animated computer characters emulate face-to-face conversation through the use of hand gestures, facial display, head motion, gaze behavior, body posture and speech intonation, in addition to speech content [30]. They personify the interface, building upon people's natural tendency to interact with machines as if they were social actors. Thus they can make interactions easier by providing more natural assistance to the user [312]. An overview of a variety of projects using different types of relational agents is given by Bickmore [29, 30]. However, most of these research and commercial ECAs were developed for desktop applications and only a few systems employed animated interface agents on mobile devices so far [70, 155, 190, 352]. Nevertheless, as has been mentioned above, animated interface agents seem to be particularly suited for mobile guide systems, since they make use of the tour guide metaphor, which most users have had prior experience with [150]. A more extensive review of ECAs for the user in mobile nature guides is given by Real et al. [266].

In a review of animated pedagogical agents (APAs) Johnson et al. [156] present a collection of functionalities, that this type of agents can offer to support the learning process of interacting students:

- Interactive demonstrations
- Navigational guidance
- Attentional guidance
- Immediate feedback
- Conveying and eliciting emotion
- Virtual companion
- Adaptive pedagogical interactions

The presented functionalities (see Appendix B.4 for more details) suggest that the employment of an APA can help to implement the instruments in a mobile nature guide system that are needed to influence environmental literacy. This implies that through displaying and eliciting emotions, the APA can motivate the active participation in EE activities and encourage direct exploration.

The conveyance of emotions can also help to transform the APA into a role model for the participants that may invoke empathy and can help to adapt responsible behavior through interactive demonstrations. The provision of immediate feedback can support the learning process but is also of importance for increasing feelings of behavioral control and the actual modification of behavior. Finally it can be assumed that the APA can serve as an adequate interaction metaphor for a human guide, since the presented functionalities match to a certain extent the traits for a human interpreter.

3.6 Knowledge base

A mobile nature guide further needs a component that stores the information presented by its context-aware services. This component should also have the capability to maintain the necessary context information specified by the context model. This functionality should be fulfilled by the knowledge base specified by this thesis (see Fig. 3.13). The knowledge base component can be subdivided into three segments:

Content knowledge base: This can be considered the main segment, since it should comprise the major part of the information to be presented by the services. For the navigator service this includes the basemaps as well as a basic description of the AOIs and OOIs (compare Table 3.3). More extensive information on these objects as well as most of the information conveyed during interpretive presentations, has to be maintained for the interpreter service. Also the encyclopedia service and the exploration tools service rely on this segment for instance for glossary entries, species descriptions or augmented reality data.

Diary and management knowledge base: Even though this segment can be essential for the services as well, it differs from the content part of the knowledge base since it stores mostly data recorded during usage. This can either be information logged by the system for the manager service or the diary tool. It further includes references to information selected by the user as well as data recorded by the user (e.g. digital photographs). Furthermore, for the manager service this segment should also store a collection of preconfigured warning messages.

Context knowledge base: Next to the content stored in the knowledge base, one of its segments is dedicated to context information. Here high-level context information can be stored in correspondence with the context model (see Fig. 3.11) and later extracted for application (see Fig. 3.12). This implies, that the context information is not solely used for the adaptation of the presentation but can also become part of the presentation itself.

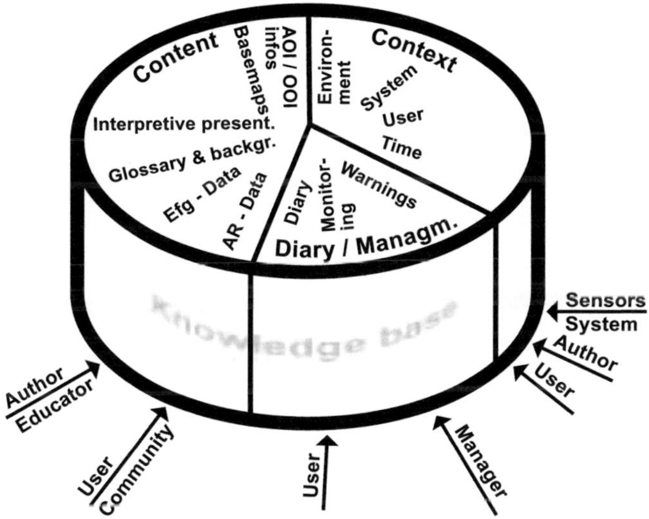

Figure 3.13: The knowledge base that stores the content for services as well as the state of relevant context features.

Fig. 3.13 further shows that the information stored in the different segments is potentially maintained by different stakeholders or sources. The data presented as content, will commonly be maintained by educators who will frequently act as authors for the system. Still, with reference to the extended mobile environmental citizen science model (see Fig. 2.4), it can also be envisioned that users and the community of naturalists can contribute to the extension and maintenance of the content knowledge base. The users are obviously also supposed to contribute to the diary segment, while the management part of this segment is to be administered by park managers. As previously described in the context life-cycle model (see Fig. 3.12), the context information stored in the context knowledge base can result from sensor readings interpreted and aggregated by the system or from the system itself. But it can also be explicitly entered by an author or the user.
Like previously discussed, it is most practical for a mobile nature guide system, to maintain the content of the encyclopedia service on the device as part of the presented knowledge base. But it should also be kept in mind, that the flexibility of the system can potentially be increased by accessing remote information resources via a network connection (see Fig. 3.6).

3.7 Control center

The system components, previously introduced, have to interact seamlessly in order for the mobile nature guide to function properly. Consequently, this calls for a central unit in form of a control center that coordinates these components. This thesis proposes such a control center unit, as presented in Fig. 3.14, that is tailored to the specified conceptual design of a mobile nature guide.
The user will primarily interact with the mobile guide system via the user interface employing different interaction modalities. This request or user input has to be processed by an interface controller that assigns the request to the appropriate service and decides if and in which segment of the knowledge base the entered data should be stored.
The interaction with the user also provides the opportunity to capture context information about the user. The system can further capture context information via a variety of sensors, which should also enter the context life-cycle, as illustrated in Fig. 3.12. The context life-cycle should, just the same, be managed by the control center. This implies that low level context is processed and the resulting high level context is assigned to the context knowledge base. Furthermore, the context life-cycle controller should take care of the reasoning and selection of context information and decides how it can be applied. Finally, this controller also needs to manage the monitoring of the context features specified by the context model. The sensors necessary for the capture of context information along with other auxiliary hardware devices, needed by certain services (e.g. networking/communications technologies, digital camera), should be managed by a hardware controller unit.

Changes in monitored context features as well as user input can constitute events that need to result in adaptations by the services. These events should be handled by the service controller. This part of the control center should further assist the services in retrieving the suitable content parts as well as context information from the knowledge base. It also needs to coordinate the interaction between the different services (e.g. switch from the navigator to the interpreter service if a user arrives at an OOI). The presentation of the respective content of a service is eventually managed by the presentation controller.

3.8 Technologies

Finally, in addition to the proposed software components a collection of hardware elements and technologies is needed to realize the conceptual design of a mobile nature guide. A comprehensive set of options for hardware ingredients was compiled for this thesis. Prior to discussing the technological components in more detail it should be pointed out that the field of mobile computing continues to evolve rapidly. Hence the proposed suite of technologies, as presented in Fig. 3.15, should be taken as a general suggestion, valid at the time of preparing the thesis.

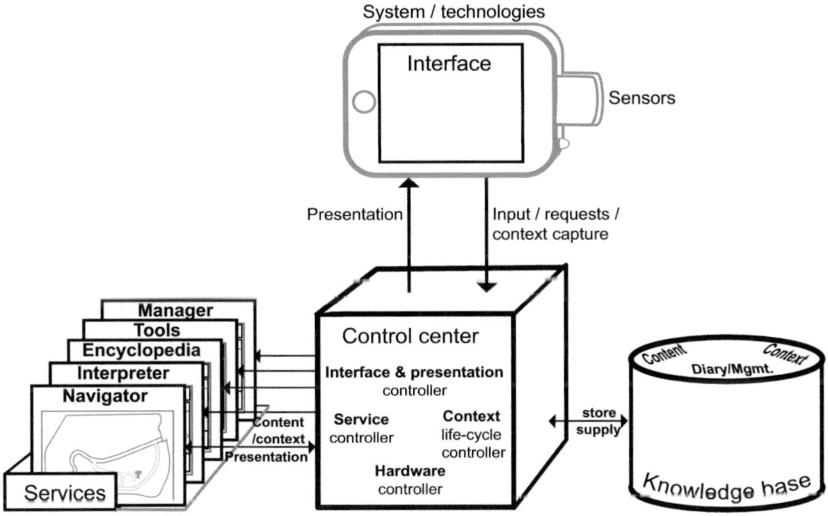

Figure 3.14: The control center is the component of a mobile nature guide that organizes the functionalities and interactions between all other components.

Mobile device The core hardware platform is a mobile computing device. In their review of mobile computational devices Sharples and Beale [311] differentiate a variety of general categories of these devices, which include wrist-worn devices, mobile phones, handheld computers/PDAs, tablet computers and laptops. A mobile nature guide device needs to be easy to carry in the field for at least up to an hour, by adults and children alike. It further needs to be easy to operate on the move and its computational power should suffice to present complex multimedia content such as animations. Their limitations in computational power and interaction capacities basically rule out wrist-worn devices and at the time also mobile phones. Laptops on the other hand have ample computational resources but are to heavy to carry for longer periods in the field and difficult to interact with while standing or walking. Thus PDAs and tablet computers, due to their light weight and easy portability, remain as the feasible options. A further benefit with respect to their application in EE institutions is their relatively low cost. The major drawback of PDAs on the other hand, is their small screen size and limitations in input modalities.
Currently there are two major operating systems that PDAs can run on: PalmOS and Windows PocketPC. Other alternatives, such as Linux, are yet only employed on a minority of devices [311]. Both current PalmOS and PocketPC (i.e. Windows Mobile 2003 SE or Windows Mobile 5.0) devices should generally constitute an appropriate platform for a mobile nature guide application. For the providers of mobile nature guides the platform that is most widely distributed and offers the best support for multimedia presentations is most likely the best choice. At the time being this is the PocketPC platform, which according to market researchers [338] holds the largest market

share in Europe and grants better support for multimedia applications [211].
It is again the multimedia content, which makes it necessary that the PDA for a mobile nature guide needs to be equipped with sufficient processing power as well as memory. Fig 3.15 lists two current alternatives for CPU technologies that can function at a speed of up to 624 MHz [319]. Flash ROM (read-only memory) up to 256 MB is available [319].
The size of typical PDA displays varies between 3.5" and 4" [319]. In order to provide a realistic presentation of the interpretive content that can satisfy the expectations of today's media spoiled public, the device should have a full-color LCD display with at least a QVGA (i.e. 320x240 pixel) or VGA (i.e. 640x480 pixel) resolution. Even though VGA displays provide greater image clarity and the ability to display more information on a single screen, they are more expensive and generally consume power faster than QVGA displays [319]. For a mobile device that is to be employed in remote natural areas, the battery life of course plays a crucial role. Taking this into account, such power hungry technologies, like the VGA display should rather be avoided. Furthermore it is important, that the battery can be exchanged and the device can be expanded with an additional battery for instance by adding an expansion sleeve. The expansion options of the mobile device (e.g. Secure Digital (SD) or Compact Flash (CF) slots) are also of relevance with regard to increasing its storage capacity or connecting it with auxiliary hardware like context sensors.

Interaction technologies In addition to these fundamental system resources, the device for a mobile nature guide also needs to comprise certain interaction technologies. For the presentation of content and feedback to user interaction, this should include a full-color display and also loudspeakers or earphones for audio output during non-visual usage. In order to realize the previously proposed non-immersive hand-held AR display it also needs to be equipped with a digital camera. Of course the interaction with a mobile nature guide also includes that the user can provide input either via selection or entering information. One form of intuitive input interaction can take place as direct manipulation via a touch-screen, using a pen-tool or finger. Since the interpretive activities will likely also require hands-free usage of the device, a microphone should also be included to enable speech input. Other potential input options for at least one-handed usage should include joystick/joypad, multifunction key or even a thumb keyboard.

Connectivity technologies A full implementation of services like the encyclopedia and the diary tool make network connectivity necessary. Network access in remote areas usually poses a considerable challenge due to the lack or sparse distribution of a respective infrastructure. Cellular radio technologies such as GPRS and UMTS appear to be the only suitable solution. Cellular radio technologies can also be used to communicate with peers for instance via cell phone. The exchange of information with other devices (e.g. mobile guide system of a peer or the MobiNaG Server) can also take place based on WLAN or Bluetooth if the distance between them does not exceed the respective range. One key connectivity technology is the vendor specific USB synchronization port that allows for the update or addition of large amounts of information on the mobile guide. All connectivity technologies can either be integrated into the mobile device itself

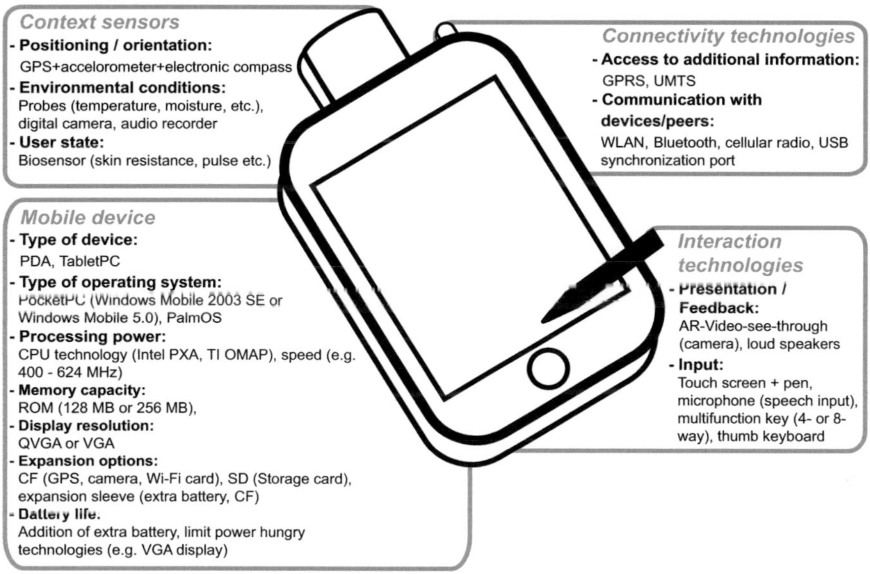

Figure 3.15: Mobile nature guide technologies subsume a number of hardware items and technologies needed to put the proposed conceptual design into practice. This includes a mobile device as well as certain interaction technologies that help to meet the MobiNaG requirements. Context sensors are needed to capture context information. The system further needs to apply a number of connectivity technologies.

or added via expansion slots (e.g. CF- or SD-cards). However, since the number of expansion slots is usually limited to one or two of each type, onboard connectivity technologies save room for the addition of other auxiliary hardware.

Context sensors These sensors are essential for a mobile nature guide system in order to capture the context as proposed in the conceptual design. This of course includes positioning and orientation technologies. GPS is the positioning technology most suitable for the usage outdoors (i.e. also remote areas), since it is a satellite-based technology that is independent of terrestrial infrastructure on site. Nonetheless, the GPS location system may occasionally fail in natural areas e.g. due to degradation of the signal (e.g. by the tree canopy). In these cases alternative positioning methods should be applied, which to some extent rely on information from additional sensors such as accelerometer and electronic compass. Next to positioning data the system should also capture environmental conditions, which requires the application of probes that can measure environmental parameters such as temperature, moisture and pH (e.g. EasySense Flash Logger and SmartQ Sensors by Data Harvest Educational [53]). Further environmental context information may be captured based on a digital camera of multiple spectra (visual and IR) or audio recorder integrated or linked to the device. Next to environmental context, user context such as attention level or emotional state [299] could be captured based on biosensors (e.g. pulse oximeter [216]). Currently some PDAs are already shipped with integrated context sensors like GPS and accelerometers. But most context sensors only exist as external hardware that has to be connected to the device. They can either be connected as CF- or SD-cards or as wireless devices via Bluetooth or WLAN. Wireless sensor devices will commonly have their own battery, which has the advantage, that they do not further diminish the power of the mobile device. The connectivity technologies can be used alike to access remote sensor networks, which may be independent of the mobile nature guide but can supply it with environmental context information.

Chapter 4

Information Technological Concepts for a Mobile Nature Guide

This chapter describes an approach for the implementation of the proposed concepts for a mobile nature guide prototype system.
As was illustrated in Chapter 1, a variety of approaches to the architectural design of mobile guide systems do exist. For the new mobile nature guide system a combination of a stand-alone system and a client-server architecture has been chosen as an adequate compromise. Particularly with respect to the deployment of the MobiNaG system in remote natural areas, a reliable network connectivity, needed for the pure client-server architecture [23], cannot be guaranteed. Thus in natural areas a stand-alone system constitutes a more reliable option that can in addition provide faster response rates to user interaction. The restrictions in extensibility and adaptability of a stand-alone system can to a certain extent be overcome by the integration of client server components for the retrieval or update of certain content.

4.1 System architecture

This thesis proposes an overall IT architecture specifically developed for a mobile nature guide. Such an architecture should be based on a frame application designed based on a three tier software architecture as displayed in Fig. 4.1. This frame application consists of a presentation layer next to a control layer as well as a data and services layer. It is designed in a modular form that allows for an easy and flexible extension as well as adaptation, by adding and removing software components.

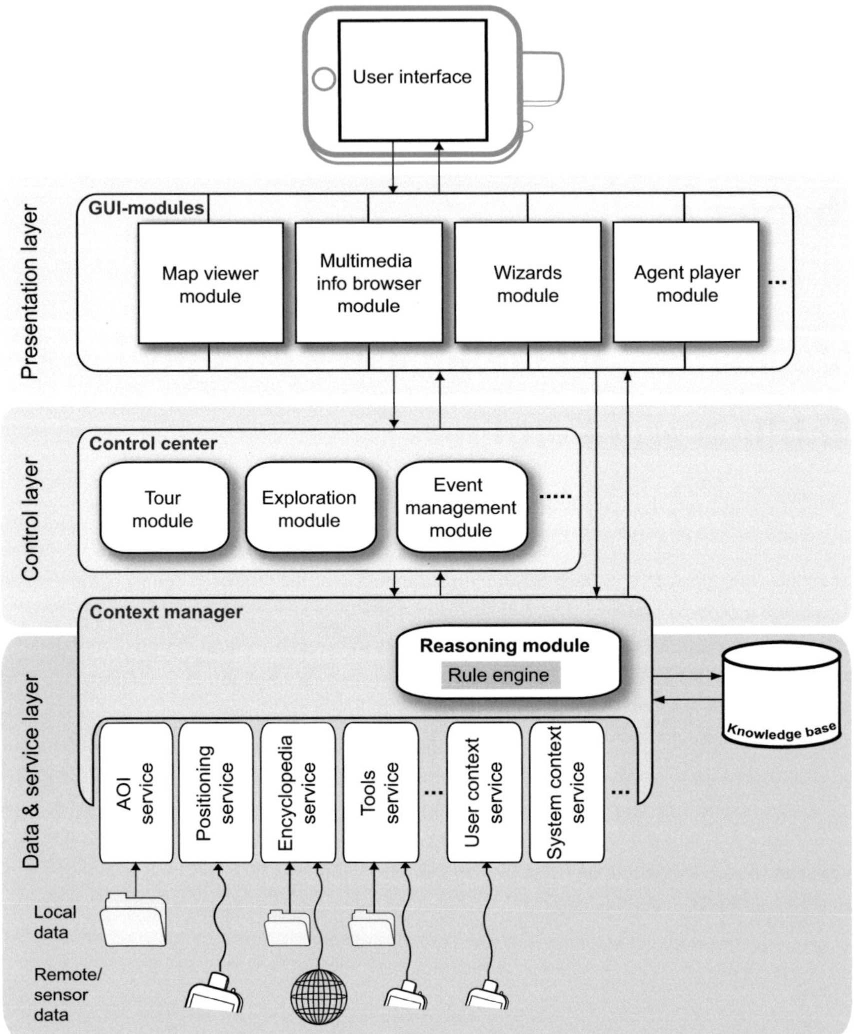

Figure 4.1: The general software design for the mobile nature guide application based on three-tier-architecture.

Presentation layer

The presentation layer incorporates a collection of graphical user interface (GUI)-modules as software components, of which each handles the interaction via the user interface for a specific func-

tionality of the mobile guide system. This includes the display of information as well as the handling of user input.
Modules for a mobile nature guide should include a map viewer component, which is responsible for the presentation of elements of the proposed navigator user service. While this module should be equipped for rendering a map representation of AOI, POIs, routes, tracking layer etc., for the interpreter user service a browser component is needed, which can display various kinds of multimedia content. Next to these core components a wizards module should also be included, which guides the user through selection dialogs for initial personalization or setting of preferences as well as the presentation of tutorials. Finally for the mobile guide system to provide a more effective interaction mechanism via an animated interface agent, an agent player module is needed, which can handle the presentation of the e-Interpreter as well as its interaction with the user. Further modules like for instance a specific augmented reality module for the tools service can be envisioned and based on the module architecture added to the system at runtime. The functionalities of a GUI module will be discussed in more detail later on.

Control layer

The frame application is to be controlled by the control center constituting an intermediate layer. It again consists of a number of modules which should coordinate and synchronize the interactions and activities of the other units of the application. Matching the different usage modes proposed for the mobile nature guide in the previous chapter, the control center should include a tour and an exploration module. These control center modules need to be aware of the content structure as well as the system status and are responsible for mediating between the GUI-modules and the context manager, next to switching between the GUI-modules as required by the content structure (e.g. switch from information browser to map viewer module after the multimedia presentation at a station has been completed). The control center also needs to include units such as an event management module that can process (overarching) events, which can not be handled by one GUI module in itself. Like for instance user requests that require switching between GUI modules. Such a module should of course also manage global functionalities of the system that lie beyond the responsibility of a single GUI module, like for example the initiation of a wizard module and the termination of the mobile nature guide application.

Data and service layer

In this software design for a mobile nature guide a service-oriented approach is proposed for accessing different types of data needed in the application. For each type of data a specific software service (as opposed to the user services proposed in the conceptual design - see Chapter 3) needs to be devised. Services can provide access to predefined content of an interpretive presentation that has been prepared by an author of an interpretive activity (e.g. the AOI service should contain the multimedia information of tours designed for an area). But services can also act as an interface to context data collected through various sensors or input modalities of the device or a remote sensor network (e.g. the positioning service collects location data from a GPS sensor). Corresponding to Fig. 4.1 this data can either be stored locally in the file system or a database or it can also be accessed remotely via a network connection as was for example proposed in Chapter 3 for an encyclopedia service. As interface to the data a service needs to primarily incorporate a representation of the data storage logic. Still the software services can also hold a number of methods which allow the system to utilize and operate auxiliary hardware to collect and incorporate data that can again be accessed via the services (e.g. the tools service should be able to activate a digital camera for the implementation of the exploration tool user services). The software services further are not only involved in capturing service specific context data from the sensors but they should also take care of the context processing corresponding to the proposed context life-cycle (see Fig. 3.12).

In accordance with the modular design philosophy the services are independent units, which can be flexibly added or removed from the system in the form of plug-ins. The services plug into the context manager component, another major unit of the frame application that mediates between the data interfaces and the controller or the GUI-modules. The context manager administers the interaction of the GUI-modules and the control center with the services. GUI modules can either communicate with the context manager directly or via the control center. For the initial access to the content of a tour, the respective control center module should notify the context manager that will retrieve the relevant content from the AOI service and provide it to the GUI-modules (e.g. map viewer and information browser). In the case of context information used for contextual adaptation and execution, the context manager can notify the control center unit of changes which will mediate these changes to the GUI or the respective GUI module can be notified directly. In case of context presentation the GUI should be served directly with the information (e.g. a change in location is transferred to the map viewer module). This notification of changes in context can either take place in form of a push or a pull mechanisms. The map viewer for example should employ a polling mechanism to periodically query the context manager for changes in the position of the user. Apart from the transaction of basic high-level context (e.g. location is x/y or brightness level is "dull" or season is "spring") that can be provided by a specific service, the context manager should incorporate a reasoning module that allows for further aggregation and abstraction of the context information from different services (e.g. positioning service provides change in location, system context service provides time - location and time can be combined to "length of stay"). As was discussed in Chapter 3 the reasoning module should also have the capacity to use such abstracted context to infer deduced context such as user motivation or generating context widgets as proposed by Salber et al. [287] that are stored in the knowledge base. Finally based on an integrated rule engine the reasoning module should be able to use the resulting context information to make decisions on appropriate reactions of the system base on certain rules. These reactions should then be enforced by the control center modules. The respective rules for the knowledge engine need to be stored in the knowledge base next to facts in the form of the context and status information. Rules should also be extendable and adaptable based on previously collected or inferred context. This context should especially include the usage pattern as well as settings supplied by the user context service.

4.2 Graphical user interface modules

GUI-modules that constitute the presentation layer of the mobile nature guide frame application are designed based on a Model-View-Controller architectural pattern (MVC) [180]. In accordance with the illustration of map viewer GUI-module in Fig. 4.2, each component encompasses its own model, view and controller unit. This three-way division entails separating the parts that represent the content as well as the operations related to the underlying application domain (i.e. the model) from the way the domain and the state of the application is presented to the user (i.e. the view) and from the way the user interacts with the model and the view (i.e. the controller) [180].

In case of the GUI modules such as the map viewer or information browser, the model constitutes a form of repository for the pertinent AOI data which is retrieved and transferred from a service, via the context manager. The controller unit handles user interactions, which can include requests resulting in a change of the status of the system as well as input that leads to an explicit change of settings or preferences. The GUI module controller principally contains an application logic that allows it to process all operations constrained to the specific GUI module and that can be based on the local model repository. It commands the view to redraw the display to render additional or updated information from the model. Besides the GUI module controller can also be triggered by a control center module of the frame application in order to synchronize the structure of the selected interpretive presentation.

Fig. 4.2 can further explain the functionality of a GUI module based on the example of the map viewer component. It can be assumed that the user has selected a guided tour and consequently the tour module of the control center has triggered the controller of the map viewer and the information browser module to request the relevant AOI data from the context manager. The tour module then sets the focus of the application to the map viewer module. In the process of inspecting the tour, the user clicks with the pen input device on one of the station icons to learn more about this POI. The controller processes the request and commands the view to obtain relevant information pertaining to station x from the model and display the corresponding short description element in an event layer on top of the map layer. Furthermore the controller can induce a change in the model that updates the user context service via the context manager, recording which information the user has requested.

The application of the MVC paradigm at the level of the GUI-modules is consistent with the modular philosophy of the overall software design for the mobile nature guide, since it increases the

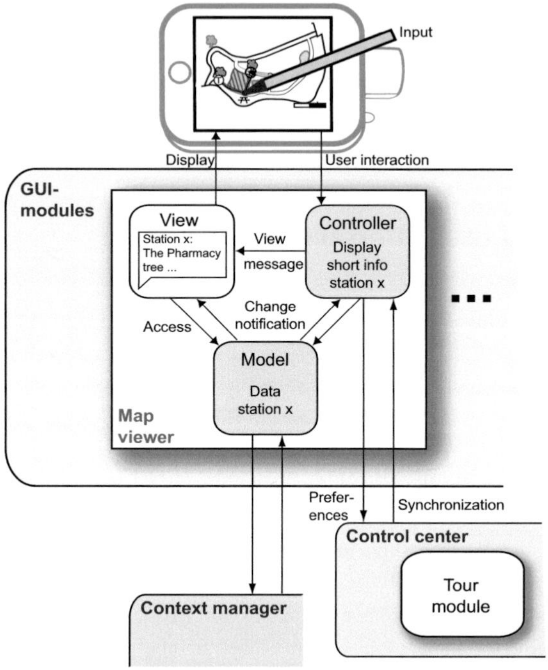

Figure 4.2: Architectural design of MobiNaG GUI-Modules based on a Model-View-Controller design pattern adapted from Krasner and Pope [180]. The functionalities of the module are illustrated based on the example of a user request to the map viewer.

reusability and "pluggability" of application components [180]. As was stressed by Düpmeier and Ruchter [94] the use of such an architecture allows for a clean separation between user interface code and application logic. Thus making it possible for user interface (UI) designers to add new views to the guide system tailored to specific target groups without needing to change the application logic.

4.2.1 User interface design

As foundation for the implementation of the GUI-modules of the system, a UI for each module was designed, in correspondence with the concepts for the mobile nature guide interface and interaction design proposed in Chapter 3. The construction of the UIs was based on the state of the art methodology for interaction design as proposed by Preece et al. [261]. Special attention was devoted to aspects related in particular to the design of handheld devices as discussed by Marcus [206] as well as Weir and Noble [354]. The prototyping of the UI designs was to a large

extent based on methods for handheld usability as proposed by Weiss [356], in particular applying the paper prototyping methodology for the production of low-fidelity prototypes [357].
The following steps were taken in an iterative design process for building the UIs for the different GUI-modules:

- Analysis of user tasks and requirements based on front-end evaluation
- Development of a low-fidelity UI prototype
- Usability tests in the field
- Redesign of the interface.

Specific aspects of the UI design process for the mobile nature guide and the results of the usability tests were described in more detail by Düpmeier and Ruchter [94]. All UI designs were founded on the overall interface metaphor proposed for the mobile nature guide in form of a combined guidebook and human guide metaphor.
Fig. 4.3 demonstrates the newly developed screen design of the UI for a mobile nature guide GUI-module. Similar to recommendations made by Weiss [356] and the Microsoft style guide for Pocket PC websites [222] the UI consists of three major components. The title and/or status bar can give the user an indication of the nature of the content being currently displayed as well as the current state of the system and functionalities in use.

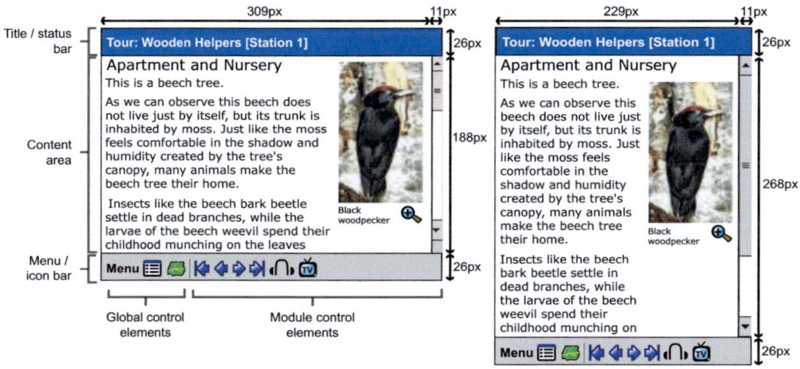

Figure 4.3: General example of MobiNaG UI (here based on Information browser module), illustrating different screen components and their dimensions as well as interaction elements. Depending on the device platform landscape layout (left) and/or portrait layout (right) should be provided. Dimensions for components are listed in pixel and founded on a QVGA resolution (i.e. 320x240 pixel) corresponding to the display of the device used for mobile nature guide prototyping.

As much screen real estate as possible should obviously be reserved for the content area directly below the title bar. For some modules such as the map viewer, the space available for the presentation of content may be maximized by sparing the title or status bar. The content area may of course also include interaction elements that can either be embedded in the content (e.g. of the information browser or of the map viewer module) and serve the direct manipulation of the content presented (e.g. the magnifying glass in Fig. 4.3 provides access to an enlarged view of the image enclosed in this page) or as part of a dialog view (e.g. of a wizard) that enables the user to set preferences or change settings.
However, next to these elements the UI of a module also needs to include permanent interaction elements for controlling the status of the module and for manipulating the content or its representation. These control elements should be incorporated in a menu and/or icon bar positioned at the bottom of the screen, since on the touch screen, interactions at the bottom of the screen can cause minimal obstruction of the user's view on the display [356]. This component can encompass menus as well as push buttons preferably represented by icons. Icons have several advantages with regard to their deployment as interface controls in small display applications, as was discussed as part of the conceptual design for a mobile nature guide UI in Chapter 3. Usability tests with an iconic UI approach for a mobile nature guide showed that regardless of some difficulties that occur with individual icons, the majority of users see iconic interfaces as an attractive way to interact with the computer [94]. However, the results of the study conducted as part of the MobiNaG project also made it clear that many users feel more comfortable interacting with a mixed iconographic/textual interface, where especially icons which the user encounters for the first time have to be supported by text [94]. Consequently the system should rely as much as possible on standard icons or apply mechanisms of text support for icons in small displays such as the "hidden labels" mechanism suggested by Heidmann et al. [136]. Next to the application of icons, commands can also be represented via menus. Menus are particularly suitable for providing access to commands which in the present module are not needed frequently or offer auxiliary functionalities [356].
As displayed in Fig. 4.3 the commands used in the mobile nature guide UIs can be categorized in global control elements and module control elements. Global functionalities will likely be consistent in all modules, representing functionalities of the frame application that are processed by the modules of the control center. Those global controls used less often such as the termination of the system and the accessing of wizards to change general settings should be grouped in one global menu. Other global elements should be kept more readily available in the form of icons. Corresponding to the guidebook metaphor these should for instance include the switching between different user services offered by the system (i.e. the navigator and the interpreter service). In order to facilitate an intuitive interaction the respective categories of functionalities should be grouped. A consistency in appearance across modules can further be sustained by placing global functions always at the left end of the bar, where they maintain their position if other elements are added or removed, and at the same time more frequently used elements are within closer reach of at least right handed users.

Two different forms of layouts can be offered for a mobile guide on a PDA, mainly depending on the operating system. Some of the current operating systems (e.g. Microsoft Pocket PC 2003 SE or Windows Mobile 5.0) have the capacity to render the display in both portrait as well as landscape format depending on the users preference or even the current orientation of the device. For the mobile nature guide UI the landscape layout turned out to be most suitable. Although approximately the same amount of content can be represented, it is especially for the representation of maps that the landscape layout will generally be more fitting.
Due to the usability study performed as part of the prototyping in this project Düpmeier and Ruchter [94] could show, that clear differences in the usability of iconic baby interfaces exist depending on the age group that the user belongs to. Thus in correspondence with the concepts proposed in Chapter 3, the presentation of the content as well as the control elements should be user adaptive (i.e. tailored to the target group) and adaptable. This requirement calls for a technology, for the implementation of the UIs, which allows a high flexibility in the presentation of control components as well as information elements.

Map viewer module

The map viewer module should realize the essential concepts proposed for the navigator user service in the previous chapter.
As was specified in Table 3.3, the service needs to support the user during the wandering, navigation and monitoring phases of his visit to a natural area. In essence this should be achieved by displaying a map of the AOI along with additional layers with different types of geographical information (e.g. position indicator, OOIs, routes etc.). The actual design of such a mobile map and the symbols for the representation of further geographic information is an issue by itself, which was discussed in depth in Chapter 3. New concepts for personalized mobile maps specifically designed for mobile nature guides have also been investigated as part of the MobiNaG project. A detailed description of these concepts can be found in Patalavitiute et al. [251, 252].

The map viewer GUI module also incorporates the three major UI components (i.e. title/status bar, content area and menu/icon bar) as proposed by the general MobiNaG UI structure. As is illustrated in Fig. 4.4, the title and status bar indicates to the user, the name of the AOI he has chosen to visit (which should correspond with the title of the map) as well as the name of the tour he is currently engaged in.
The content area serves for the display of the mobile map of the respective AOI along with the other map layers and components. In accordance with the general UI approach for the MobiNaG system, the menu and icon bar comprises two groups of commands. The global functionalities on the left side include the menu as well as an icon representing a list, which allows the user to switch to a table of contents, offering a listing of the stations contained in the specific tour. This textual overview of stations and corresponding content parts allows the user an additional means to "nav-

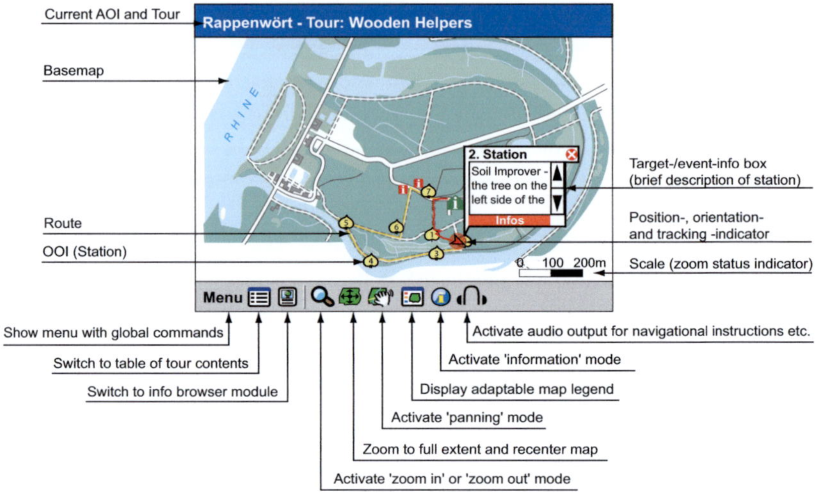

Figure 4.4: MobiNaG map viewer UI - including the presentation of a sample mobile map as content and an explanation for the proposed UI components and control elements.

igate" between stations and access their multimedia content next to the geographical access via the map. The next global command is represented by an icon depicting a page with multimedia information and allows the user to switch from the map viewer module to the information browser module rendering the last information unit selected by the user.

The module control elements include functionalities for the manipulation of and interaction with the map and the map viewer module. The collection of icons includes control elements that upon user interaction result in an immediate change of the map display and those that activate a specific "interaction mode". This actually changes the mode of the pen (or stylus) as input device. For example if the user presses the magnifying glass icon, he will be presented with an additional choice between "zoom in" mode and "zoom out" mode. In "zoom in" mode any tap on the map with the pen will lead to an enlargement of the the map (i.e. a reduction of the scale) around the selected area. For reasons of performance and usability a zooming mechanism based on three zoom steps is proposed for the mobile nature guide, as opposed to a continuous, seamless zooming mechanism. Which zoom steps are provided essentially depends on the dimensions of the AOI as well as the tours provided. As it was described by Patalavicute et al. [251] in case of the mobile nature guide prototype system it seems reasonable to start with the an overview scale that yields the display of the entire AOI (e.g. scale of 1:25.000). This should be followed by an intermediate step that still offers an overview of the entire tour (e.g. scale of 1:10.500) and eventually a detailed view of the map, showing details of the tour, that should ideally match the range of direct (visual) perception of the user (e.g. scale 1:4.500). As was also discussed by Patalavicute et al. [252] it is recommended to add a graphical scale to the map that adapts to the zooming interaction and

provides the user with a reference to judge distances more easily based on the map. For a quick and easy orientation the system also provides a shortcut to reset the map to the overview scale by means of a button that zooms directly to full extent and simultaneously recenters the map.
Another common function for controlling a map is according to Reichenbacher [269] the panning mode. This mode can be activated via the icon represented by a hand that skids the map sideways. Once activated the user can change the position of the map by tapping with the pen on the map and simultaneously sliding the pen into the direction the map should be moved to, which results in a change of the visible section of the map. As was pointed out by Reichenbacher [269] mobile maps should be adaptive to changes in context as well as adaptable by the user to suit his preferences. An opportunity for users to at least partially modify the appearance of a map as well as the content displayed is to provide them with an interactive legend like it is commonly used in modern geographic information systems (GIS) [314]. A map legend generally fulfills a crucial role regarding the legibility of the map, since it provides an explanation of the symbols, codes, names given to variables and other information appearing on the map [251, 269, 314]. Also the mobile nature guide map viewer UI provides access to an interactive legend of the displayed map via an icon representing a map accompanied by color code symbols. Due to the restrictions in screen size the interactive legend is presented in a separate screen. Next to viewing the explanation of the map components the user can add or remove geographic information via adjacent check boxes.
Apart from the interactive legend, information on map objects can also be accessed via direct manipulation in an specific "information mode". After this mode is activated by the user via the standard information icon, the pen can be used to access additional information on specific components of the map if available. The information mode can for one serve as an additional form or short cut to an interactive legend but it should rather be employed to provide quick access to more specific information on POIs. The user can for example tap on one of the station icons placed on the map, which should result in the display of an information box as displayed in Fig. 4.4. This info box is labelled with the name of the specific station and encompasses a short description. The brief textual description may be scrolled up or down and the box further includes a link via an "info-button" to the detailed content on the station presented in the information browser. The info box can however also be closed without proceeding to more detailed information by means of a standard close button. The screen shot in Fig. 4.4 yet also illustrates that this info-box dialog is not only displayed following direct user interaction but is also triggered by context-based events such as the matching of the users position with the location of the specific station. In the information mode the pen can further be used to tap on other POIs than stations to receive a brief description of these objects (e.g. opening hours for a visitor center [252]).
Finally the headphones icon to the far right of the icon bar allows the user to activate an audio output functionality. If turned on, the mobile nature guide can provide context-based announcements or warnings and navigational instructions as audio files. In general the user should be made aware of, any kind of context-sensitive automatic notification of information presentation by some kind of audio signal. Such a functionality, can serve to provide additional opportunities for direct ex-

perience, since the user does not have to focus on the display at all times to navigate or to become aware of phenomena in his environment.

In accordance with the concepts previously discussed the maps utilized in a mobile nature guide need to be adaptable as well as readily adaptive to changes in context. It was also discussed earlier, that map applications on mobile devices have to meet additional requirements. Thus next to adaptation key issues are linked to performance and readability, calling for a technology which supports scalability of visualized geographic information along with small storage demands and dynamic adaptation. A variety of research projects illustrated the particular suitability of the XML-based scalable vector graphics (SVG) format for the implementation of maps on the web as well as on mobile devices [31, 133, 253, 268]. The scalability of a vector graphics format guarantees that no information is lost while zooming, because SVG based map objects, such as text labels, can be easily scaled to stay readable at other zoom steps, an essential feature on a device with a small display. Furthermore, scripting SVG using the Document Object Model (DOM) makes it possible, to dynamically change maps in nearly every aspect imaginable, like adding or deleting map objects and layers, changing visual attributes of the map objects etc. Further advantages of the XML-based format include, that the files size of even complex maps is generally small, further SVG is being established as an open-source vector graphic standard for web applications by the World Wide Web Consortium (W3C) [103] and also as a standard for generation of maps in conjunction with the OpenGIS standardization process [71].
The specific advantages of SVG for mobile nature guide UIs in general and mobile nature maps in particular are discussed in more detail by Düpmeier and Ruchter [94], Sobek [316] as well as Patalaviciute et al. [251]). This also includes the realization of concepts proposed specifically for mobile maps in natural areas such as the context-based adaptation of map features to changes in environmental conditions (e.g. seasonal changes in color schemes etc.).

Thus the map viewer GUI module for the mobile nature guide should consist of a software component that allows the display of SVG to be embedded in the frame application. Due to the relative novelty of the SVG technology the number of implementations for SVG viewers that run on Pocket PC platforms is still limited to date. As part of the MobiNaG project a number of "SVG-engines" that are platform compatible and at the same time can be embedded as a software component into a frame application were analyzed. A more detailed review of the implementations can be found in Sobek [316]. Based on the status of SVG implementations at the time, a "SVG-engine" was chosen (i.e. Intesis eSVG), that allows for the rendering and dynamic manipulation of SVG objects as well as user interaction with these objects. Prerequisite for this is that the SVG component grants the frame application access to the Document Object Model (DOM) of SVG documents and allows the controller to process user interaction with the UI rendered by the "SVG-engine".
In order to provide any form of speech-based acoustic support like for audio based navigational instructions (when audio support is activated), the module needs to have the capacity to replay or generate audio messages. A straight forward solution appears to be the integration of an audio

player component into the module that in case of an event can be triggered by the module controller to replay a prerecorded sound file. This would however make it necessary to produce an extensive library of spoken audio messages, which is likely laborious to manufacture and difficult to maintain [266]. Instead, as was pointed out by Real et al. [266], speech based audio output of an application is commonly generated by a Text-to-Speech (TTS) component. Real et al. [266] also provide a review of TTS systems that can be employed on a PDA platform.

Multimedia information browser module

The multimedia information browser module should realize the essential concepts proposed for the interpreter user service in the previous chapter.
As is specified in Table 3.5, this presentation of interpreter components primarily requires the display of sets of multimedia content. Such elements of an interpretive presentation are built on textual content, photos or other images as well as animations or videos and audio files. A discussion of aspects regarding the actual design (development) of multimedia content parts for a mobile nature part will follow in a proceeding section.

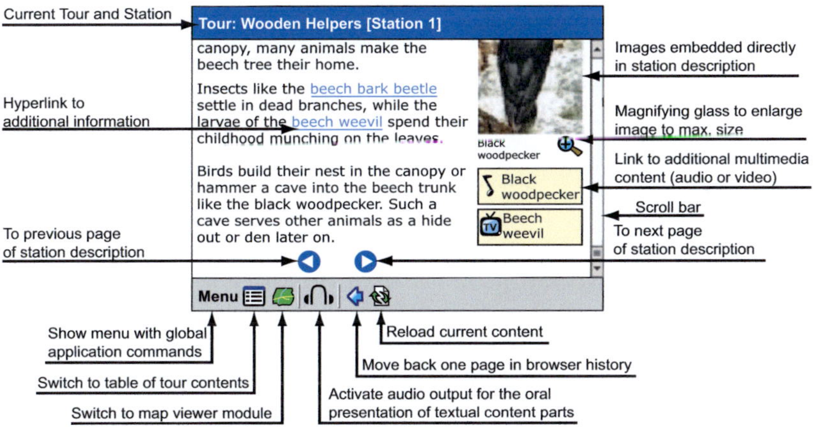

Figure 4.5: MobiNaG multimedia information browser UI - presenting the long description of a station as a sample multimedia content page along with an explanation for the proposed UI components and control elements.

Corresponding to the map viewer module the information browser module is constructed of the three major UI components: title/status bar, content area as well as menu/icon bar.
It can be observed in Fig. 4.5 that the title bar informs the user about the type of interpretive presentation (i.e. also the usage mode of the mobile nature guide e.g. guided-tour) in this case a thematic tour. By means of the status bar the user should also be able to readily recall the name

or theme of the interpretive presentation (e.g. "Wooden Helpers"). He should further be able to determine which part of the unit the presented content belongs to (e.g. "Station 1"). The presented example already indicates, that the options for the textual representation of this status information are fairly limited by the restrictions in screen size. This either has be to defined as a guideline for authors or should potentially be circumvented by using iconic representations for this status information.

The content area is reserved for the presentation of the actual multimedia information that is produced by authors/educators for the interpretive presentation. The core element of each content unit will commonly consist of a textual message, which can be accompanied by different forms of additional visual, audio-visual or acoustic media. In principal the specific layout for one or a collection of interpretive presentation for an AOI can be determined by the actual content provider or environmental education institution. Still, the overall layout of such a multimedia presentation should follow the basic design principles as promoted by Preece et al. [261]. Also the main usability principles as recommended by Nielsen and Søndergaard [236] and in particular the recommendations given by Weiss [356] for handheld systems should be taken into consideration. Fig. 4.5 presents one example for a content layout. In this sample layout the textual content is left aligned and about 2/3 of the screen are assigned to this part. The remaining section of the screen on the right is reserved for the other media elements that should further illustrate the information presented in the accompanying text. This can include graphs or photos, further characterized by an image caption, like the photograph of a black woodpecker in the presented sample station description. Next to such media, which are directly embedded in the content, the media section can also encompass interaction elements such as buttons, that offer a link to additional multimedia content presented in a separate screen (i.e. pop-up window) or player. Different types of additional media like audio files (e.g. a recording of the black woodpeckers call or drumming pattern) or videos (e.g. beech weevil larvae mining a leave) but also computer animations, can be represented by specific icons. All of these additional media will require a generic or specialized (e.g. Macromedia Flash) player component that provides the user with specific UI elements to control the presentation of the media file (e.g. "stop", "replay"). If the player is presented in its own screen, this should of course include the option to return to the info browser UI.

As was pointed out previously also the embedded multimedia elements may come with additional control elements also enclosed in the content area, like for example the magnifying glass icon. Based on a hypertext paradigm the textual content itself can also include further interaction elements that provide links to additional information. Such links may connect to other content parts within the same interpretive presentation or may serve as a reference to more detailed information on specific terms (e.g. "beech weevil" explanation beetle species) and thus initiate the glossary service. Corresponding to the concept for background information a link may, however, also lead to external information resources (e.g. web-based EIS or wiki systems). In addition to these control elements for multimedia information another category of control elements can be included directly in the content area. These are navigation elements, which are specific to this content unit (e.g. station description). The example in Fig. 4.5 illustrates an example for such navigation but-

tons (i.e. "previous" and "next"), which provide the user with an option to move back and forth between pages of the unit.
If the amount of content exceeds the visual dimensions of the content area, another standard control element is added to the UI, the scroll bar. It should work analogous to its functionalities on a desktop system.

In case of the information browser module the menu and icon bar consists of three groups of commands. In consistency with the general MobiNaG UI approach the global functionalities, including the menu as well as the list of contents icon, maintain their position of the far left of the bar. The multimedia content icon is replaced by a map icon, which correspondingly allows the user to switch back to the map viewer module. The number of module control elements is fairly limited, since the functions needed for the manipulation of the content are basically restricted to the navigation between pages and scrolling within the content of one page. Both of these functionalities can principally already be covered by control icons embedded in the content area itself. The headphone icon remains. It can initiate the audio output functionality. If the audio output is activated in the information browser UI, the complete textual content contained in the content area should be converted by a TTS and presented acoustically to the user. This is an important feature that can improve the accessibility of the system and at the same time may reduce the need of the user to focus on the device while meant to explore the natural environment.
Finally the last group of icons in the bar also represents module specific commands. They include standard web- or information browser functionalities resembled by the "back" (i.e. move back one entry in browser history) and the "reload" (i.e. renders the data of the displayed unit anew) icons. Due to the flexibility the module offers the author with respect to the design of content units and the inclusion of embedded navigation instruments these standard commands should remain as a basic control options for any content part.

The example shows, that a lot of the functionalities of the multimedia information browser unit match those of a standard web browser (also available for Pocket PC devices in form of the Microsoft Pocket Internet Explorer). Thus one option for the technical realization of the information browser UI would be the integration of a web browser component into the system. A further approach would however be to continue to rely on SVG as an implementation technology (compare also Düpmeier and Ruchter [94]). According to the current SVG standard established by the W3C [103] it should be possible to describe or embed all forms of multimedia content discussed above also with SVG itself. In comparison to plain HTML or XHTML content displayed by the Pocket PC webbrowsers, SVG would bring the advantage that also dynamic (i.e. animated) content could be presented in the browser or engine, without the need to make use of an auxiliary player application. It needs to be taken into account however, that in SVG there are not yet predesigned standard control elements like for instance a scroll bar that can be used out of the box in an application, increasing the amount of functionalities which have to be implemented from scratch.

Wizards module

For certain functionalities of the MobiNaG system, especially those requiring user input (e.g. changing of settings) an additional GUI-module is needed. This wizard module needs to have the capability to display and manage wizards or forms that offer a variety of user input modalities. For the display of tutorials, the module should, however, also be able to present interactive multimedia content.

Wizards may for example be offered to guide users through an initial personalization process which implies the initial collection of user context to build a user profile. Further a wizard should be made available, that allows the user to change basic settings of the system (e.g. activation of GPS unit) or that guides him through an initial setup process of the software. Settings for the MobiNaG system can for instance include the selection of a language for content presentation as well as commands or the selection of an AOI to visit along with the loading of the relevant data provided for this particular area, which is a necessary step prior to engaging in interpretive activities on site. Based on this scenario (i.e. a user selecting an AOI) Fig. 4.6 illustrates two sample screens of the wizard module.

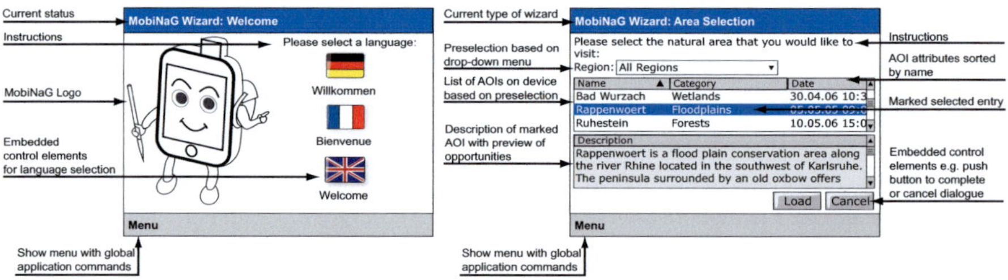

Figure 4.6: MobiNaG wizard UI - The module displays wizards or forms that either guide the user through a personification process, present tutorials or allow for the changing of settings. This Fig. depicts to sample wizard UIs, the welcome screen allowing the user to select the system language (left), followed by the wizard that allows the user to select an AOI he wants to visit (right).

Corresponding to the general UI layout, the screen for the wizard UI is also constructed from three components. The title/status bar reminds the user of the type of wizard he is currently using and potentially also of the step of a multi part wizard or tutorial that he is involved in.

In the welcome screen example with the "language selection" scenario, the user is greeted by the MobiNaG project logo placed in the content area next to a collection of embedded control elements. These push buttons represented by flags of different countries allow the user in an initial step to choose a language (e.g. German, French, English) for the subsequent usage of the application. Fig. 4.6 also displays one of the wizard screens following the welcome screen. In this sample scenario "AOI selection" the wizard unit presented in the content area shows an AOI specific file browser, similar to a standard "open file" dialog on Pocket PC systems. At first the user is provided with an instruction on the task that he is expected to perform with this wizard unit (e.g. "select the desired natural area from a list of available areas"). In order to fulfill this task the user is presented with a list of AOIs whose data can be accessed by the system. In order to reduce the length of the list and make it easier to comprehend, the user can pre-filter the AOIs by geographic region (e.g. state, bioregion etc.), choosing an entry from a drop-down menu. Next to the name of the AOI, the list entry can encompass a number of attributes of a respective area, such as a category it belongs to (e.g. type of ecosystem it primarily represents, such as flood plains) or the date that the data package for the AOI was generated. Corresponding to standard file lists, the user should be able to sort the records by each of the attributes. Users can choose an AOI with a tap of the pen on an entry. As feedback for the selection the AOI record is marked with a blue underlay. In order to grant the user a preview of the chosen AOI a brief textual summary of the marked entry is presented in an associated description box below the list. Finally corresponding to dialog screens there are additional control elements embedded in the content area. These control elements visualized as push buttons allow the user to complete the dialog by either finalizing the selection of a specific AOI via the "load' button or by terminating the entire wizard procedure via the "cancel" button. In case of this wizard the menu bar is solely devoted to the menu including the collection of global functionalities as previously described.

Since wizards and forms encompass mostly plain textual components next to standard input elements similar to other Pocket PC software, it seems reasonable to implement this module employing the same technology as for the frame application (i.e. using .Net and C++). Tutorials of course pose an exception as they should display a variety of multimedia information.

Agent player module

Next to the guidebook metaphor the interaction paradigm concepts for a mobile nature guide also envision the application of a human guide metaphor in form of an animated interface agent. The agent player module should constitute the presentation environment for the novel "e-Interpreter". Previous work conducted on an affective pedagogical agent as part of the MobiNaG project (see Real et al. [266]) provides a detailed discussion of the conceptual design of this module. According to these concepts, the agent player serves the presentation of scenes, which are pre-compiled by the controller unit of the module. A scene will commonly consist of an animation of the embodied agent character (also referred to as puppet) accompanied by an oral (i.e. speech output) and/or textual presentation as well as various interaction elements (e.g. lists of check boxes) (see Fig. 4.7).

Similar to the other GUI modules, the agent player UI also consists of three major UI components. Fig. 4.7 presents sample screen shots of two different forms of scenes.
The content area should be dedicated to the presentation of scenes. Core component of a scene is the embodied presentation agent or puppet itself, which in accordance with the definition for embodied conversational agents (ECA) by Bickmore [29] is animated to emulate the experience

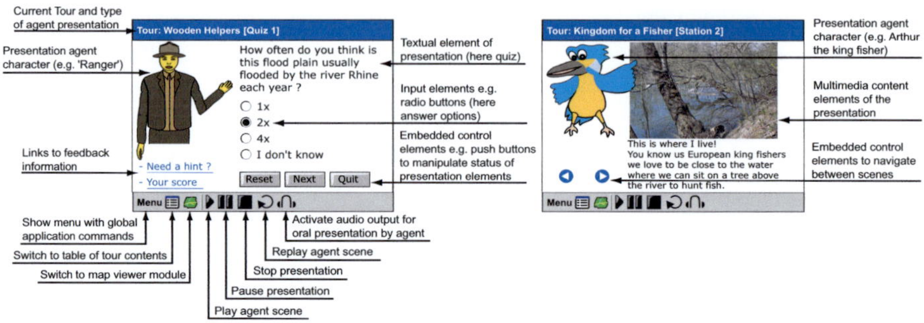

Figure 4.7: MobiNaG agent player UI - allows for the application of an additional interaction paradigm based on an animated presentation agent. A variety of characters can be used and different kinds of animated scenes can be played. This Fig. illustrates the presentation of a quiz by a "ranger" character (left) as well as the presentation of an interpretive unit by a bird-like animated character (right).

of human face-to-face conversation with the user, based on non-verbal (e.g. facial expression, gesture, posture) and verbal (e.g. speech, intonation) modalities. Isbister and Doyle [150] note that each character represents some form of specific personality. It was further determined, that the appearance and interaction style of the character influence its believability [151], which in turn can make a character more sensorially engaging and can facilitate or motivate interaction with the agent. In addition Krenn et al. [185] advocate, that users generally tend to trust ECAs more, which exhibit similar socio-cultural traits and thus make it easier for them to identify with, suggesting an agent's graphical appearance should be tailored to the target group or user preferences as well as the application domain. Consequently in a mobile nature guide system the characters (i.e. the appearance of the ECA) for an animated pedagogical agent in form of an e-Interpreter should in principal be exchangeable too suite the target group as well as the respective content of the presentation, like it is illustrated by the two screen shots in Fig. 4.7.

The agent character representing a park ranger should for instance serve as a realistic metaphor for the human guide that visitors are familiar with. The personality commonly associated with a ranger should help to increase trust in the system and support the believability of the agent and the system on the whole. Further a ranger is likely considered as a role model even by adults, which should help in influencing the participants environmental literacy. However the application of an APA that can represent different characters offers the opportunity to go beyond the mimicking of traditional metaphors for guided tours, for instance by making use of representations of natural phenomena such as plants, rocks or animals. These representations of natural phenomena acting as guides can potentially help especially children to immerse (i.e. to fully engage in direct exploration) in the natural world that is to be explored but may initially not be familiar or seem accessible. By presenting the natural environment in a first person mode and from his personal angle, a natural phenomenon character such as the king fisher, depicted in the sample screen shot, may further facilitate the perspective taking of the participant and thus the generation of emotional affinity with the character and his environment.

Next to the animated presentation agent, the content area should be used to display additional textual or other multimedia content as well as interaction elements. The combination of these elements depends on the unit presented (i.e. interpretive unit, quiz). In case of an interactive unit like a quiz or a feedback form, the respective question may be posed via speech output as part of the APA animation but should simultaneously be displayed as textual element. This redundancy grants the accessibility of the system to hearing impaired users and generally allows the repeated evaluation of the question at one's own pace, without having to review the entire animation. Essential parts of a quiz or form unit are various types of input elements. If for instance a query is intended as an open question, a plain text entry field could be used. Due to limited input modalities of handheld devices in particular with respect to text entry, it is however recommended to employ standardized questions with predefined answer options. These sets of options may be implemented for instance as check boxes for non-exclusive options or radio buttons for exclusive sets [356]. In order to manipulate the status of the specific unit, additional control elements should be embedded in the content area. For a quiz or form unit these may for example include push buttons to reset the given input elements, move to the previous or next presentation part or quit the entire unit.
Based on the APA concepts, the agent should interact with the user beyond the presentation of the question or form. This includes especially feedback to the user's selection of an answer option but also the provision of further hints or information on the user's performance (e.g. current quiz score). Such feedback or additional information may be offered either automatically based on system or user context (i.e. previous input, time passed since initiation of unit) or in form of links to feedback information that can be triggered by the user as displayed in Fig. 4.7.
If the agent player module is rather used for the presentation of an interpretive unit, the textual content may again be displayed redundantly to the oral presentation of the APA for accessibility reasons. The animation by the agent can further be illustrated by additional visual media such as photographs that the APA animation can incorporate in form of a demonstration, e.g. by using deictic gestures to point out certain features. A further option with reference to the APA functionalities of giving demonstrations or acting as role model, is the immersion of the agent character in a more complex animation, that may represent a virtual replication of the given natural environment, in which the agent can explain particular natural processes.

For the agent player module the menu and icon bar is composed of two groups of commands. The global functionalities, positioned again on the far left of the bar, match those of the multimedia information browser module. Since the presentation given through the agent player module can frequently be considered a supplement for the multimedia information presented in the information browser, the respective command was not included in the icon bar but the user will return to the information browser upon completion of or quitting of a unit. The synchronization between

the two modules is managed by the control center.
The module control elements are primarily dedicated to the manipulation of the agent animation. Similar to a standard multimedia player and in accordance with the conceptual design proposed by Real et al. [266], this includes icons for "play", "pause", "stop" and "replay" of the animation sequence including synchronized speech output. The headphone icon is still included as command to deactivate the, in this case as default, activated audio output. This allows those users who potentially do not want to be "distracted" by acoustic presentation to only watch the animation of the agent along with the rest of the content elements.

The implementation of the agent player module is described in detail by Real et al. [266]. The animation of life-like characters on mobile devices can best be achieved by employing a 2D character model based on a vector graphics format that supports tweening animations like for instance Macromedia Flash or SVG. Similar to maps SVG offers a viable solution due to its, dynamic adaptability based on the DOM, its small file size and the possibility to integrate other multimedia data. Since for the map viewer module a SVG-engine is to be employed it seems even more reasonable to rely on SVG as graphical display technology for the character animation and share the same component.
The application of a TTS system constitutes a suitable approach for the realization of the speech output [266]. A specific challenge composes the synchronization of the different presentation elements in particular animation and speech output. This task has to be managed by a specialized controller unit (i.e. the "director"), which is comprehensively portrayed by Real et al. [266].

4.2.2 Services

The software services provide an access to the different data types needed for presentation in the GUI modules. This section will present a "walk through" of the core functionalities of the MobiNaG based on the presented UI modules and elaborate on the respective service interfaces used for the retrieval of data.

When the application is launched a configuration object is created from a configuration file, which contains default values for settings of the system like for instance the default language or ports used for auxiliary sensor devices. This configuration object can then be incorporated by the various services upon initiation, where it provides access to the required attributes.
The first GUI module displayed following the start up of the application is the wizard module showing a "welcome screen" as illustrated in Fig. 4.6. In this first screen the user is asked to choose a language for the information to be presented in. The selected language is recorded by the configuration object. The next wizard screen allows the user to choose a natural area (i.e. AOI) that he wants to visit. In order to provide the user with a list of available AOIs (as displayed in Fig. 4.6) the wizard module needs to call upon the AOI service.

AOI service

At this point a thorough description is given of the AOI service, and its application programming interface (API), which offers the methods to access the core data needed in particular for the display of geographic data as well as multimedia content related to an interpretive presentation. Fig. 4.8 shows its main interface, the "IAreaService" interface along with the methods to access respective data and its interrelatedness to other interfaces. This representation of the API admits a more detailed description of the service functionalities.

At first the service has to be initialized and configured (i.e. init(conf:Configuration):void) with the attributes encompassed in the configuration object, supplied via the dependency on the "ConfigurationService". Once the service is initialized, various functionalities can be used. One of these functionalities of the "IAreaService" interface is the provision of AOI attributes to a wizard. As previously described the wizard for the AOI selection, following the initial welcome and language selection screen, needs to retrieve a list of AOIs available for a particular region (compare Fig. 4.6). The information needed for the dialog, the list of regions and a corresponding list of AOIs, can be accessed via methods of the "IAreaService" interface (i.e. getAreaDescription(region:String):List ≪IAreaDescription≫ and getAreaDescription():List≪IAreaDescription≫). The relevant attributes for the AOI description (i.e. name, category, date and AOI description) are retrieved via the "IAreaDescription" interface (i.e. getName():String etc.) connected to the "IAreaService" interface. If the user confirms his selection by pressing the 'Load' button, the ID of the selected AOI is sent back to the "IAreaService" interface where the corresponding AOI data is loaded (i.e. load(id:String):void).
Succeeding the selection of a natural area (i.e. AOI) the user is supposed to choose a desired usage mode (i.e. "guided tour" or "self-determined exploration") based on a matching wizard screen. The user's choice effects the further wizard steps as well as the kind of data that is loaded for the interpretive presentation. If the user decides to take a guided tour, he will be presented with a further wizard screen for the tour selection. This wizard will correspond to the AOI selection wizard shown in Fig. 4.6. The user should be able to choose from a list of tours available for the chosen AOI, which he can pre-filter by their attribution to specific target groups and sort out by attributes like "theme", "general topic" and "length". When a tour entry is marked the user will be presented with a brief overview description of the corresponding tour. The data needed to offer the selection dialog in the wizard, can also be accessed through the "IAreaService" interface by means of the "TourService" object. This sub-object is retrieved via the "ITourService" interface (i.e. getTourService():IAreaTourService). This interface offers the methods to supply the list of target groups as well as tours (i.e. getTourDescriptions(userClass:String):List≪ITourDescription≫ and getTourDescriptions():List≪ITourDescription≫). The relevant attributes can again be gathered based on a connection to the "ITourDescription" interface based on the tour ID (i.e. getTour(id:String):ITour), granting access by means of tour specific methods (i.e. getLength():double). This selection wizard is again completed when the user confirms his choice with the "Load" but-

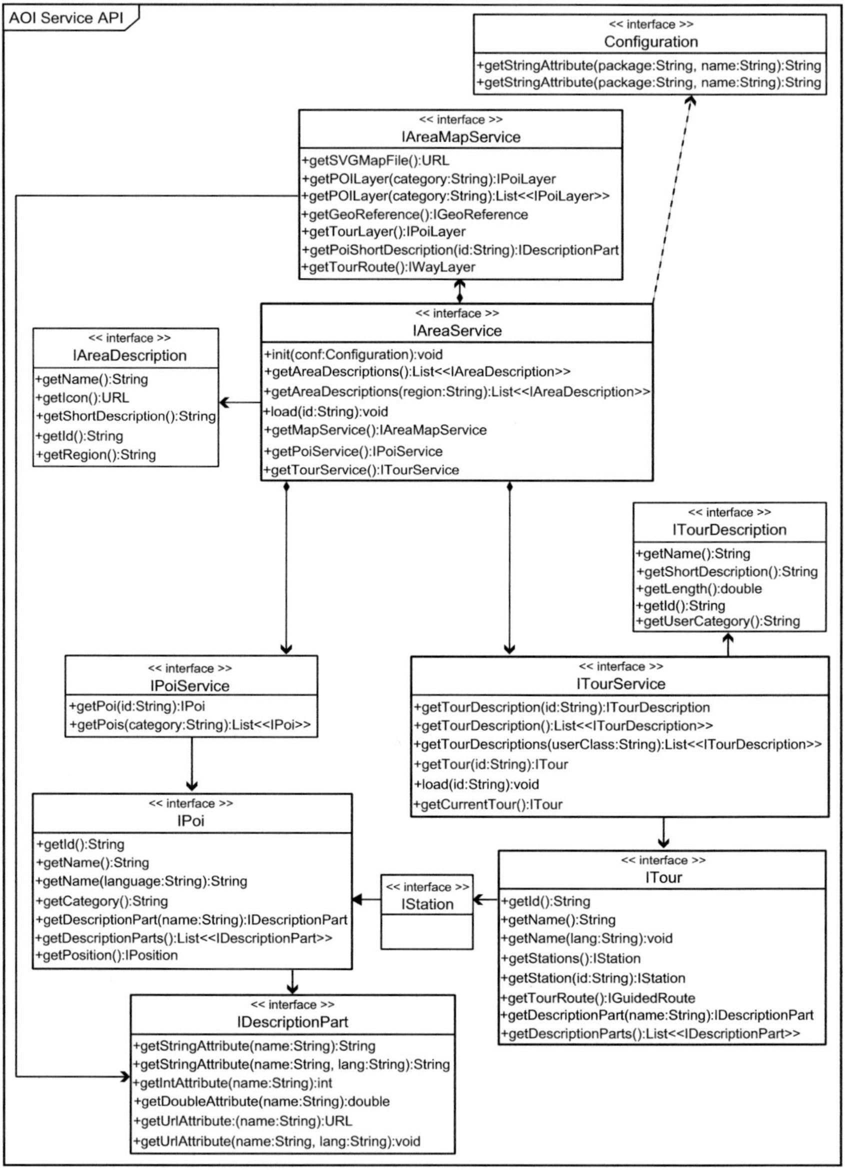

Figure 4.8: The diagram illustrates the application programming interfaces of the AOI service to the respective data models.

ton. This interaction is passed on to the "ITourDescription" interface, resulting in the loading of the selected tour based on its ID (i.e. load(id:String):void).

At this point the initial selection steps administered by the wizard modules are completed and the AOI service can prepare all data needed for the interpretive presentation based on the user's choice of AOI, usage mode and potentially also tour. In case the user chooses the "self-determined exploration" as usage mode, the tour selection wizard is omitted and he will directly be presented with a map of the chosen AOI including a default set of POIs (e.g. tourism infrastructure) in the map viewer. If a guided tour is chosen, the user will first be presented with an introduction to the tour in the multimedia information browser unit, before being presented with the map of the AOI in the map viewer, which then includes not only the default POIs but also geographical representation of the tour, including the respective stations and route. Thus in order to provide the information required for the interpretive presentation, the "IAreaService" interface has to employ specific methods to retrieve the data for the map, POIs and the tour.

The map data needed for the map viewer module can be gained via the connection to the "IAreaMapService" interface (i.e. getMapService():IAreaMapService). The map viewer GUI should display a base map of the AOI along with OOIs, routes and landmarks among other geographic information. The base map should be loaded for example in form of a SVG file, which is referenced via a universal resource locator (URL) (i.e. getSVGMapFile():URL). The base map is then overlayed with layers encompassing particular groups of map components such as POIs, routes or stations (i.e. getPoiLayer(category:String):IPoiLayer). In order to place the layers including the positioning indicator correctly over the base map, all map components need to be georeferenced. The necessary data for this step are also retrieved via the "IAreaMapService" (i.e. getGeoReference():IGeoReference). The display of certain layers of map components like for instance certain categories of OOIs, should be subject to user control by means of an interactive map legend. For the purpose of generating such an interactive legend the "IAreaMapService" interface offers a method to retrieve a list of OOI or in particular POI layers (i.e. getPoiLayers(categories:String): List≪IPoiLayer≫).
POIs as one specific type of OOI are characterized by a number of attributes which are also of importance for their representation in the map viewer GUI as well as the presentation of the associated location-based content in the information browser module. The "IAreaService" interface can gain access to the corresponding data via the "IPoiService" interface (i.e. getPoiService():IAreaPoiService). The "IPoiService" interface supplies methods to either get single POIs and their attributes based on an ID or can return lists of POIs based on a specific category (e.g. rest areas) (i.e. getPois(category:String):List≪IPoi≫). The actual POI attributes such as name, category and location are handled by the 'IPoi' interface (i.e. getCategory():String). A POI can further be described by any number of description parts, constituting the actual multimedia content. These data are administered through the "IDescriptionPart" interface. Each description part can be based on a variety of attributes of different types (e.g. string, integer, double) and it can

also reference further content like for instance images or videos through an URL reference (i.e. getUrlAttribute(name:String):URL). It needs to be pointed out, that in coherence with the language selection through the respective wizard module, the description parts, name as well as content should be provided in the different language and also need to be accessed specifically (i.e. getStringAttribute(name:String,lang:String):String).
In accordance with the proposed concepts, tours incorporate a combination of elements, including a route, stations along with additional description parts such as an introduction and a summary. The matching data can be accessed via the "ITourService" and through the "ITour" interface (i.e. getStation(id:String):IStation, getTourRoute():IGuidedRoute). Stations are essentially a specialized form of POIs, with equivalent attributes. They may match the location of a registered POI but have their own content parts and are accessed by the "IStation" interface.
Description parts of POIs as well as stations or the tour are generally displayed via the multimedia information browser GUI (see Fig. 4.5), which is initialized simultaneously to the map viewer module. Description parts belonging to the tour as a frame structure (e.g. introduction) can be directly retrieved via the "ITour" interface (i.e. getDescriptionPart(name:String):IDescriptionPart). Analogous to the described architecture for GUI modules both map viewer and information browser need to be linked to the control center. It is in particular the relationship to the event manager module of the control center that should be mentioned here, since it is responsible for switching between GUI modules. Once a user arrives at a station along a tour route and requests further information through interaction with the "target-info box", the event manager should switch from map viewer to the information browser, which displays the associated description part.

Positioning service

Another service which is crucial especially for the provision of location-based services by the MobiNaG system, is the positioning service. This service administers the access to the data collected by any form of positioning sensor technology (e.g. GPS sensor) utilized by the system. The API for the positioning service, as illustrated in Fig. 4.9, will be portrayed in more detail in this section.

Any component of the system that wants to employ positioning data (e.g. the map viewer requests the current position in order to adjust the positioning indicator) needs to go through the "INavigationService" interface. In a first step, the service is initialized and the configuration object will be integrated (i.e. init(conf:Configuration):void). Configuration parameters read by the "INavigationService" interface may include settings of the port, which the sensor is binding to, as well as the polling interval etc. Following its initialization the "INavigationService" interface provides methods to control the acquisition of sensor data (i.e. start():void, stopp():void) or to free the device again for usage by other applications (i.e. close():void). If another service wants to access position data, it needs to trigger the "INavigationService" interface to retrieve positioning data

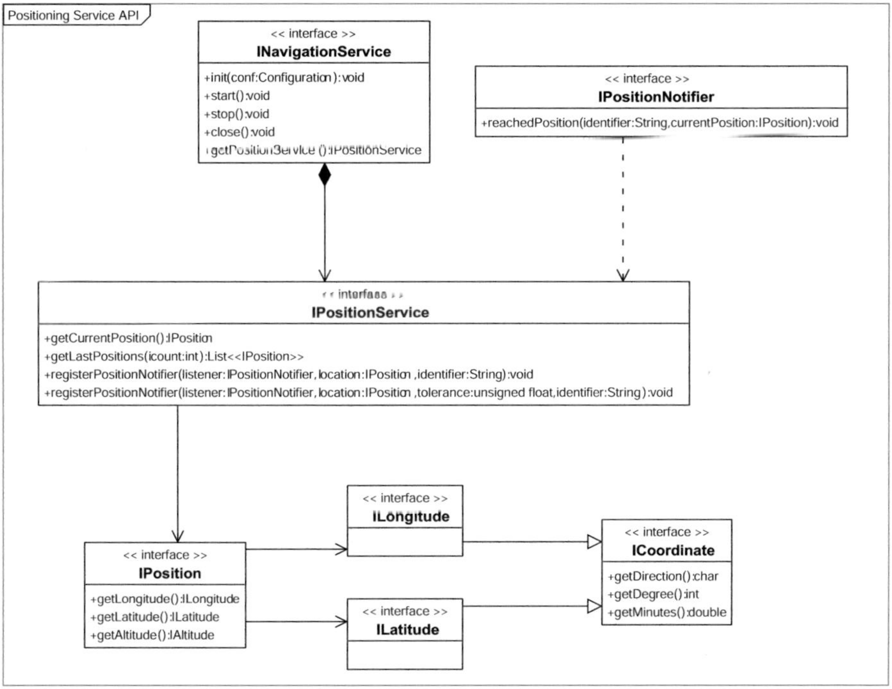

Figure 4.9: The class diagram illustrates the interfaces of the Positioning service, managing the access to relevant positioning data for the mobile guide to generate location-based events.

(i.e. getPositionService():IPositionService). The necessary method for the retrieval of these data (i.e. getCurrentPosition():IPosition) is supplied by the "IPositionService" interface. The actual position data set, which is enclosed in the "Position" object should include at least a longitude and a latitude attribute and may be completed by an altitude measure. As is shown in Fig. 4.9, these geographic features are accessed via the "IPosition" interface (i.e. getLongitude():ILongitude). Both "Longitude" and "Latitude" can be described as coordinate objects, consisting of a direction (e.g. N), a degree value (e.g. 48°), a decimal minute value (e.g. 59.795')(or minutes and seconds). The "IPosition" interface can also be shared by other services such as the "IPoi" interface of the AOI service, which also requires a position data set for each POI.
Next to the current position the "IPositionService" interface provides an option to retrieve a number of previous position data sets, with the number being specified by the programmer (i.e. getLastPositions(count:int):List≪IPosition≫). Based on the "getLastPositions" method, previous records can be used to perform alternative positioning strategies such as dead reckoning as well as certain error control procedures. If no current position data set is available, the previous records can be used for the interpolation of the users current position.

The fundamental functionality for location-based services is, however, furnished by the Position notifier. This unit can be used through the "IPositionNotifier" interface. With the "reachedPosition" method (i.e. reachedPosition(identifier:String, currentPosition:IPosition):void) the position notifier compares the current position data set with the position of a specific geo-object such as a POI or station, distinguished by an identifier. Consequently, if the current position matches that of the identified object (i.e. the user reached a station as illustrated in Fig. 4.4), a service is notified and triggers a location specific event like the presentation of an info-box. In order to utilize this functionality, the service needs to register with the position notifier through the "IPositionService" interface (i.e. registerPositionNotifier(listener:IPositionNotifier,location:IPosition,identifier:String): void). As shown in Fig. 4.9 this listener mechanism can next to the geo-object identifier be supplied with a tolerance setting. This parameter characterizes a circular tolerance zone around the target object that will allow the notifier set off when the user reaches its vicinity even if, due to potential inaccuracies of the sensor, the positions due not match up perfectly.

4.3 Implementation of a mobile nature Guide prototype

As part of the MobiNaG project a prototype mobile nature guide system was implemented that could serve as a foundation for an evaluation of the feasibility of the proposed conceptual design and IT concepts. The prototyping implementation was further the prerequisite for an assessment of the applicability and impact of a MobiNaG system in the EE domain.

4.3.1 Graphical user interface prototyping

For the implementation a prototyping-oriented incremental development approach was taken, similar to Pomberger et al. [258]. Following an initial assessment of user requirements, the prototyping cycle involved several stages of GUI prototypes of different fidelity, constructed in the course of this thesis.

Fig. 4.10 demonstrates the sequence of UI prototypes developed for the core GUI modules (i.e. map viewer and information browser). The figure shows the original prototypes designed in the course of the MobiNaG project, therefore the content and textual interface elements are presented in German. The first set of prototypes was produced as sketch-like mock-up. This low fidelity paper prototype was employed to evaluate the initial MobiNaG concept derived from the first requirements analysis. It was further utilized for a quick and low-cost field test of the proposed interaction paradigm along with the identification of the major usability issues, documented by Düpmeier and Ruchter [94]. The next set of the series was still generated using a paper-prototyping methodology but the graphical representation yielded a higher fidelity, being drawn with a vector-graphics

software tool. Similar to the low fidelity prototypes, this medium fidelity set was used for the re-assessment of user requirements next to the reevaluation of usability issues especially with respect to adjustments that were based on previous test results.
The experiences gained from the two mock-up based GUI prototyping cycles were then put to use in the development of the first functional prototype. This first implementation of the MobiNaG system again served as the basis for the impact evaluation of a mobile nature guide, which will be presented in the following chapter. Founded on the findings of the evaluation the functional prototype again is undergoing a redesign including improvements to the GUI.

4.3.2 Technologies and development environment

The MobiNaG technologies encompass a mobile device including interaction and connectivity technologies as well as context sensor. For the implementation of the prototype, Pocket PC based devices were used primarily the T-Mobile MDA II (next to HP IPaq 5450) with touch screen and pen, microphone and multifunction key as well as a camera as input technologies (see Fig. 4.11). Further the device is equipped with connectivity technologies, specifically GPRS and BT as well as a proprietary synchronization port. With regard to context sensors, the functional MobiNaG prototype was equipped with a GPS receiver to determine location. Specifically Bluetooth GPS receivers like the Rikaline^{TM} GPS-6031-X7 (see Fig. 4.11) as well as Compact Flash models like the Fortuna^{TM} PocketXtrack were employed. A more extended review of the technologies used in the implementation is given by Sobek [316].
Since a Pocket PC based device was chosen as a hardware platform the application was developed for an, at the time state of the art, Windows Mobile 2003 (Phone Edition) operating system running on an Intel PXA263 processor.

The frame application of the MobiNaG prototype, encompassing the major components of the control and the service layer were programmed in C++. Mainly for reasons of performance and processing power on the mobile device C++ was chosen over Java [316]. Consequently Microsoft eMbedded Visual C++ 4 was utilized as a development environment. Corresponding to the preceding discussion on the the various GUI modules, the implementation of these components of the presentation layer required the utilization of a technology, which can represent different forms of textual and other multimedia content, while the graphics should be scalable to different sizes and dynamically adaptable to changes in context. The suitability of SVG for this purpose was previously pointed out. Because of the promising qualities of SVG for presentation of multimedia content on small display devices the decision was made for the prototype system to implement the core GUI modules (i.e. map viewer and multimedia information browser) including all UI elements in SVG. A "SVG-engine" was chosen, which could be embedded in the frame application and allowed for the rendering and dynamic manipulation of SVG objects as well as user interaction with these objects. Prerequisite for this is that the SVG component grants the frame application

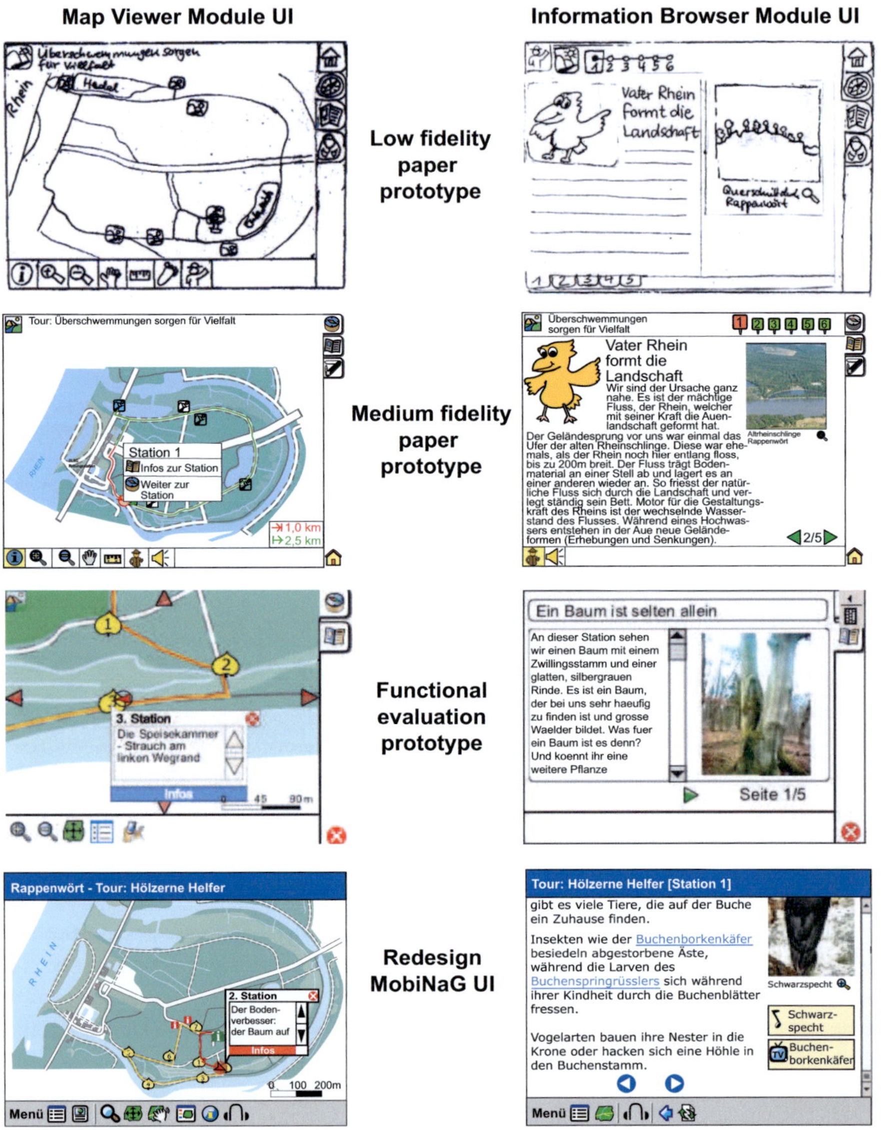

Figure 4.10: Sequence of UI prototypes of different fidelity levels, for map viewer and multimedia information GUI modules representing different stages of the prototyping process.

access to the Document Object Model (DOM) of SVG documents and allows the controller to process user interaction with the UI rendered by the "SVG-engine". SVG-DOM manipulations

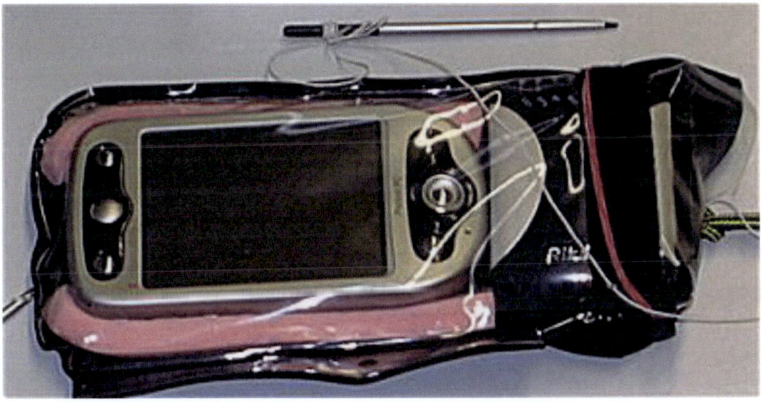

Figure 4.11: Hardware technologies employed for the MobiNaG prototype include a Pocket PC based PDA (left: here T-Mobile MDA II and stylus) and a BT GPS receiver unit (right: here Rikaline^{TM} GPS-6031-X7) enclosed in protective casing for field test.

as well as callback-functionalities for the communication between the embedded component and the frame application needed to be handled with JavaScript (i.e. ECMAScript). The Intesis eSVG was employed for the implementation since at the time it best fulfilled these requirements.

4.3.3 Prototype architecture

Next to the decision to use C++ for the development of the application as well as SVG for the realization of the GUI modules, it also became clear early in the development process, that it would be important to achieve a clear separation between UI design elements and content as well as the application logic [94]. This rationale was based on the results of the requirements analysis and the paper prototyping procedure, which yielded that the different target groups needed different styles of interaction and control elements. Thus similar content should be presented with different types of UI design. In addition, the scenarios developed for a mobile nature guide system clarified that content and UI elements will likely be maintained by different factions (i.e. EE educators and UI designers respectively) likely non being IT specialists. With this in mind, an IT-architecture was devised for the MobiNaG prototype (see Fig. 4.12) following the MVC pattern introduced above.

In accordance with the MVC architectural pattern, the C++ application is separated into a controller component, a view generating template engine and different components which define the model logic of the application. The user interface of the application is rendered in form of different SVG screens by a C++ SVG rendering component (i.e. embedded SVG-engine), which also accepts the interactions of the user with the UI elements and forwards them to the central

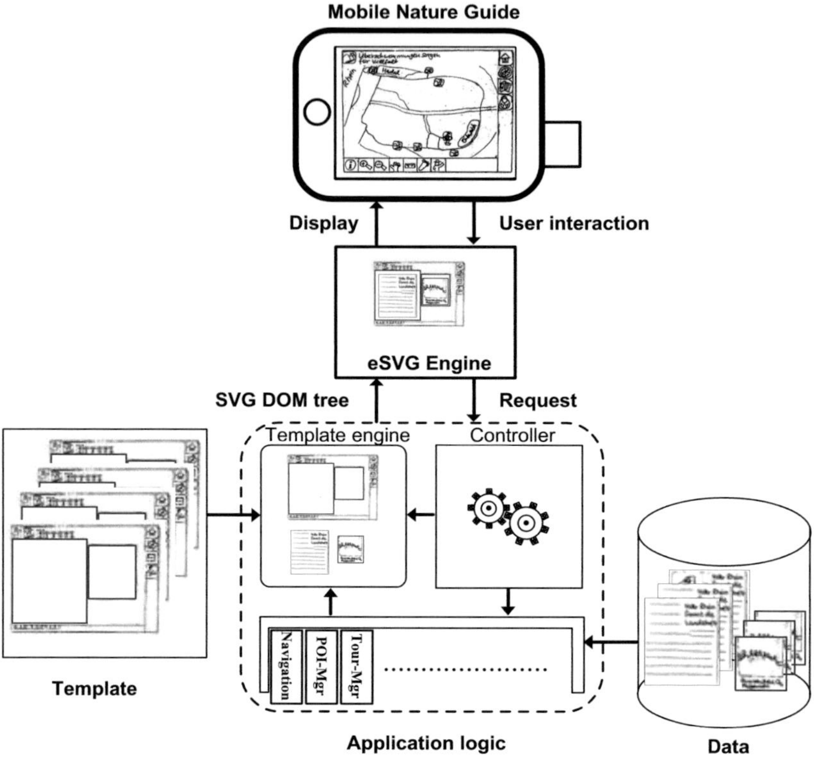

Figure 4.12: The first functional prototype was based on an earlier concept for the overall system architecture - also based on an MVC pattern (adapted from Düpmeier and Ruchter [94]).

controller component. The application logic components, such as the navigation manager, POI manager or tour manager register with the controller to intercept relevant SVG interactions, and process them based on context data like AOI features or positioning data.

The components use a well defined SVG generation interface to perform changes in the SVG DOM tree. These DOM changes can be used to deliver information to the user by changing parts of the SVG based user interface, such as the map viewer or rendering a popup window with a textual notification for the user. This SVG generation interface is referred to as "template engine" because it can use templates of SVG code, which are completed by application data and are then transformed into DOM tree changes resulting in potentially complex changes of the GUI, like adding new images or a complete layer of symbols to the map, after being rendered by the SVG-engine.
The embedded SVG-engine does not only allow direct changes to the DOM tree of a displayed

SVG document via its eSVG component API, but it also grants the execution of JavaScript functions, defined in the SVG document itself. This makes it possible to define entire user interface elements, such as a popup info box or a tooltip label, as JavaScript implemented UI objects. These objects can be stored in a UI library linked to the SVG document and called from the C++ application by executing JavaScript code fragments, which will be interpreted by the eSVG component's JavaScript interpreter. Some basic user requests, like zooming of the map, can be handled more easily without the intervention of the frame application, by just calling a JavaScript function within the SVG document itself.

The overall functionality of the prototype can be described in a brief scenario. For instance if a user requests detailed information on a station by interacting with a push-button (i.e. "Details") in the SVG screen, the SVG-engine forwards the interaction to the central controller component in the main application logic of the MobiNaG system. The controller decides, which data objects are needed to fulfill the user's request (e.g. textual description, image etc.), and initiates the retrieval of these objects by the corresponding application logic module (e.g. the tour module). The controller thereafter decides, which SVG screen should be used to present the response to the user's request and calls the template engine component to choose the corresponding SVG template file. The template engine either assembles an entirely new screen from template and data objects or merges the data objects into an already loaded DOM tree representing a SVG template in memory. The template engine then calls the SVG-engine to display the new or modified SVG DOM tree to the user.

4.3.4 Implementation issues

Several of the technical issues that were encountered throughout the implementation relate to the decision to rely entirely on SVG for the construction of all GUI components. In accordance with the proposed advantages of SVG for adaptive visualization especially on mobile devices [103,268] the MobiNaG implementation provided an opportunity to examine if a user interface solely based on SVG technology could lead to a more natural user experience. Despite these promises a number of difficulties were encountered while developing the MobiNaG prototype:

Lack of standard GUI elements In contrast to classical GUI applications, which rely on GUI libraries (e.g. Qt, JSF) for their UI elements, SVG did at the time not provide any standard GUI widgets. Consequently in the process of developing the UI for the MobiNaG prototype all GUI elements had to be created from scratch, including SVG object and JavaScript functionality. Next to MobiNaG specific buttons this also included the design of standard interaction elements such as scroll bars and pop-up windows, making it a time consuming endeavor. Moreover, this became a usability issue given that in some cases the visual or temporal characteristics of the standard Windows desktop GUI elements, that most users are familiar with, could not be exactly matched (e.g. feedback animation of scroll bar), resulting in initial confusion and to some degree frustration of the test users.

Limitations of the SVG specification The present SVG 1.1 specification, recommended by the W3C [103], has certain limitations regarding the presentation of multimedia content apart from SVG objects. In particular it does not support advanced text features such as text flow (i.e. wrapping continuous text onto multiple lines), but only text labels. In order to present non-vector-graphics textual content in flow text, the wrapping feature had to be added. But this text wrapping procedure reduced the performance of the SVG-engine, increasing the time for loading new textual content.

Incomplete implementation of SVG specifications The eSVG component implements only a subset of the SVG 1.1 and SVG Mobile specifications. As a result not all the SVG objects (e.g. object with a color gradient) created in an SVG editor (i.e. Inkscape) could be rendered properly by the eSVG-engine. A more severe constrain however was the lack of support for URL references by means of the "xlink:href" attribute. In conjunction with the '<use>' element, the "xlink:href" attribute can serve to reference (i.e. embed) external SVG or other multimedia files, which is the foundation for the construction of true SVG templates. For example the map viewer module is supposed to consist of a frame template encompassing all control elements which are not part of the map itself, while each map component, or group of map components united in one layer, is maintained in a separate SVG file and reference via a "<use>" element. Since this feature is not realized in the utilized SVG-engine each UI, in particular the map viewer had to be constructed as one complex SVG file. The display of specific SVG objects, was then controlled via the "visibility" attribute. A further disadvantage of this restriction was, that the complex and large SVG "frame template" was difficult to maintain and restricted the performance of the GUI modules. Even though the layout of the SVG UI could be controlled via an external CSS stylesheet, a proper MVC pattern could not be established.

Limited performance on Pocket PC The performance of the MobiNaG prototype system turned out to be one of the key issues. The more or less cumbersome communication mechanism between C++ frame application and the SVG GUI limited the reactivity of the application considerably. Because of the overall limitations of processing power and thus also performance of the eSVG-engine on the Pocket PC device, SVG animations would not run smoothly and could therefore only be put to limited use in the prototype. This constrained in particular the implementation of the animated interface agent (i.e. the e-interpreter), which is discussed in more detail be Real et al. [266]. For the prototype system the decision was made to present the entire GUI in a landscape layout. This was a further challenge for the development because the operating system utilized on the test PDAs (i.e. Windows Pocket PC 2003) did only support a portrait layout by default. Therefore the entire SVG

GUI had to be adjusted with a rotation transformation to realize the landscape layout. This transformation step further reinforced the performance issues.

Restricted context-awareness The first implementation of a functional MobiNaG prototype was focused on the development of the frame application along with the main GUI modules (i.e. map viewer and multimedia info browser) and the core services needed to support the participation in a guided nature tour (i.e. navigation and tour). During this process context-awareness was also realized only in a rudimentary form. In essence the system made use of the two context features location and target group. Location was captured explicitly via the GPS sensor and applied to adapt the position indicator on the map as well as for presenting location-based content on natural features. The target group feature was captured explicitly through user selection of predefined tours for families or students, resulting in the presentation of content tailored to the requirements of the respective group. But the prototype did not include a unit like the context manager that would allow context aggregation and abstraction. In particular context features crucial to a mobile nature guide such as aspects from the categories environment and time (e.g. season, status of phenomena) could not yet be considered.

The redesign for a future MobiNaG system needs to address these issues identified during the development of the first functional prototype. The IT-concepts proposed in the preceding sections have already been formulated with the experience of the prototype development in mind. The revised system architecture (see Fig. 4.1) offers a potential resolution for most of the identified issues.
Several of the discussed issues are tied to the implementation of the entire GUI in SVG. The new concept for GUI modules suggests that the frame elements of the GUI, including title/status bar and menu bar should be implemented as part of the frame application (i.e. in C#). Next to these native UI elements, the GUI Modules will also allow for the embedding of various software components that can be employed for the presentation of multimedia content. In this way SVG can still be employed where it is most suitable, ergo the presentation of graphic-rich content that needs to be dynamically adaptable and scalable (i.e. presentation of maps and animations). Thus the SVG-engine software component can for instance be embedded in the content area of the map viewer module. In those cases where the current SVG specification and/or its mobile implementation for SVG-engines are not suitable for building an adequate GUI (i.e. standard interaction elements and presentation of multi-line textual content), other components, such as an internet browser software component can be embedded into the application. This approach should resolve several of the issues related to the performance of the application and the usability of the UI elements. It should further facilitate the overall UI design process since standard GUI elements, like scroll bars will be provided by the software component (i.e. internet browser) or, for instance in case of wizards, can be built with standard GUI widgets utilized by the frame application itself.

The redesign further promises a more consistent implementation of the MVC pattern as presented in Fig. 4.2.

Next to the improvements to the presentation layer, the revised architecture in particular provides the opportunity for an extended use of context. This includes the implementation of specific context services that capture and process for instance user or system related context features. A key component with respect to context-awareness of the application, which has to be added to the architecture, is the context manager. It administers the access to context data and offers notifications about changes in context. But in conjunction with the incorporated reasoning module and the knowledge base, the context manager also has the capacity to go beyond the aggregation and abstraction of context to inferring context and storing context states as context widgets. Finally the utilization of the proposed rule engine should allow the MobiNaG system to perform complex adaptations based on the state of its context model.

Also the controller has to be enhanced in the extended system architecture. The clearly structured modules for example allow the management of interpretive presentations tailored to the usage mode of the system. It should also be noted that in the redesigned MobiNaG system the data model for multimedia content services (i.e. AOI services) will support the presentation of multimedia content in multiple languages.

Further progress regarding system performance and flexibility of use can be made by using novel mobile devices with more processing power. Also the newer versions of Pocket PC operating systems (i.e. Pocket PC 2003 SE, Windows Mobile 5.0) have the advantage of layout rotation being already built into the system. In order to improve the efficiency of the development the programming has to be shifted to the Microsoft .Net Compact Framework employing C# for programming language.

Chapter 5

Evaluation of the New Mobile Nature Guide

5.1 Motivation and related work

Following the implementation of a mobile nature guide prototype, based on the devised concepts, it remains to be assessed if such a system can actually contribute to solving the discussed EE challenges.

A fair amount of evaluations of mobile guide systems has been conducted, as indicated by the review by Kray and Baus [182] (see also Appendix A.3, Table A.4). However, most of these studies focus on the usability of the respective systems [172] and to some extent investigate their support for domain specific tasks [61, 301].
Among the evaluated mobile guides are also a few mobile environmental guide systems. The WebPark system [84, 85] was evaluated in the field but the evaluation of the system also focuses on the satisfaction of user needs and does not include an outcome evaluation or performance measures. Of the evaluation studies on nature guide systems, Rogers et al. [276] (i.e. Ambient Wood project) and Okada et al. [243] (i.e. DigitalEE II project) made important contributions regarding the potential of nature guides for scaffolding environmental learning in general. Next to usability related aspects the study by Abe et al. [1] (i.e. Forest Education Support System) also addresses the educational efficacy of the nature guide system for forest education. However, educational efficacy in this study is only based on a measure of self-reported increase of interest in nature. Thus, none of the studies on existing nature guide systems actually included an impact assessment, commonly used in EE research. Impact assessment is defined by Rossi and Freeman (Rossi and Freemann 1999 cited in [32]) as an evaluation study that answers questions about program outcomes and impact on social conditions it is intended to ameliorate. Thus it can be considered a specific type of summative evaluation or outcome evaluation [32]. Leeming et al. [195] as well as Gunderson et al. [123] provide extensive reviews of such outcome research of

"out-of-class" EE interventions. Only two of these studies, however, include a comparison of the effects of different media-based interventions (Fortner 1985 and Simmons 1984 cited in [195]). This lack of empirical comparisons between different media can also be observed for the research on mobile guides. Even though, a number of studies point out differences between mobile guides and traditional guidance media [11, 50, 62, 91, 203, 342], very few empirical comparisons have actually been conducted in the field. Apart from an ethnographic survey by Woodruff et al. [366], those studies mostly focus on the comparison between electronic and paper-based maps and their effects on navigational performance as well as user preferences [117,273]. Some work comparing the effects of different interpretive media and techniques was conducted in the field of visitor studies mostly in the tourism and museum domain [226, 308].

Thus, studies comparing mobile guides to traditional media or approaches (i.e. paper-based brochures/guide books or human guides) are generally overdue, since they constitute the bases for further improvement as well as for the judgement of the economic viability of such systems in the various domains. Especially in the EE domain the evaluation of new computer-mediated approaches, such as mobile nature guides, is essential to select those systems that will yield a true benefit over traditional media and approaches and can thus help to meet new challenges of the domain.

5.2 Evaluation objectives

The goal of this evaluation is the assessment of the effectiveness of the mobile nature guide prototype along with a comparison to traditional environmental interpretation activities offered at EE institutions. These include tours along an interpretation path either as self-guided tour employing a brochure or tours given by a human guide[1].

The following hypothesis were tested as part of this impact evaluation:

Hypothesis 1 *Intervention effect:* The conducted EE intervention has an effect on the environmental literacy components of the participants.

Hypothesis 2 *Media effect:* The media employed to administer the intervention cause a difference in the state of environmental literacy components.

Hypothesis 3 *User satisfaction:* The satisfaction of the users with the tour and thus the success of the interpretation will differ between the groups employing different media.

[1] For the purpose of simplicity, the human guide will hereafter also be referred to as an interpretive medium.

5.3 Materials and methods

Corresponding to the front-end evaluation this remedial evaluation was conducted in collaboration with the MobiNaG project partner NAZKA.

5.3.1 Participants

The evaluation focused on students and families with children as two of the major target groups identified during the front-end evaluation. The third group (i.e. adult nature lovers) could not be included in the assessment, due to constraints in time and personnel. Both families and school classes were recruited through an information leaflet or articles in a newspaper and on the web.

Families Among the families responding to the call for participation only those were accepted who had at least one child of the age of seven or older. This age barrier was made a prerequisite, since children were expected to have the ability to read texts in order to fully interact with the interpretive media. Furthermore, according to Beringer [25] the ages of 5-12 years characterize a period of childhood, in which significant and transformative encounters made with nature can imprint on a child's mind for life. This makes the age group a critical target group for EE activities. A total of 34 family groups participated in the study, including 78 parents (i.e. adults including parents and other relatives or friends) and 76 children of the target group. Of the parents 53.8% were female and on average 40 years old. For male parents the average age was 41.5 years. Among the children being evaluated 51.3% were girls with an average age of 11 years. Boys were on average 10 years old.
Families did not constitute a true random sample, but were subject to self-selection based on their interest in engaging in an interpretive activity at the test site. The family clusters were further randomly assigned to the three different treatments (i.e. interpretive media) on a "next medium available" bases. Table 5.1 gives a more detailed representation of the basic demographic characteristics of the adult family members. About 33% of the parents were assigned to the intervention with the mobile nature guide, while 27% were delegated to the brochure group and approximately 40% took part in an intervention with a human guide. The gender was basically evenly distributed in all groups, except for a slight offset in the group assigned to the human guide that included 58% women. The average age of 41 years was the same in all groups. Adults with an academic degree and those without an university education were close to evenly distributed in the group assigned to the mobile nature guide. The human guide group sample included a slight excess of 8% of adults with an advanced education, whereas the group assigned to the brochure only consisted to 33% of adults with a higher level of education. In all groups the majority of adult participants had previously visited the visitor center, especially the mobile nature guide group only included about 35% of first time visitors.

Table 5.1: Basic demographics of adults as part of families assigned to media groups.

Media Groups	No. of subjects	% of all parents	Female	Average age (Median)	Academic education	First time visit
Parents MobiNaG	26	33.3%	50.0%	41	53.8%	34.6%
Parents Brochure	21	27.0%	52.4%	41	33.3%	42.9%
Parents Human Guide	31	39.7%	58.1%	41	58.1%	41.9%

Table 5.2 presents the basic demographic characteristics of the children, who were also part of the family groups. The participating children were on the whole evenly distributed among the three treatment groups. It is noticeable, though, that girls and boys are not quite evenly scattered. More girls were part of the human guide group, whereas, it was vice versa for the other groups. It is further apparent that children in the mobile nature guide group are slightly older and 50% attend the most advanced type of high school, while most of the remaining children are still in elementary school. In the two other media groups more than half of the children were still at the elementary school level while most of the others attended also the advanced school form. In comparison to the parents, all of the media groups included fewer first time visitors. The highest proportion of first time visits (i.e. 39%) was determined for the brochure group.

Students The target group "students" was sampled in clusters of school classes. Only classes starting at fifth-grade were accepted for participation. A total of five secondary school classes, comprising 109 students, could be recruited for the study. These classes ranged from fifth- through seventh-grade (i.e. 2x5th, 2x6th, 1x7th grade) and represented all three tracks of the German school system. It can be assumed that the participation was not the students' choice. Still, the teacher willing to engage his class in an EE activity made the selection and thus, the classes also have to be considered a self-selected sample. Each class was, however, supervised by a different teacher and was evaluated on separate days. In a second stage cluster sample, students were assigned to the treatments in sets of three students. At the same time it was attempted to balance gender and performance level within treatment groups, resulting in a disproportional stratified sampling. None of the school classes received prior instructions on the intervention content in preparation for the participation.
Table 5.2 displays the basic demographic characteristics of the student groups assigned to the different treatments. Similar to the children, also the students are nearly evenly scattered among the media groups, with only slightly more attending the intervention with the human guide. The

Table 5.2: Basic demographics of children (family groups) and students (school classes) as assigned to media groups. (Types of school corresponding to German 3-track school system: Basic="Hauptschule", Medium="Realschule", Advanced="Gymnasium".

					Type of school				
Media Groups	No of subjects	% of all children/stud.	Female	Average age (Median)	Elementary School	Basic track	Medium track	Advanced track	First time visit
MobiNaG (Children)	26	34.2%	42.3%	11	42.3%	0.0%	7.7%	50.0%	23.1%
Brochure (Children)	23	30.3%	43.5%	10	56.5%	8.7%	4.4%	30.4%	39.1%
Human Guide (Children)	27	35.5%	66.7%	9	59.3%	3.7%	3.7%	33.3%	29.6%
MobiNaG (Students)	33	30.3%	48.5%	12	0.0%	54.5%	27.3%	18.2%	30.3%
Brochure (Students)	36	33.0%	47.2%	11	0.0%	50.0%	25.0%	25.0%	38.9%
Human Guide (Students)	40	36.7%	45.0%	11	0.0%	50.0%	17.5%	32.5%	52.5%

same is true for gender, there are only slightly more boys than girls represented in all groups. The average age was 12 for those taking part in the mobile guide group and 11 for all other groups. In all treatment groups about 50% of the students attended the basic school track. With the exception of the brochure group, more first time visitors were identified among the students than the children. In particular the human guide group stands out with about 50% being new to the site.

5.3.2 Study design and procedure

It is the objective of the evaluation to assess if the mobile nature guide, developed for mobile EE, can be effectively employed in comparison to traditional EE instruments. Mobile, interactive computer systems are commonly evaluated in lab studies [171]. However, since the conditions given in natural areas cannot be realistically replicated in the lab, a mobile nature guide should be tested in the field [94]. According to Kjeldskov and Graham [171] field studies are characterized by taking place in "the real world" under realistic conditions, drawing upon qualitative as well as quantitative approaches. Consequently, a combination of an ethnographic field study and a field experiment was conducted attempting to maintain a high degree of realism with regard to the application domain.

With reference to Bittner [32] this study was designed as a quasi-experimental pretest/posttest design with comparison groups as displayed in Fig. 5.1. As was previously pointed out, subjects were sampled as clusters in the form of families or school classes. The fact that no true random

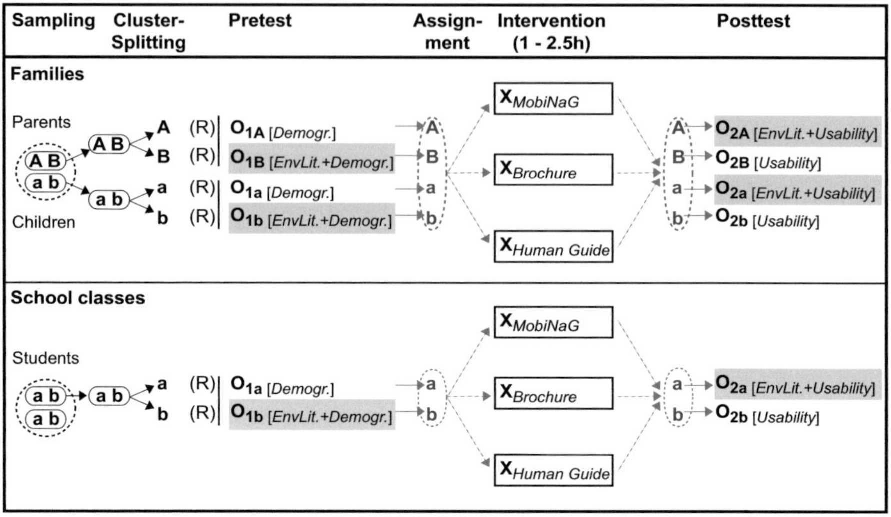

Figure 5.1: Quasi-experimental pretest/posttest design with comparison groups employed in this study to evaluate the effectiveness of the mobile nature guide prototype in comparison with traditional interpretive media. In the pretest participants were asked to complete either a questionnaire combining the environmental literacy instrument (EnvLit.) with the socio-demographic elements (Demogr.) or a questionnaire including only the socio-demographic elements (Demogr.). In the posttest the participants had to vice versa answer the EnvLit. in combination with usability questions (Usability) or solely the usability questionnaire (Usability).

samples could be taken but given population segments or clusters were employed to test hypothesis about differences between these groups, renders the study a quasi-experiment [32, 294]. A key issue regarding this methodology, is that the investigated groups may not only differ in the treatment but also concerning several other characteristics [293, 294]. As a precaution Schahn and Bohner [294] suggest an investigation of the comparison groups, in particular by means of a comparison between pretest values for dependent variables. The pretest/posttest design is further commonly used in outcome research in the field of EE, as shown in a review by Leeming et al. [195], in order to determine an effect of the intervention. This typically means that groups take a test measuring the dependent variables before participating in the intervention and take the same test again at some time following the intervention [195]. Still such a pretest/posttest design potentially involves additional issues, as discussed by [293]. A danger with repeated-measures is the existence of carry-over effects [121]. Carry-over effects can be positive (i.e. lead to an increase in performance) e.g. through practise or negative (i.e. cause a decrease in performance) e.g. due to frustration or fatigue. Especially in the case of a short-term intervention, with at most 2.5 h between observations, a practise or learning effect is likely. To avoid this, it would be necessary to take the posttest measurement on a different cluster of subjects. However, as Bittner [32] points out the pretest and posttest groups should be as similar as possible with respect to crucial characteristics, in order to make a valid comparison between the groups. As a solution to this dilemma Bittner [32] proposes an alteration of the classic pretest/posttest design in the way that a cluster is split into ideal halves, which are then assigned respectively to pretest or posttest.

A similar approach has been taken in this study (see Fig. 5.1). Following their arrival at the visitor center and a brief introduction, the families were asked to take a first test before engaging in the EE activity. Prior to the pretest the family cluster was split and parents were randomly assigned to a group A or B. The same procedure was performed with the children assigning them randomly to a or b. Thereafter, all subjects were asked to take the pretest based on a semi-standardized self-completion questionnaire, without information on the subsequent type of intervention that they would be assigned to. As described above, only half of each sub-cluster (compare Fig. 5.1 group B and b) received a questionnaire at this point that included the instrument (i.e. questions related to environmental literacy variables) next to socio-demographic elements. The other half of the cluster was also administered a questionnaire at the same time that, however, only included the socio-demographic elements. This procedure is recommended, since it gives all subjects the impression of equal treatment, avoiding unnecessary frustration or bewilderment in either the pre- or posttest group [32]. Following the pretest the entire family as a group was randomly assigned to one of the three treatments including a self-guided interpretive tour with the mobile nature guide prototype ($X_{MobiNaG}$), a self-guided tour with a brochure ($X_{Brochure}$) and a guided interpretive tour administered by a human guide ($X_{HumanGuide}$). All EE interventions had the same content and essentially, differed only in the presentation medium. The families assigned to self-guided media were granted as much time as they needed to complete the tour at their own pace. All groups were accompanied by an experimenter making observations.

The intervention lasted for 1 - 2.5 hours. Immediately after the tour, participants were asked to take the posttest. In a kind of "measurement counterbalancing," the other half of the adult and child cluster (compare Fig. 5.1 group A and a) was now served with the environmental literacy instrument plus a usability questionnaire. The other participants were vice versa asked to fill in only the usability questionnaire. For either pre- as well as posttest, participants took approximately 15 to 25 minutes for the completion of the questionnaire.

Essentially the same procedure was performed with the school classes participating in the evaluation, also displayed in Fig. 5.1. The students were randomly assigned to group a or b before taking the pretest. Following the pretest, sets of students from both groups were joined and assigned to one of the three treatments. For organizational and pedagogical reasons, sets assigned to a self-guided tour could not include more than three students.

The field study was conducted April through May of 2005. Evaluation sessions for families were scheduled on weekends, whereas school classes were invited to attend on weekday mornings.

5.3.3 Environmental education intervention

For the EE intervention, administered during the study, a guided nature tour was used. The EE activity was conducted at the flood-plain conservation area, chosen as a test site, and was designed in collaboration with NAZKA, the EE institution located on site. The intervention materials were created based on the environmental interpretation principles previously discussed and the concepts for guided tours presented in Chapter 3 of this thesis. The resulting tour with the theme "Wooden Helpers" guided the visitors along an easily accessible 1.5 km trail, starting and ending at the visitor center. Along the trail 7 natural phenomena, in this case extraordinary trees, were selected as Points of Interest (POI). At each of the 7 stations the visitors were presented with a brief text and images, telling them something about the functions trees fulfill in an ecosystem and which services they perform for man kind (see Fig. 5.4). Two different versions of the tour were created, adapted to the target groups of families and students, respectively. Contents for school classes were mainly kept shorter and easier to comprehend, while the essential information remained the same. Corresponding to the objectives of the thesis and the developed concepts, the intervention went beyond the transfer of environmental information and incorporated constructive elements to foster direct experience with the interpreted natural phenomena. Similar to the approach taken by Bogner [38], this included cognitive activities (e.g. searching and recognizing plants), emotional elements (e.g. trying to empathize with a squirrel in order to help it choose a den location) and also performing activities (e.g. embracing a tree together to measure its circumference) (see Fig. 5.2). These hands-on activities were integrated as games into the interpretive presentation at each station.

Still, the study did not focus on the tour itself, but on its effect on the participants based on the form of interpretive medium employed. Consequently the intervention was administered through different media corresponding to the three treatments in the study design, encompassing both personal- as well as media-based forms of interpretation. The attempt was made to design the different interpretive media in a form such that the treatments would only differ with respect to the different presentation media. The content of the tour, including geographic elements, remained the same across media.

Mobile Nature Guide The sets of participants in the group with the MobiNaG prototype were equipped with the mobile guide application running on a PocketPC-based PDA (here a T-Mobile MDAII or a HP iPaq 5450). As positioning technology, GPS was utilized, with the PDA being connected to an external GPS receiver via Bluetooth. The deployed prototype systems (see Figure 5.4) provided the user with two basic services. For one the navigation service, displaying a map of the natural area as well as the current position of the user. The trail that should be followed was highlighted and the system tracked and displayed the route taken by the user. Furthermore a tour service included in the MobiNaG, offered location-based information on the POIs that the

Figure 5.2: Group of participants engaging in performing activities during evaluation tour.

Figure 5.3: Family group using the MobiNaG prototype during the intervention.

user encounters throughout the guided tour. The information associated with a station was divided into several parts based on the environmental interpretation concepts. The participant could flip back and forth between these parts using the "next" and "previous" arrow-buttons. If a content part included a long text the user had to scroll down to read the entire text. Each content part was completed by an image that could be enlarged and minimized using a magnifying glass icon. In accordance with the guidebook metaphor, the user could switch between the map and the tour-information, employing the taps on the right hand side (see Figure 5.4). In order to protect the mobile device and the GPS receiver unit from weather as well as physical impacts during the test tours, they were enclosed in a light-weight transparent casing that the users could wear around their neck (see Fig. 5.3).

Brochure The paper-based guidebook was realized as booklet containing 17 pages of text and color images as well as a foldout paper map of the area (see Figure 5.5). The route for the tour was highlighted and the map included references to the stations in the form of a photograph as well as the name of the station and page reference to its content within the booklet.

Participants of both interpretive media groups had to read the presented texts. Usually the person carrying the device or the booklet had to read the text out loud for the rest of the group (see Fig. 5.3). With respect to navigating through the tour it should be mentioned, that neither stations nor junctions were equipped with any signage related to the test tour. While the groups using the brochure had to rely on the map and the pictures of the respective trees, the groups using the

Figure 5.4: Presentation of the route and the stations by the navigator service (left) and presentation of the content to a phenomenon by the interpreter service (right). Screenshots of original MobiNaG prototype with German content for the evaluation.

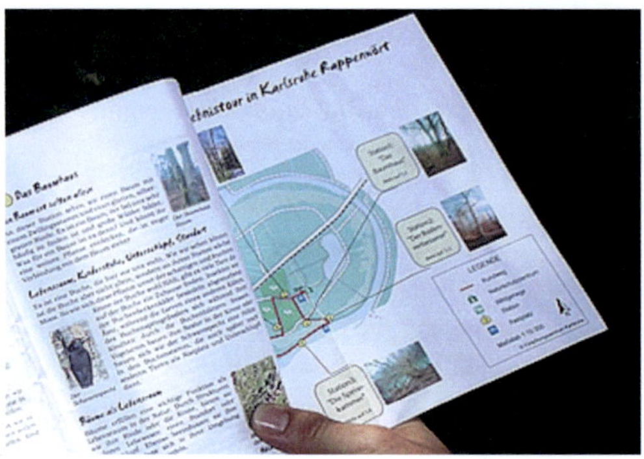

Figure 5.5: Brochure presenting the content to natural phenomena along with a foldout paper map.

MobiNaG system were notified by an audio signal, when they penetrated an area approximately 10-15m around a station. The audio signal was accompanied by a message box announcing the station and allowing the user to view more detailed location-based information on the respective POI (see Fig. 5.4). In accordance with the paper map the mobile device did not provide additional navigational cues at intersections. It should also be pointed out that neither of the two groups received additional instructions or training regarding the usage of the guide media.

Human guide The third treatment group did not engage in a self-guided tour but was guided by a person. Overall two human guides assigned to give the tours. But all sets of one target group were always attended by the same guide, since the continuity of instructions helps to keep the variation between sessions low [38]. Obviously, in contrast to the self-guided groups, the human guide groups did not have to worry about navigational aspects due to the guidance of the educator.

The human interpreter gave oral presentations at each station, based on the same textual content used in the other media (see Fig.5.6). At the same time the guides followed the principles of environmental interpretation, which implied personalization and trying to engage participants in the tour as well as fostering direct experience. Personal-based tours were supposed to be given under realistic conditions which also implied that for organizational reasons the sets guided by the human guide commonly included more participants (i.e. 2-3 Families or 9-12 students) than those guided by the brochure or mobile nature guide.

Figure 5.6: Human guide giving an interpretive presentation to students at one of the stations.

5.3.4 Instruments

The evaluation of effects of interventions requires a valid instrument [38]. In this study the impact on the participants environmental literacy was assessed by measuring a number of environmental literacy components from the proposed environmental literacy model (see Fig. 2.1). These components include:

- Knowledge
- Attitude
- Values and concerns
- Behavior (based on reported behavior) as well as behavioral intention (based on verbal commitment)

Corresponding to these components, scales were developed to obtain a pretest and posttest measure of the participants' environmental literacy. Two different versions of the instrument were designed, one for children/students and one for adults. A total of 21 and 20 items were distributed among the four scales of the children/students and those of the parents respectively. In accordance with the recommendation by Leeming et al. [195], the instrument for this study was not generated entirely from scratch but items from existing instruments were employed. This should make comparisons with the results of related studies more meaningful. At the same time it allowed for the integration of items that were constructed based on psychometric techniques, despite considerable time constraints for this study.
The knowledge scale was formed from a combination of items evaluating system and action related knowledge adapted from Frick [109] as well as knowledge items assessing the retention of facts related directly to the content of the intervention. The attitude scale is composed of statements related to the utilization or exploitation of nature, adapted from Bogner et al. [39] as well as Bogner and Wilhelm [40]. Furthermore, items regarding the interest and consideration for conservation were used based on Bogner and Wilhelm [40] as well as Bittner [32]. The values and concerns scale was built from items concerning the sense of responsibility taken from Kaiser et al. [160] and Bogner et al. [39]. In addition statements were included, assessing empathy, in particular perspective taking with respect to other organisms, adapted from studies by Schultz [305] and the Interpersonal Reactivity Index (IRI) by Davis [78]. Finally the Behavior Scale includes items expressing verbal commitment by Bogner and Wilhelm [40] as well as Bittner [32]. Furthermore, statements of reported behavior, based on the "General Ecological Behavior scale" by Frick [109] and items from the "Actual Behavior scale" from Bogner and Wilhelm [40], are included.
As is described above, the instrument was administered as a self-completion questionnaire (see Appendix C.1). Similar to related studies [32, 37, 39, 40, 196], the questions of the attitude, value/concern, and behavior scales were presented in a bipolar 5-point Likert response format (i.e. "strongly agree" to "strongly disagree", "always applies" to "never applies"), always including an "undecided" category. Items were generally accredited 5 to 1 points with the most proenvironmental response receiving 5 points and the least proenvironmental 1 point. Four of the items on the children/student questionnaire and five items on the parent questionnaire were negatively connoted and reverse scored in order to reduce the likelihood of response sets [196]. Only the knowledge scale items were not presented as rating statements but as questions with multiple-choice answers similar to Frick [109].

As is described in the study design, the questionnaires further included semi-standardized questions on demography (i.e. in the pretest) and usability (i.e. in the posttest). Next to basic socio-demographic characteristics of the participant (i.e. age, gender, class level, occupation etc.), the demography section also assessed additional factors that may have an influence on environmental literacy. With reference to studies by Bittner [32] and Brämer [45], variables assessing the individual state of nature experience, the engagement of family and friends in conservation, and

questions relating to the environmental state of the hometown, were included. Additionally questions relating to prior experience with environmental interpretation, specifically the test site and mobile computing devices, were posed.
The usability part of the questionnaire was designed to evaluate the users overall satisfaction with the EE activity, as well as the perceived quality of guidance and assistance by the respective guide medium. This section was composed of semi-standardize as well as multiple choice questions and rating statements. Next to items particularly designed for this study, building on the principles of successful interpretation discussed by Ham [129] and Moscardo [226], questions were included that were adapted from Bittner [32]. Some of the questions relating particularly to the usability of the mobile system were adapted from Abe et al. [1]. The questionnaires for parents as well as for children/students were pilot tested in two test trials and consequently revised, resulting in the rewording or removal of inefficient items.

In addition to the measurements taken during the pre- and posttest, the performance of the groups employing self-guided media (i.e. mobile nature guide or brochure) was monitored throughout the intervention. Each set of participants was accompanied by an observer who completed a semi-standardize observation protocol. Principally an observation approach was taken that relates to the concept of "user shadowing" as proposed by Dias et al. [85] as well as Semper and Spasojevic [310]. Some of the measures used, were adapted from suggestions made by Goodman et al. [117]. In this protocol, data were recorded on navigation behavior, such as the success in locating the stations, as well as the time of arrival at and departure from each station. In addition to these measures, navigation decisions at junctions were documented. The accompanying observer further surveyed the utilization of content, by noting which content parts were read out loud to the group.
Furthermore, the observer was to record usability problems encountered by groups using the mobile nature guide. Due to the social group context within the field study, the observer could benefit from a form of "natural talk aloud method" based on visitor shadowing, to detect difficulties with the use of the device. Listening in on the groups' conversations about the utilization of the device instead of asking the participants explicitly to follow a talk aloud procedure, may compensate for some of the restrictions of the talk aloud procedure as discussed by Nielsen and Yssing [237]. The observers were instructed to always remain in the background, in order to avoid influencing the subjects especially with respect to navigation and orientation. The observer only intervened in case a group got seriously lost or the mobile guide system malfunctioned. Nonetheless he could, to a certain extent, observe the interaction with the device. Major difficulties like for instance a system crash would be registered by the observer while assisting the group to resume the tour.

Statistical methods For testing the hypotheses one and two, the following statistical methods were applied. Initially the pretest scores for the respective environmental literacy scales were tested for differences in between media groups using a one-factorial Analysis of Vari-

ance (ANOVA). If no statistically significant differences were determined, the pretest scores were pooled to form the control group mentioned in the study design. Hereafter, t-tests for unpaired sample groups were performed to test the posttest scores of each media group against the control group. A further set of t-test was conducted to check for differences in posttest scores between the mobile guide groups and each of the traditional interpretive media groups. Further tests were carried out to examine the influence of additional factors on environmental literacy scores. Fisher's exact test was used to check for differences in distribution between binary sociodemographic factors (e.g. gender). A one-factorial ANOVA was employed to check for differences in scores of sociodemographic subgroups within the pretest and the posttest, while t-tests were conducted to test for changes between pretest and postest scores. Furthermore, a two-factorial ANOVA was administered to investigate interaction effects between sociodemographic factors and media. In order to test the third hypothesis, the Mann-Whitney U test was deployed to check for media effects in ordinal scale answer sets. Binary data were again examined with the Fisher exact test. Furthermore t-tests were utilized to compare the arithmetic means of media groups. In the figures presenting the results in the following section, the $*$-sign is employed to indicate statistically significant test results for the respective group, either at the ($p<0.01$) or the ($p<0.05$) level.
The statistical analysis was performed with the statistics software package STATISTICA, StatSoft [325].

5.4 Results

5.4.1 Intervention and media effects on environmental literacy

As was previously described, the study was designed to compare the effect of the mobile nature guide on participants environmental literacy with the effect of two traditional forms of environmental interpretation. Thus the hypotheses for this section is, that the intervention has an effect on environmental literacy and that this effect differs between the three interpretive media.

Knowledge

Prior to the analysis of intervention effects, a comparison of knowledge scores of all adults that took the pretest was conducted across media groups that they were subsequently assigned to. Since the one-factorial ANOVA did not result in significant differences between pretest knowledge scores of media groups (see Appendix C.3.1, Table C.1), these data were pooled to form the control group.
T-tests for unpaired sample groups between the pretest control and the posttest media groups of

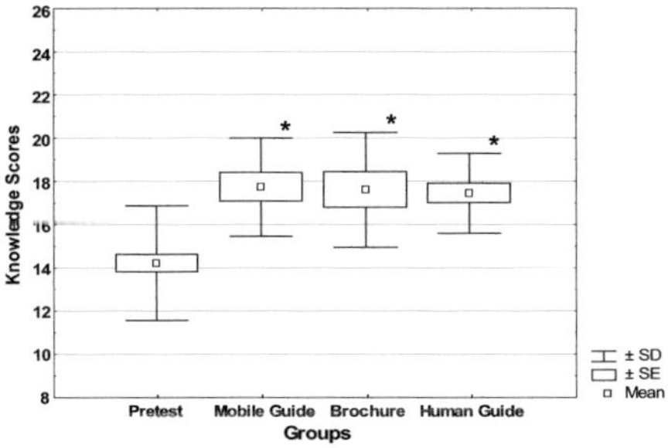

Figure 5.7: Mean knowledge scale scores of parent groups, including: pooled pretest (n=40), mobile nature guide (n=11), brochure (n=10) and human guide (n=16). $*$ = significant difference to pretest (p<0.01).

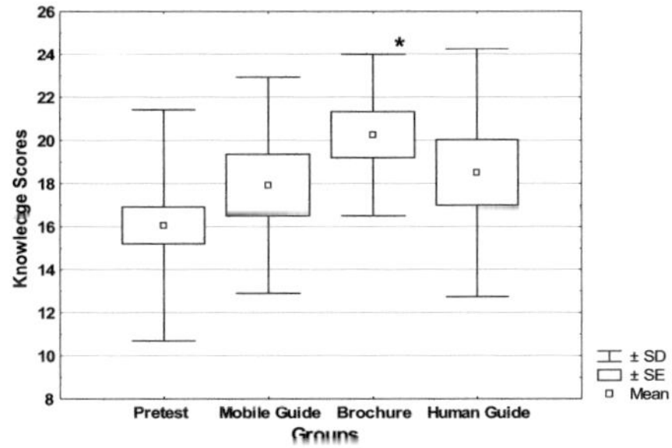

Figure 5.8: Mean knowledge scale scores of the children groups, including: pooled pretest (n=38), mobile nature guide (n=12), brochure (n=12) and human guide (n=14). $*$ = significant difference to pretest (p<0.05).

the parents yielded a significant increase in knowledge scores for all media (see Appendix C.3.1, Table C.3), the mobile nature guide (t=-4.009, df=49, p<0.01), the brochure (t=-3.618, df=48, p<0.01) and also the human guide (t=-4.450, df=54, p<0.01). It can be observed in Fig. 5.7 that all groups exhibit a similar increase in knowledge and a t-test could not determine a significant difference between the mobile nature guide and the traditional interpretive media for the parents.

Also for children the one-factorial ANOVA resulted in no significant difference between pretest knowledge scores (see Appendix C.3, Table C.1) and consequently all pretest participants were treated as one control group. A significant increase in knowledge due to the intervention could, however, only be determined for the group of children whose family took a tour guided by the brochure (t=-2.516, df=48 p<0.05). The mobile nature guide group (t=-1.065, df=48, p>0.05) actually showed the smallest gain based on mean posttest knowledge score, followed closely by the human guide group (t=-1.431, df=50, p>0.05) (compare Fig. 5.8). Still, no significant posttest difference could be found between the mobile nature guide and the other media for the children.

In case of the students, there was also no significant difference between the subjects taking part in the pretest (see Appendix C.3.1, Table C.1). Similar to the adults, the t-test between the pooled pretest scores and those for the posttest media groups showed a significant knowledge gain for all groups. The group administered by a human guide (t=-4.329, df=71, p<0.01) exhibited the highest increase. It is followed by the group attending the test tour with a mobile guide (t=-2.253, df=68, p<0.05) and the brochure group (t=-2.044, df=70, p<0.05), exhibiting little difference in means (compare Fig. 5.9). Nonetheless, also for the students no significant difference between the media could be determined following the intervention.

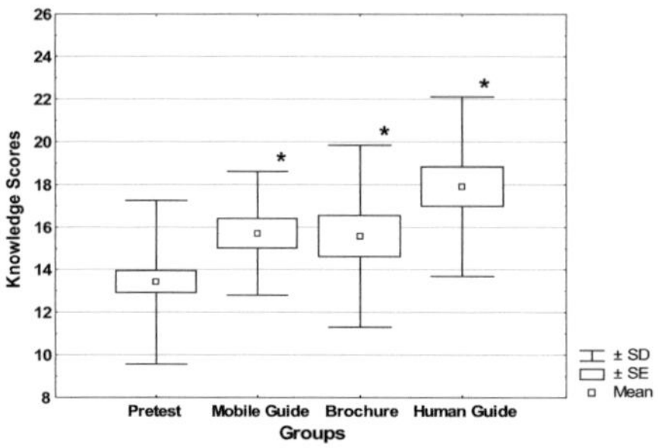

Figure 5.9: Mean knowledge scale scores of student groups, including: pooled pretest (n=53), mobile nature guide (n=17), brochure (n=19) and human guide (n=20). $*$ = significant difference to pretest ($p<0.05$).

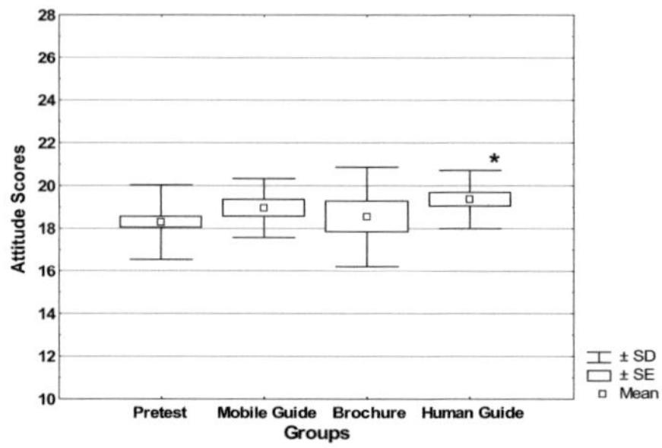

Figure 5.10: Mean attitude scale scores of parent groups, including: pooled pretest (n=40), mobile nature guide (n=11), brochure (n=10) and human guide (n=16). $*$ = significant difference to pretest ($p<0.05$).

Attitude

As there was no significant difference detected between the scores of pretest adult participants, they were again pooled to form the control group for the attitude scale (see Appendix C.3.1, Table C.1). A significant change in the attitude scores of the parents, following the intervention, could be found for the human guide group ($t=-2.192$, $df=54$, $p<0.05$) (compare Fig.5.10). The mobile guide group ($t=-1.160$, $df=49$, $p>0.05$) displays a slightly smaller increase, constituting no significant difference to the pretest scores. The smallest increase in the environmental attitude was observed for the brochure group ($t=-0.380$, $df=48$, $p>0.05$). Also for the attitude scale of the parents, the different posttest media groups do not differ significantly from each other.

With respect to the children's attitude scale no significant differences could be found in the tests. There is only a trend that indicates that, in comparison to the pretest group, the brochure and the mobile nature guide group show a slight improvement in environmental attitude, due to the intervention. Whereas the human guide group shows a decline in the attitude score (see Appendix C.2, Fig. C.1).

Similar to the children, no significant intervention or media effect was detected for the students' attitude scale. The brochure group's attitude scale remains basically unchanged, whereas the mobile guide as well as the human guide group show a decline in attitude score following the tour (see Appendix C.2, Fig. C.2).

Values and concerns

For the values and concerns scale none of the tests showed a significant difference. Thus, neither an intervention nor a media effect was found for the adult participants (see Appendix C.3.1, Table C.1). The comparison between the media showed for all parent groups a slight increase in value and concern scores.

With respect to the value and concern score for the children, a significant intervention or media effect could not be found for this target group either. The mobile nature guide and the human guide group increased slightly compared to the pretest comparison group while the value and concern scale of the brochure participants remains practically unchanged.

The picture for the students' value and concern scores is slightly different. While mobile nature guide and brochure group increased slightly the human guide group lies marginally below the pretest mean. Still none of the changes and differences turned out to constitute a significant effect.

Behavior

Also for the behavior scale no significant effects could be found for the parents groups. All groups showed an increase in their mean score compared to the pretest comparison group. The mobile guide and the human guide groups achieved higher behavior scores than the brochure group (see Appendix C.2, Fig. C.3). This did, however, not result in a significant difference between the media groups.

In case of the children's behavior scores, the ANOVA resulted in a significant difference between children participating in the pretest ($F(2,34)=4.258$, $p<0.05$). As a consequence, the test for an intervention effect was not performed between the pooled pretest control group and the posttest media groups but the individual pretest media groups were used for the comparison (see Appendix C.2, Fig. C.4). However, the t-tests performed for these within media group comparisons did not yield a significant intervention effect for the behavior scale. The mean score of the mobile nature guide group increased slightly, while both, the brochure and the human guide group exhibit a decline in their scores. Also the comparison between the posttest scores of the media groups did not result in a significant difference (see Appendix C.2, Fig. C.5).

A similar trend was observed for the posttest scores of the student media groups. For the student groups no significant differences were found between the pretest participants, nor were significant intervention or media effects determined for the behavior scale. The scores of all media groups increased slightly in the posttest with the mobile nature guide group showing the largest increase (see Appendix C.2, Fig. C.6).

5.4.2 Effects of additional factors on environmental literacy

Other studies like the one by Bittner [32] documented, that additional factors including sociodemographic variables have an influence on the environmental literacy components in addition to the treatment. An ANOVA revealed a number of statistically significant effects of some of the assessed sociodemographic factors on the environmental literacy (i.e. knowledge, attitude and behavior) scores of participants. Here only those factors will be discussed which exerted an influence on attitude and knowledge, the only scales that changed significantly due to the intervention (see Appendix C.3.1).

Gender effects Based on a Fisher's exact test the distribution of female and male participants differed between none of the media groups significantly, in all target groups (see Appendix C.3.4, Table C.9, Table C.10). In contrast to the findings of Unterbruner and Unterbruner [347], no gender effect was found with respect to the knowledge scores following the intervention. Neither in the pretest nor in the posttest knowledge scores an ANOVA detected a significant difference between male and female participants of all target groups.
The findings for the attitude scores of the parent groups did not yield a significant gender effect either. Furthermore a 2-factorial ANOVA was conducted to test for interaction effects between gender and interpretive media used on the knowledge or attitude score but no such interactions could be identified (see Appendix C.3.1, Table C.2). Thus it can be assumed, that for all target groups the media intervention worked similarly for female and male participants.

Prior knowledge Especially for children and students the questions on additional factors included an assessment of the state of nature experience. Among other aspects this entailed questions on general prior knowledge about plants, animals, and food sources. Similar to gender prior knowledge seemed to be equally distributed between the media groups of children and students in the pre- and post test. For children no significant effect of prior knowledge could be observed for the posttest knowledge scores. The analysis of the student data, however, yielded an interaction effect between the prior knowledge of the students and the medium they used on the posttest knowledge scores (see Appendix C.3.1, Table C.2). Students who had the highest prior knowledge, had significantly higher knowledge scores when they attended a human guide group than with the mobile nature guide ($t=-3.375$, $df=10$, $p<0.01$) or the brochure ($t=2.231$, $df=15$ $p<0.05$).

Education With regard to the level of education, two groups of parents participated in the test, those with a college/university degree and those without a college education. Still, based on a Fisher's exact test, no significant differences could be found between the media groups in the pre- or posttest, with respect to the distribution of education levels (see Appendix C.3.4, Table C.12). Adults holding an academic degree scored significantly higher on the knowledge score of the pretest (F(1,38)=8.262, p<0.01). Still, both levels of education exhibit a significant gain in knowledge from pre- to posttest without a significant difference between the posttest scores. This indicates, that those without an university education gained comparatively more knowledge. A 2-factorial ANOVA did not result in a significant interaction between education and media (see Appendix C.3.1, Table C.2). Thus it can be assumed that parents of both educational levels are equally effected by the media.

Furthermore, children and students attended different types of schools, due to Germany's 3-track school system. Thus knowledge scores where analyzed for these target groups to test for an influence of the type of school.
Among the students, the basic track school made up the largest fraction of participants. No significant intervention effect nor a media effect on the knowledge scores of this group could be determined. The comparison of knowledge scores of medium track students yielded significant intervention effects in the mobile nature guide group (t=-2.195, df=15, p<0.05) and the human guide group (t=-3.218, df=14, p<0.01). But also for this student group no significant differences between media groups were found in the posttest. Knowledge scores of advanced track students showed a significant intervention effect for the human guide group (t=-4.072, df=18, p<0.01) only. Furthermore, a significant media effect was found in the posttest results. The advanced track students attending the human guide group scored significantly higher than both the mobile nature guide group (t=-2.415, df=7, p<0.05) and the brochure group (t=-2.904, df=9, p<0.05) (see Appendix C.2, Fig C.7).
Among children those attending elementary school and those visiting the advanced track school made up the largest fractions. For elementary school children significant intervention effects were found for the brochure group (t=-2.925, df=23, p<0.01) and the human guide group (t=-2.542, df=25, p<0.05). Still, there was no media effect in form of a difference between groups in the posttest. For children in the advanced track schools, no significant intervention effects nor media effects were determined.

5.4.3 Media effects on the success of the interpretation and the usability

Following the completion of the test tour all participants, independent of assignment to pre- or posttest, had to answer questions relating to their emotional state after the tour as well as their overall satisfaction with the tour. In addition they were asked to rate statements or give grades for aspects addressing the usability of the media as well as the success of the interpretation unit.

Emotional state

Regarding the emotional state of the participants after the intervention, statistically significant differences could not be found between the media groups based on U-test results for any of the target groups (see Appendix C.3.3, Table C.7). All media groups for parents and children as well as for students displayed similarly high median ratings, indicating that they felt mostly content and relaxed or motivated and excited as opposed to bored or frustrated and angry. Even though statistically not significant, mean scores for parents indicated, that in the group following the human guide more of the participants were in a positive emotional state than in the brochure group, which lay also ahead of the mobile nature guide group. Among the children, the brochure group was ahead of the human guide group, which is followed by the mobile nature guide group. Finally among the students those groups exhibited a higher mean emotional state, that attended the self-guided media.

The following sections will only present the results for those items, which yielded a significant difference between the media groups for at least one of the target groups.

Perceived knowledge gain

Only among the parents a significant difference could be found with respect to perceived knowledge gain. Based on a U-test, significantly more of the parents taking the tour with a human guide rated the gain of new knowledge higher than those participating in the tour with the mobile nature guide ($U=218$ $Z=2.83$ $p<0.01$). A similar trend could be observed for the comparison between the brochure and the mobile nature guide group, but the difference is not statistically significant. Neither for children nor for students any of the comparisons between media groups yielded a significant difference in perceived knowledge gain. The medians for the child groups only indicated, that the mobile nature guide and the brochure group more consistently stated that they had gained new knowledge through the tour. Whereas the students group did not differ in their perception of gaining additional knowledge (see Appendix C.3.3, Table C.7).

Grades for the interpretive presentation

The study subjects were asked to grade six aspects of the experienced interpretive presentation. Grades to be assigned, were based on the German system of school grades (1=highest to 6=lowest).

Entertainment qualities Only children and students were asked to grade the entertainment qualities of the interpretive presentation. Since today's children are used to an extensive consumption of entertainment media, this aspect was considered to be of more relevance to them then opposed to grading the content, as the parents were expected to do. The entertainment quality grades were actually one of the aspects that differed significantly between the media groups.
The children guided by a person, graded the entertainment qualities of their tour significantly

higher than those taking the tour with a mobile nature guide (U=208.5 Z=2.018 $p<0.05$). The students group on the other hand showed the opposite result. Students participating in a tour with a mobile nature guide experienced the tour as significantly more entertaining than those that took the tour with the human guide (U=420.5, Z=-2.176, $p<0.05$).

Games The only significant media effect for the grading of games experienced during the tour, was found for the parents. Games were valued higher during tours with the mobile guide than during those with the human guide (U=128, Z=2.287 $p<0.05$) (see Appendix C.2, Fig. C.8). For children and students no significant differences were found with respect to the grade for games. But in both cases, grades for games tended to be higher in the brochure groups than in the other groups.

Guide media Again only the media groups of the parents differed significantly in their judgement of the performance of their guide. The human guide was granted significantly higher grades by his audience than the mobile nature guide by his (U=165, Z=-2.362, $p<0.05$) (see Appendix C.2, Fig. C.9).
Children and students groups did, even though not statistically significant, exhibit differing trends with respect to the grading of the guide. Children guided by a person and attending the brochure group, tended to give their guide higher marks than the mobile nature guide group. The students group using the mobile nature guide seemed to give their guide medium the best grades.

Overall tour When asked about their overall impression of the tour, students employing a mobile nature guide assigned significantly higher grades to their tour than those following the human guide (U=389.5, Z=-2.119, $p<0.05$) (see Appendix C.2, Fig. C.10). For the media groups participating in the family tour the differences concerning the overall impression of their tour did not reach statistical significance. Still, for children a similar trend was found as for the students, while parents showed the tendency to mark the tour with the human guide higher.

Perceived interactivity

Parents taking part in an interpretive activity with a brochure perceived the tour to be more interactive and providing opportunities for self-determined exploration, than those parents using the mobile nature guide (U=165, Z=2.105, $p<0.05$) (see Appendix C.2, Fig. C.11). For children and students the perceived interactivity did not differ significantly between the media groups. Still, children demonstrated the tendency to value the interactivity of the tour with a brochure or a human guide higher than the one with the mobile nature guide. The data collected from students show again an opposite trend.

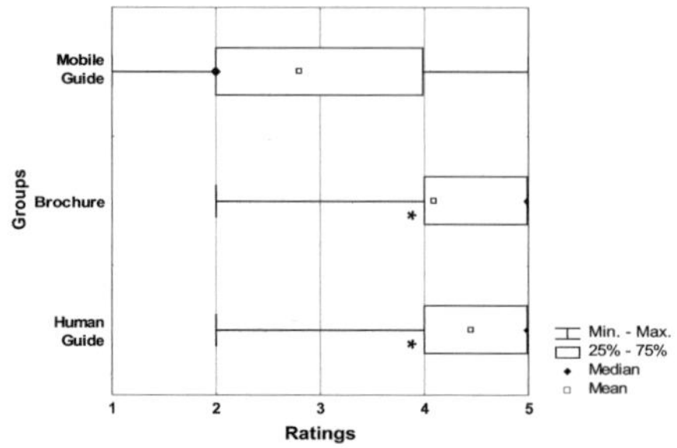

Figure 5.11: Rating of distraction from nature due to the the guide medium given by parents using mobile nature guide (n=26), brochure (n=21) and human guide (n=29). $*$ = significant difference to mobile guide (p<0.05).

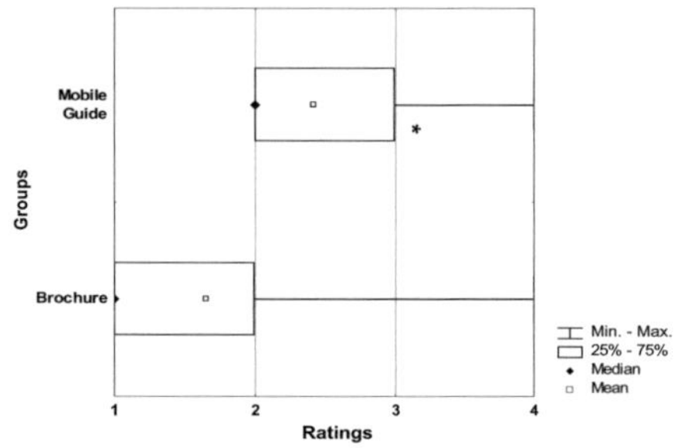

Figure 5.12: Rating of perceived involvement with self-guided media by children using mobile nature guide (n=24), and brochure (n=23) $*$ = significant difference to brochure group (p<0.05).

Perceived distraction

The scores related to the perceived distraction, indicate that both, parents participating in the tour with the human guide (U=152.5, Z=3.785, p<0.01) and the brochure (U=144 Z=2.76, p<0.01), felt significantly less distracted from the actual experience of the natural environment by their tour, than the mobile nature guide group did (see Fig. 5.11, p. 188). Neither the analysis of the children's data nor that of the students', resulted in a significant difference in distraction scores between the media groups (see Appendix C.3.3, Table C.8). A trend among the student groups shows that students taking part in the intervention with the brochure felt less distracted than the other students.

Perceived excitement

Further, children and students were asked if they felt, that the tour had been exciting for them. As it turned out, significantly more of the children taking the tour with a human guide could agree with this statement, than among the children guided by the mobile nature guide (U=236, Z=-2.046, p<0.05). Whereas among the students there was a stronger tendency of those in the mobile nature guide group to confirm this statement than in the other groups.

Perceived involvement in media-based interpretation

Furthermore, children and students, who took part in a self-guided tour with one of the interpretive media, were asked how much time of the tour they had spent "operating" the guide medium themselves. The result shows that children taking a family tour in a mobile nature guide group reported a significantly higher involvement in using the medium than those whose family took the tour with a brochure (U=121.5, Z=3.288, p<0.01) (see Fig. 5.12). For the students, however, the usage of different interpretive media did not result in a difference in the perceived degree of involvement with the medium.

Open visitor feedback

Next to the collection of standardized questions, test subjects where also asked to name the two things they liked best as well as the two things they liked least about the tour. Answers to these open questions where categorized and coded before analyzing them for comments about the guide media.

Positive comments The results of Fisher's exact test of the parents' data show, that positive comments about their guide medium were made by a significantly higher proportion of the mobile nature guide users than of the brochure users (Fisher's exact p one-sided p<0.01) (see Appendix C.3.4, Table C.10).
The comparison between the children's media groups shows, that participants of the mobile nature guide tour listed their guide medium significantly more often among their favorite aspects than either the human guide group (p<0.01) or the brochure group (p<0.01) (see Appendix C.3.4, Table C.11).
In the case of the students, there was also a significantly larger proportion of the mobile guide users that made positive comments about their guide medium and the tour than those attending the tour with the human guide (p<0.05). The same trend could be observed for the comparison between mobile guide and brochure group, but Fisher's exact test did not result in a significant difference (see Appendix C.3.4, Table C.11).

Negative comments The analysis of the list of aspects liked least by the parents showed, that the mobile nature guide group also gave significantly more negative comments about their guide medium than the brochure (p<0.01) or human guide group (p<0.01) did. The same is true for the children, where the mobile guide group also exceeded the brochure (p<0.05) and human guide group (p<0.05) in negative comments.
For the students, however, no significant differences could be found between the media groups with respect to the frequency of negative comments. Still, the proportion of negative comments by the mobile nature guide group only exceeds those of the human guide group but is slightly smaller than the proportion identified for the brochure group.

5.4.4 Navigation success

In contrast to a tour administered by the human guide, a tour with the brochure or the mobile nature guide constitutes a self-guided interpretive unit. Thus, participants taking a tour with the brochure or mobile nature guide had to find the correct route and stations on their own and could choose themselves, which content they would actually want to consume. As was discussed previously, navigational assistance and the presentation of interpretive content are the core services of interpretive media. Consequently, an evaluation of their effectiveness is of particular interest for the comparison between mobile nature guide and traditional interpretive media. For this purpose the observer made recordings of the navigation behavior and content consumption for each group participating in a self-guided tour. The following results are based on the analysis of the observation protocols recorded for seven family groups and nine student groups using a brochure as well as 11 family groups and eight student groups employing a mobile nature guide.

Navigation behavior

Participants of the self-guided tours were confronted with a number of navigational decisions including taking the correct turn at junctions and deciding if they had reached a station of the tour. As was described above, both media offered support for the navigation tasks in form of a map of the area of interest (see Fig. 5.5 and Fig. 5.4). On the paper-based map as well as the mobile map, the route of the tour was highlighted and stations were marked by specific symbols, completed by numbers indicating their chronological order. The two types of maps differed in the following aspects. The paper-based map included a legend to its symbols. Due to the small screen size the legend was spared on the mobile map. But the navigator service of the mobile nature guide offered basic navigational support with a position and tracking indicator, based on data from the GPS unit connected with the device.

Junctions Fig. 5.13 presents the results of the analysis of navigation decisions at junctions. Family groups using the paper-based map of the brochure made 97.9% of possible navigation decisions correctly, compared to the family groups using the mobile map who made only 87.5% correct choices. Based on a Fisher exact test, family groups with the brochure performed significantly better at junctions than family groups using the mobile map ($p<0.05$) (see Appendix C.3.4, Table C.13).
Student groups were generally less successful navigating the route than family groups. Student groups using the paper-based map took 68.8% of decisions correctly, whereas student groups using the mobile map made the right decision in 73.4% of the cases (see Fig. 5.13). Thus in case of the student groups, the mobile nature guide groups show the tendency to perform better at junctions than the brochure groups. This difference was, however, found to be not statistically significant.

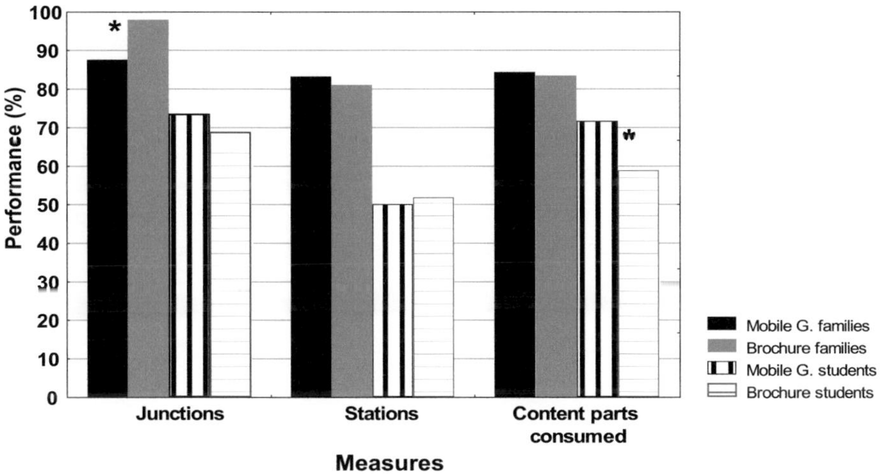

Figure 5.13: Navigational performance and consumption of content by families with mobile nature guide (n=11) and those with brochure (n=7) as well as student groups with mobile nature guide (n=8) and with brochure (n=9). $*$ = significant difference to other media group ($p<0.05$).

Stations With respect to finding the stations correctly without assistance from the observer, family groups with the mobile map performed slightly better than those with the paper-based map, but no significant differences could be determined between the two (see Fig. 5.13).
The student groups did not differ significantly either. In this case the brochure groups performed marginally better. Fig. 5.13 further indicates, that overall family groups performed apparently better than student groups.

Temporal navigation behavior Navigational performance was further evaluated by measuring the time that the groups spent on the trail sections in between stations, assuming that most of this time was spent walking along the trail and making navigational decisions such as locating the next station.
A t-test comparing the overall means of the groups for all trail sections did not result in a significant difference between families employing the brochure and those using the mobile guide. Such a difference could neither be found for the student groups. Still, a test for differences on the level of the individual trail section indicates that families using a mobile guide took more time navigating section 1, 2 and 4. A t-test shows a significant difference for trail section 2 ($t=-2.627$, $df=16$, $p<0.05$). A similar pattern can be observed for the student groups. Student groups guided by the mobile nature guide take considerably more time to navigate trail sections 1, 2 and 3, with section 1 resulting in a significant difference ($t=-3.265$, $df=15$, $p<0.01$) (see Appendix C.3.2, Table C.6).

Perceived navigation success

In the posttest the participants were asked if they felt that the map they had used was of assistance in solving the navigation tasks. The analysis of the responses based on a U-test did not result in a significant difference between mobile nature guide and the brochure users (see Appendix C.3.3, Table C.8). Still both parents and children using the brochure tended to value the support by the map more than those using the mobile map. Again it can be observed that this trend is reversed in the student group, where the mobile map was judged as slightly more helpful.

5.4.5 Utilization of content

Observed content utilization

The observation protocol also documents the content consumption of the participating groups, that could be perceived by the observer.

Content parts consumed The observer recorded which content parts were read out loud to the group throughout the tour. The results illustrate, that both family groups, using the mobile nature guide and the brochure, essentially consumed the same percentage of content (see Fig. 5.13). Whereas among the student groups utilizing the mobile nature guide and those employing the brochure a significant difference could be observed according to the Fisher exact test ($p<0.01$). The mobile nature guide groups were observed reading 71.6 % of the content, while the brochure groups read only 58.8% of the content parts (see Fig. 5.13).

Time spent at stations The time spent at each station was recorded as an indicator of the intensity at which the participants engage in the content and the proposed environmental education activities. The mean time spent at each station was calculated for the families and students using the paper-based guide as well as for the ones using the mobile guide. What has to be taken into account, though, when comparing the time spent at a station is that the groups visiting a station with the mobile guide, will not only take the time to read and explore the natural phenomenon but also to interact with the device. This "interaction time" includes activities like scrolling and moving between pages but also time taken by the system to load station data. The interaction time was determined for the two test tours by timing four testers during a "walkthrough", solely navigating through the tour without actually reading the content. The mean interaction time was then subtracted from the mean time spent at the stations by the mobile guide groups.
A t-test comparing the overall means of the groups for all stations did not result in a significant difference between families employing the brochure and those using the mobile guide, even though family groups with the mobile guide tended to stay some more time at the stations. For student groups a significant difference between the two types of media-based tours was detected, with

student groups using the mobile nature guide spending significantly more time at the stations than those with the brochure ($t=-2.458$, $df=12$, $p<0.05$) (see Appendix C.2, Fig. C.12).

Perceived content consumption

Following the completion of the tour the participants were also questioned on whether they had read or listened to all content parts (i.e. texts) offered by the self-guided medium. The results show for the parents, that significantly more of the mobile guide users reported to have consumed all of the content than the ones taking the tour with the brochure ($U=176.5$, $Z=-2.065$, $p<0.05$) (see Appendix C.2, Fig. C.13). For both, the children and the students, no significant difference could be detected with a U-test. In each case both media groups claimed to have consumed about the same amount of content.

5.4.6 Main usability issues

Next to monitoring the navigation behavior and the utilization of content, the observer also recorded usability problems encountered by 22 groups, including both families and students using the mobile nature guide. Fig. 5.14 displays the main usability issues for family as well as student groups along with the total for all groups. The overall performance and stability of the prototype system was a major issue that more than 70% of the groups had to deal with. Most of these groups commented on the system being relatively slow in comparison to the computer programs that they are used to. Some also experienced an application crash throughout the tour. Furthermore, 68% of the groups had some difficulties interacting with the application. This included aspects related to interface elements like buttons. The functionality offered by certain buttons was for instance not intuitively understood, leaving the user puzzled if and how a functionality like the switching between navigator and interpreter service could be achieved. In some cases users complained about a lack of feedback or reaction to the use of buttons or a long delay between input and reaction. Particularly the scrolling of textual content components in the interpreter service caused a problem experienced by 55% of the groups.

Another set of usability problems was associated with malfunctions of context sensors and unfavorable environmental conditions. Thus, some of these issues were associated with the GPS-based functionalities of the navigator service. During 55% of the tours with the mobile nature guide participants reported problems with the positioning indicator, tracking or location-based information presentation. Errors appeared to be mostly associated with insufficient GPS coverage in the forest habitat or the inaccuracy of the supplied position. Next to malfunctions of the navigator service, 50% of the groups also complained about restrictions in the readability of the information on the display. They either experienced the font size as being to small or were bothered by reflections on

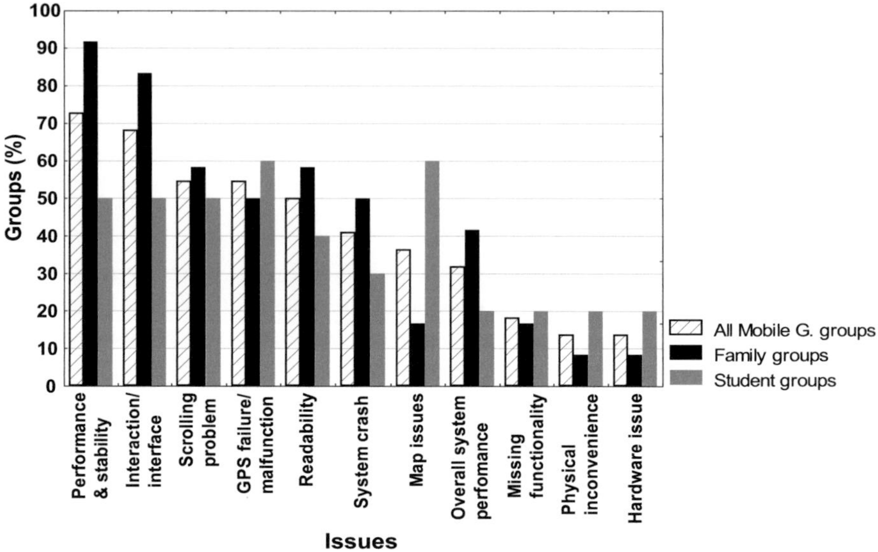

Figure 5.14: Usability issues documented for proportions of family groups (n=12) and student groups (n=10) by the observer.

the display due to the lighting conditions.
Less than 40% of the groups had trouble interpreting the map elements. Finally 27% of the groups encountered difficulties with the hardware (e.g. low battery power) or complained about physical fatigue due to carrying the device.

5.5 Discussion

5.5.1 Environmental literacy

The results show that of the investigated environmental literacy components essentially only knowledge was effected by the conducted EE interventions.

The results related to the effects on environmental literacy, illustrate that the interventions led to an increase in knowledge in both parent and student groups and with certain restrictions also in the children groups. This corresponds to the findings of several other studies [38, 74, 123, 195]. Bogner [38] for instance reported significant changes in knowledge both in 1-day and 5-day ecology programs. With respect to their review of the outcome of EE interventions, Leeming et al. [195] stress, that the reviewed studies focusing on knowledge all found significant effects. Also Gunderson et al. [123] found in their review of wilderness education studies several programs that resulted in an increase in knowledge. As was discussed in conjunction with the environmental literacy model in Chapter 2, knowledge can not be assumed to have a direct influence on environmental behavior [138]. Nonetheless, since knowledge can be considered a prerequisite for other components such as attitude, an increase in knowledge is still an essential step towards the achievement of environmental literacy [38, 138]. Bogner [38] further stresses the importance of knowledge acquisition in any kind of education, and that it can as such be seen as a kind of basic responsibility of EE interventions.

As no significant difference was found in the posttest knowledge scores between the different media groups, it can be concluded that all media were able to serve this basic purpose of EE in a similar fashion. This is in particular noteworthy with regard to the mobile nature guide. Consequently it can be assumed to function as an effective alternative to the respective traditional interpretive media for the enhancement of environmental knowledge through EE interventions.

Still, the analysis of the influence of additional factors for students shows that, despite the generally effective applicability of all media, students with a high amount of prior knowledge gained more knowledge from a tour with a human guide than from a self-determined tour with the other guide media. One reason could be that the human guide was perceived as a teacher-like authority that managed to encourage the students to focus on the content of the tour. It is further possible, that those students with high levels of prior knowledge, who can be assumed to be the ones more interested in nature, benefited in particular from the more personalized form of presentation of the human guide. The assumption is further underlined by the discovery that advanced track students only achieved a significant knowledge gain when attending a tour with the human guide. Still, the results show that, regardless of the influence of additional factors, the students in the human guide group display the highest increase in knowledge. This stresses the power of interpersonal communication in conveying information, which has been propagated by Ludwig [199] as key strengths of the human guide.

It is remarkable, though, that these findings are not matched by the results of the children. This may be explained by the fact that the children participated in family groups. During these interventions, the human guide had to find a compromise between a presentation for children or adults and could thus not tailor the tour as much to the needs of the children. The children's results involve a further peculiarity. Children exhibit a similar overall trend of an increase in knowledge but only the brochure group achieved a significant gain. However, since the posttest scores of the media did not differ significantly, one can not speak of a true media effect. A possible explanation would be that in comparison to the students, children entered the pretest with an already high level of environmental knowledge and thus could gain relatively less new information, which matches findings of other studies on environmental knowledge [74] as well as attitudes (Horsley (1977), cited in [195]). This may have been especially the case for the children of the mobile and human guide group. Although pretest results did not vary significantly between media groups, the brochure group started out with lower knowledge scores than the other groups. A further explanation for the constrain of the children's significant gain in knowledge to the brochure group could be that, in contrast to the mobile guide group, the participants were less distracted by the technology. In addition, in the brochure group the parents commonly acted as the facilitators for the tour and made it more of a point to engage their children in the interactive activities and could assist them individually in the comprehension of the tour content. This also corresponds to the increased interactivity ratings, of children, for the brochure and human guide.

An influence of additional factors on knowledge was also discovered for parents. Such effects of education on knowledge as well as other components of environmental literacy were also documented in the literature [166, 178, 244]. The analysis of the present study reveal a significant influence of education on the pretest results, with the parents holding an academic degree scoring significantly higher on the pretest than those with less advanced education. This is confirmed by Kollmuss and Agyeman [178] who state that the longer the education, the more extensive is the knowledge about environmental issues. The difference in environmental knowledge between educational levels is, however, balanced by the intervention and does not prevail in the posttest results. It can thus be assumed that the parents without an academic degree were able to gain more knowledge from the intervention, regardless of the environmental interpretation medium used. A reasonable explanation for this finding could be, that the design of the contents for family tours required a compromise between comprehensibility for the children and adaptation to the prior knowledge of the parents. For the parents with a lower level of education (i.e. less prior knowledge) the presented content was apparently more suitable, whereas academics could profit less from the presented information, with regard to knowledge gain. This stresses the need for the provision of target group specific presentations, which has also been advocated by Milfont et al. [223] and can be considered a strength of the human guide but in particular also of the mobile nature guide.

With regard to environmental attitude the effect of the intervention was generally not as evident as in the case of knowledge. The only significant increase in attitude was documented for the adults participating in tours with the human guide. With his capability for interpersonal communication the human guide can make use of particular EE mechanisms, like persuasion [49, 259] and direct feedback, to specifically address the attitudes of the participants. Thus with his more personalized presentation, the human guide may have induced a shift at least in specific attitudes of the parents. But even though the self-guided media, including the mobile nature guide, lack this capability at the time, the parents attending the respective media groups also showed a slight increase in attitude. Moreover, since a significant media effect could not be found, this observed difference between the media groups should be interpreted with caution.

The children and students attending a human guide group did not match these results either but to the contrary even showed a slight decline in attitude following the intervention. This posttest decrease in attitude scores is, however, not restricted to the human guide groups but can also be observed for the students using the mobile nature guide. There are a number of studies that found no effect of EE interventions on attitude [49, 51, 123, 195] but no reference to negative effects of EE on attitude could be identified in the literature. It is thus likely that this decline was not caused by an actual negative effect of the intervention itself but is rather linked to constrains in the evaluation methodology. As Gotch and Hall [118] pointed out, many of the weaknesses of survey/questionnaire research are magnified when working with children. In spite of a thorough target group specific design and pilot testing, the questionnaires used in this study did potentially not suite the cognitive development of the children and students. One possible cause could be limitations in reading comprehension [112, 118]. Especially the younger children and the students attending the basic track school form may have had difficulties to understand the questions, resulting in random responses. Another key issue is the attention span of children [118]. It can be assumed that the children and students taking the pretest could still focus sufficiently on the whole questionnaire. But the experimenters could observe that the attention span of the children and students did sometimes barely last for the entire intervention. Thus during the post test following immediately after the intervention some of the participants likely experienced fatigue and may have even been demotivated, resulting possibly in responses that were less considered or even deliberately negative. Family tours with the human guide or the mobile guide lasted on average longer than those of the brochure groups, which may explain some of the observed differences. For the students the tours with the mobile guide were presumably most challenging due to the need to familiarize with the technology. Whereas the tour with the human guide may have induced more fatigue based on more intense social interaction and maybe also a certain amount of demotivation due to the more formal setting. Furthermore, as was suggested by Gotch and Hall [118], the attitudes of younger participants are frequently not well developed yet, which reduces the reliability of their responses. In combination with the lack of attention this phenomenon may also explain the negative trends to some extend.

Still, it should be stressed, that none of the discussed trends resulted in a significant change of attitude for children or student groups and a significant media effect was not be detected either. As a consequence it can be concluded that there was no overall effect on attitude, which corresponds to the results of previous studies reviewed by Leeming et al. [195] and Gunderson et al. [123] in which interventions induced changes in behavior but failed to affect attitudes. Still, there remains a certain controversy over the effect of EE interventions on the attitudes of the participants.

Leeming et al. [195] also cited studies that did result in at least short term changes in attitudes (e.g. Fortner 1985 and Simmons 1984). Further, a study by Marynowski and Jacobson [212] documented changes in attitudes due to mass media educational approaches. However, based on their research, Bogner [38] as well as Bittner [32] stress that short term informal EE interventions have only limited potential to induce permanent changes in attitudes. Bogner [38] points out that the process of changing attitudes is inevitably slow and that fostering a shift in attitudes demands intensive and long term interventions exceeding at least 5 days. Thus it can be deducted, that the overall lack of a significant effect on attitudes in this study can be also attributed to the short term nature of the interventions, which lasted 1 - 2.5 hours, as well as the general difficulty to influence attitudes through EE activities.

Nonetheless, the requirements analysis, performed as part of this thesis, revealed that the respec-

tive target groups commonly spend at most 3-4 hours on site. Hence, short term interventions constitute a realistic form of informal activity that can be offered by EE institutions like the NAZKA. Consequently, subsequent research should also explore alternatives for the development of content, which enables also short term interventions to effectively target emotions and beliefs, as was recommended by Pooley and O'Conner [259] for achieving changes in attitudes.
Finally, Reid [271] provides a comprehensive discussion on the deficits of current attitude measurements. It suggests, that the instrument may not be sensitive enough to detect subtile changes in attitudes or may not address the relevant aspects of environmental attitudes. The improvement of the instrument should be address prior to future studies.

Almost all media groups of the three target groups showed a slight increase in their values and concerns scores following the intervention. Only for the student group administered by the human guide a small decline in values and concerns scores was detected following the intervention. Possible explanations for such a decline that are related to methodological constrains have already been discussed for attitude outcomes but do principally also apply to the results of the values and concerns scores. In particular, it can be assumed that similar to attitudes, values are not well developed among the children yet. The larger sets of students that participated on average in the human guide groups, may contribute to the explanation of the negative trend. A larger group size implies also an increased level of social interaction that does not directly pertain to the EE activity and thus leads to an increase of distraction and those also less engagement in the interactive tour.
Nonetheless, alike the environmental attitude component, the values and concerns scale, constructed for this study, did not measure a significant treatment nor media effects. This is the case despite the fact that all treatments included elements to promote direct experience as well as perspective taking, aspects which were suggested by Kals et al. [163] and Schultz [305] to further emotional affinity as well as biospheric concern. Also for the children taking part in family tours, no significant change in the values and concerns measure was detected, regardless of the opportunity for sharing nature experiences with significant others, which according to Kals et al. [163] is another potential amplifier for the impact of stays in nature. Most likely the human guide should have succeeded in effectively employing such mechanism for changing values and concerns based on his capacity to act as a true role model. In particular with respect to the students group, the guide may have not been accepted as a role model but rather perceived solely as a teacher like authority. But the human guide may have not succeeded in general because of deficits in the design of the intervention, which did not include explicit discourse on values or moral ethics as proposed by Kals et al. [163] as well as Fransson and Gärling [107].
Still, it has been noted in the literature [38, 210] that, in correspondence to attitudes, values and concerns are difficult to influence. Thus it remains difficult to draw definite conclusions on the actual impact of these media on environmental values and concerns, since it must be assumed that changes to this component will likely also require long term interventions and may not be detectable immediately after the intervention.

As was stated previously, the ultimate aim of EE interventions should be to motivate participants to engage in responsible environmental behavior [38, 195]. Since none of the evaluated direct antecedents of behavior (i.e. attitudes, values and concerns) were effected significantly, it is not surprising that the statements of all participants about their present and future environmental behavior did not change significantly due to the intervention.
Nonetheless, all of the parents and students display, independent of the respective media group, at least a slight increase in behavior scores during the posttest. This positive trend appears to be most pronounced in the mobile nature guide group. The thorough utilization of contents with the device and the novelty of the experience may have motivated the participants slightly more to agree to pro-environmental behavioral intentions. However, this positive trend was not detected for all media groups of the children. The significant differences between the pretest behavior scores of the media groups indicates that certain additional factors, which were not evenly distributed between the media, potentially influence the behavior ratings. As a consequence, the brochure group shows a positive trend of behavior scores on the posttest, while the other media groups exhibit a negative trend compared to their respective pretest results. Again, the negative trend can potentially be explained by a certain amount of fatigue and demotivation of these children. This trend may have been amplified by relatively high pretest scores in these groups that can result from socially desirable answers by the participants, a phenomenon that will be discussed in more detail below.
Regardless of the described trends, no significant intervention or media effect on behavior was detected by this study. This contradicts several of the examinations documented in the literature. Studies reviewed by Gunderson et al. [123] on K-8 level wilderness education programs resulted in short term behavioral gains but no long term behavior changes. Most studies reviewed by Leeming et al. [195] measuring observed or reported behavior, found a strong effect. However, a key difference between these studies and this work is that the majority of them involved training the students in relevant behaviors. In the presented study, the participants were encouraged to engage in activities to directly experience the natural environment but they were not explicitly trained in skills or specific environmental responsible actions for everyday life as suggested by Fransson and Gärling [107]. Neither did the intervention include specific incentives or other reinforcers nor was the attempt made to induce cognitive dissonance as discussed in [43, 138, 335]. Furthermore, there are also studies similar to this one, which investigated the idea that nature experience fosters action, revealed only small effects on behavior [38]. Also a study by Knapp and Poff [175] yielded only limited impacts of an environmental interpretation program on students' behavior.
Thus the lack of an effect on the behavior measure can likely be accounted to the fact that the intervention did not imply training in specific responsible behavior. Besides, based on data by Bogner [38] it can be assumed that the duration of the intervention was not sufficient in order to influence behavior.

One potential issue, relevant to all studies that attempt to assess intervention effects on self-reported variables, is that of socially desirable answers. According to Schahn and Bohner [294], this phenomenon occurs especially during the assessment of environmentally relevant behavior or attitudes leading to a shift in answers that are perceived as "socially desirable", which causes a distortion of the results. But Schahn and Bohner [294] further stress that even for self-reported behavioral data, at least a conscious distortion does not appear too plausible, since respondents are typically not motivated to distort responses in the context of a research study. The emphasis of anonymity and the scientific purpose of the study are normally adequate measures to reduce social desirability. Nonetheless, in this study the phenomenon of socially desirable answers may lead to elevated pretest scores, particularly concerning self-reported behavior and commitment, which would contribute to a balancing of pretest and posttest scores.

5.5.2 Success of interpretation and user satisfaction

The evaluation of the success of the interpretation as well as overall user satisfaction suggests that the different target groups experience the guidance by the specific guide media differently. This difference is particularly obvious between the participants of family tours and those taking part in the student tours.
The general difference between family groups and students regarding the satisfaction with the guide media can potentially be attributed to the differences in social interaction. Within the class context students are commonly also engaged in student-student interaction, which can have beneficial (i.e. motivational) as well as detrimental effects (i.e. distraction) on a student's performance [154, 256]. As was discussed previously, groups of students administered by the human guide were commonly larger in size, potentially resulting in more distraction due to social interaction that does not relate to the tour. Furthermore, the human guide tour represents a form of EE that displays certain parallels to formal educational settings, with the human guide as a teacher-like figure. Even though here the human nature guide tried to facilitate a self-directed exploration of the natural environment, the students still tend to judge the actual self-determined media, in particular the mobile guide, as more interactive and seem to rate them overall higher than the human guide. This underlines the assumption, that the mobile nature guide constitutes a novelty that, with its gimmickry qualities, tallies to the fascination of students for ICTs [13, 147], resulting in an overall higher valuation of this medium.

Parents commonly tend to be less driven by the fascination for technology. A few adults even explicitly stated their dislike of mixing ICT with the experience of nature. The analysis of requirements and assessment of expectations yielded that parents to the most part primarily seek the immediate and recreational experience of nature and want to receive new expert information on the respective natural environment (compare Chapter 3). This may explain why parents overall tend to be more satisfied with the human guide as opposed to the self-guided media. This state of mind is in particular reflected by the fact, that despite identical information content as well as the lack of a significant difference in knowledge gain of media groups, parents administered by a human guide perceived their knowledge gain as higher than those using a mobile nature guide did.

It is further emphasized by the explicit rating of the tour guide, where the human guide receives higher grades than the mobile nature guide. Furthermore, corresponding to their focus on direct experience, parents generally felt more distracted from the direct experience of nature, when using the mobile nature guide in comparison to taking a tour with the brochure or the human guide. This may partially explain why they were less satisfied with the mobile guide. The parents did, however, favor the self-guided media over the human guide when it comes to overall interactivity, with the brochure receiving the highest interactivity ratings. The accompanying experimenters observed, that the parents participating in a tour with the self-guided media, were frequently more involved in the tour and did also engage in the activities. Especially in case of the brochure groups they repeatedly took on the role of a facilitator for the overall tour and its activities. This observation may provide an explanation for the recorded differences in interactivity ratings. Likewise it may explain the grading of parents with respect to games offered during the tour, where the self-guided media were again valued higher and the mobile nature guide received the best grades. Despite attempts of the guide to involve the parents in the tour also during interactivities, the parents acted more as bystanders during tours with the human guide and consequently seemed to assign lower grades for games.
Even though the detected differences were not significant, children as well as students also tended to value the self-guided media higher with respect to games. The brochure received the highest mean grades for games. This generally confirms the findings of the requirements analysis that revealed that children and students favor particularly the self-determined exploration aspects of environmental interpretation activities. One reason for lower grades for games in the mobile nature guide may be, that the children and students were more distracted by the novel technology and were not able to focus entirely on the games. Whereas the young participants that used the brochure were familiar with the medium and could unrestrictedly enjoy the games.
In general children seemed to be equally satisfied with all guide media. Whereas students tended to be more pleased with the tour administered by a mobile nature guide.

Finally regarding the statements about the favorite aspects of a tour, the mobile nature guide exceeded the two other media, not only for the students but also in case of the children and parents. At a first glance this seems to contradict the previously discussed results. Still, it may also be an indicator for the proposed novelty effect associated with employing a mobile device for a nature tour. Even though it is most pronounced among the students, it does also apply to some degree to the family groups.
However, it also needs to be taken into account that among the least favorite aspects of the tour, the mobile nature guide was also the guide medium listed most frequently by parents and children. This makes it clear that even though intrigued by the novelty of the approach, the family groups were not entirely satisfied with the performance of the mobile guide medium. Next to general issues related to the guide, the mobile guide groups also experienced certain difficulties with the usability of the system resulting in additional negative feedback. These aspects will be further clarified by a subsequent discussion on the usability issues documented in the following section.

The students appear to constitute an exception in this respect, since no significant differences were found regarding the dismay with the guide media. Still again a tendency can be observed that they were even less satisfied with the brochure than with the mobile guide. It can thus be assumed that the novelty of the approach and the fascination for ICT may have counterbalanced the experienced difficulties with the usage of the system.

5.5.3 Navigation success

The observations on the navigation success of brochure and mobile guide groups did reveal a difference in wayfinding performance. Families using the brochure made fewer mistakes during spatial decision making (i.e. at junctions) than those with the mobile nature guide system. This may again be explained by the novelty effect of the mobile nature guide. Since none of the groups received prior training, families using the mobile guide commonly first needed to get accustomed to the device, including handling and reading the map on the small display. Further, the detected usability issues, comprising occasional malfunctions of the GPS and with it the failure of navigational support, may have resulted in a certain frustration and anxiety. According to Saucier et al. [288], this anxiety may have a negative impact on navigational ability by reducing attention to features in the environment or impairing the ability to encode these features. Families using the paper-based map, on the other hand, could apply a type of instrument that they were already familiar with. In addition, the accompanying experimenters frequently observed that while in the brochure groups the parents were regularly in charge of the map, in the mobile guide groups the children were allowed to operate the device and were thus also more involved in navigation decisions. This may have added to the difference in navigation success between the media groups, since children have usually less experience in using maps, which will be discussed in more detail below.
Still, these results contradict to some extent the findings of a previous study by Goodman et al. [117] comparing an electronic pedestrian navigation system to a paper map. Here, the electronic pedestrian navigation aids could increase the navigational performance of users compared to a paper map. It needs to be considered, though, that the mobile nature guide prototype, used in the evaluation, did not offer any additional navigational support apart from a position indicator, plus location-based notification of a station close-by, including basic instructions to locate the station exactly. When it comes down to the plain decision between paper-based map and electronic map, Reilly et al. [273] showed in an exploratory study, that most participants used plain electronic maps very little. They commonly reasoned that the display was to small for the map. Furthermore, a crucial difference to other studies is based on the fact that the MobiNaG system is employed in natural areas, which generally lack additional navigational cues like signs or prominent buildings and thus make the navigation tasks generally more difficult for participants, as was stressed by Soh [317].

For students, however, the novelty effect did not result in a decreased wayfinding performance. On the contrary, even though no significant difference was determined, a tendency can be observed that students using the mobile guide made slightly fewer errors at junctions. It can be assumed, that students were generally less experienced in reading maps. Thus the brochure group did not have the advantage of familiarity with the instrument. Moreover, students with the paper-based maps likely had to deal with the uncertainty of navigational decisions due to a lack of confidence in their navigation abilities, while those with the mobile guide potentially put more trust in their instrument. Further, driven by the fascination for the technology, the system might have even helped them to focus on the map and the navigational task.
This difference between families and students is further substantiated by the observation, that students were overall less successful with respect to spatial decisions. This can be explained by findings of Uttal [348] who states, that the acquisition, mental representation and use of spatial information evolves throughout childhood and with it improves the children's understanding of maps. It is thus evident, that the family groups, where navigational decisions were commonly performed or at least monitored by adults, generally outperformed the student groups, in which all decisions were taken by the children.

Concerning the target recognition task (i.e. finding stations), no significant differences were found between the media. Still, a trend could be observed that mirrors the results for spatial decisions at junctions. In this case the familiarity advantage of paper-based map was apparently balanced by the navigational support of the mobile nature guide in form of the location-based notification by means of an audio signal which was played upon reaching a 5–10 m diameter around the station. The fact that despite this technological assistance in target recognition the mobile guide groups did not outperform the brochure group may potentially also be ascribed to the detected usability issues including the inaccuracy and malfunctions of the GPS technology.
Again families clearly outperformed the students groups with regard to the target recognition, which can most likely be also attributed to the students' lack of experience with the task and ongoing development of their spatial abilities [348].

Despite the discussed differences in wayfinding performance, the two media groups did not vary significantly concerning their overall temporal navigation behavior (i.e. mean time taken to travel trail sections). This was true for families as well as students. Still an in depth analysis of individual sections shows that families as well as students using the mobile nature guide had the tendency to take more time navigating the initial section (i.e. trail sections 1–4). This further strengthens the assumption that mobile nature guide users had to go through an initial period of familiarization with the electronic device and the guide application (see also Ruchter et al. [286]), which resulted in additional time needed for the navigation of the initial sections. This familiarization effect apparently lasts for about one to two stations and trail sections. Familiarization and improvement throughout the course of a related study have also been documented by Reilly et al. [273].

The observed difficulties and usability issues with the mobile guide were expected to result in a difference between the groups regarding their perception of the navigational assistance by the guide medium. This difference could, however, not be clearly detected. Only trends confirm the preference of families for the brochure and the students' inclination towards the mobile guide for the purpose of navigational assistance. It can thus be assumed that the observed difficulties and differences in navigation success were not so pronounced that they could have effected the user's satisfaction with the navigational support.

Summing up, the findings show that, due to usability issues and restrictions in the reliability of the GPS technology in natural areas, as well as the limited provision of true navigational assistance, the mobile nature guide does not provide pronounced benefits over the traditional brochure with a paper-based map. Still, despite these constraints that existed in the prototype system, most of the disadvantages may be compensated by prior training and with increasing familiarity with mobile guides in the future. And especially for students, who are the participants with limited experience and at the same time constitute a critical target group, the mobile guide in some aspects tended to work better and was preferred over the traditional medium.

There remains a certain amount of ambiguity in the explanation of the documented results, which should be subject to further investigation. Specific evaluation of wayfinding performance and tasks in natural areas, comparing the assistance by paper-based maps and mobile nature guides, which however needs to provide state of the art navigational support. Further, additional questions include for instance, if differences between the media will prevail if participants receive prior training with the instruments? The evaluation should also control for additional factors that are reported to exert influence on wayfinding in spatial cognition literature (see [60, 204, 317]). In particular an assessment of the applicability and suitability of the proposed classification of landmarks for natural areas should be conducted.

5.5.4 Utilization of content

Student groups using the mobile guide consumed more content than those employing the paper-based guidebook. The same trend, even though not significant can be observed for the families. Schwarzer [308] cites similar findings for users of audio guides in museums who spend more time on the tours than those with other programs did.
Considering the overall consumption of content parts, families again outperformed the students. A crucial difference between the two target groups, which may explain this variance is that in the case of families the parents typically initiated the participation in the study in order to expose their children to an environmental education experience. School classes on the other hand were signed up by the teachers. Hence, usually parents were highly motivated to learn more about the natural phenomena, which could explain that families generally read more of the content parts.

It is striking, though, that students using the mobile guide, even though not a truly non-captive audience, clearly spent some more time at the stations and were also observed to read more of the content. There could be two possible explanations for this result. For one, this may be related to the interaction modalities of the mobile nature guide interpreter service. Once the students entered the information related to a station, they were more or less directed through all content parts until they could exit the station information again. The students using the brochure on the other hand could more easily skip whole parts of the information. Another reason for the result could be a stronger motivational effect that the mobile guide system had on the students. This may be supported by the result discussed previously, that students using the mobile guide exhibited a slightly higher tendency to feel motivated and content following the tour. This may of course be attributed to the general fascination of children of this age with technical and especially mobile devices of all kinds [13, 147].

Nevertheless, this could be a basis for employing the mobile nature guide as an anchor to motivate especially students or children to engage in EE activities and eventually learn more about their natural environment through direct experience.

Although for families the measures of content utilization did not differ significantly between the groups, more of the parents using the mobile guide still had the feeling they had used the content to a large extent. This impression could be an effect of the interaction modality of the device as well. This also includes the aspect that if they did not hold the device themselves and another member of the group read the content to them, they had the experience of spending more time and also likely presumed that they had consumed the entire content. This finding may further indicate, that a mobile nature guide could result in a more intensive consumption of the content also by families.

5.5.5 Usability issues

Next to aspects relating to the impact of a mobile nature guide in comparison to traditional guide media, the study also yielded crucial results regarding the usability of the prototype system. As was pointed out before, one of the crucial issues regarding the effectiveness of a mobile nature guide for EE interventions is the increased distraction from the direct experience of nature, that participants criticized. This matches findings by Rogers et al. [275] made with students in the "Ambient Woods" project. Based on the observations during this study it can be concluded that a fair amount of this distraction was caused by usability issues that test subjects encountered while operating the device. Considering the identified issues, it has become clear that next to the implementation of further functionalities proposed as part of the mobile nature guide concept (compare Chapter 3), adjustments need to be made to the core services implemented so far. The following issues should be primarily addressed:

Overall performance and stability of the system It can be assumed that restrictions to the performance and stability have caused a certain amount of frustration for the mobile guide users and thus also limited its effectiveness as an instrument for EE activities. The fact that complaints about these limitations were documented for a considerably higher proportion of families than students, may indicate that families had higher expectations regarding the device, expecting it to perform similar to an office or home PC. Whereas students again seemed to be less annoyed by these issues, either being more intrigued by technology in general or taking it as a welcome change to their standard formal school setting. Potential solutions could include changes to the architectural design and the deployment of more efficient technologies.

Interface and interaction design issues SVG can still be considered a promising approach for the design of user interfaces on small display devices as proposed by Düpmeier and Ruchter [94]. But the experiences of this study, relating to the constraints immanent to the current implementations for SVG-engines on PocketPCs, make it clear that it is still a fairly new technology which is yet restricted in its applicability of interactive mobile systems to be employed by non-expert users. Due to its scalability and scriptability (plus layering etc.) SVG proved an adequate technology for the display of mobile maps. The user interface for the interpreter service, however, was also rendered in SVG and caused a variety of difficulties. For instance neither standard text flow nor scroll bars had been implemented as standard elements in the employed eSVG engine. In addition slow response times caused a certain amount of anxiety among the participants and resulted in the high percentage of complaints about readability. As a consequence, many users called for additional output modalities like audio. A study by Woodruff et al. [366] showed that the presentation of content using acoustic modalities can also help to reduce the fixation on technology by freeing the user from focusing on the display. But then again, acoustic presentations can of course also distract the user from experiencing sounds of the natural environment especially when consumed via headphones. Similar concerns were raised by Cheverst et al. [62] with respect to city guides and traffic. Further they may also distract other visitors of the natural area. Possible solutions to these issues were discussed as part of the implementation issues in Chapter 4.

GPS and navigational assistance As was mentioned afore, another major usability issue listed by participants was that of malfunctions of the GPS component, resulting in inaccuracies or wrong information presented by the navigator. This made the wayfinding task additionally challenging. It should be pointed out that this is an issue that was criticized by more students than families. The same is true for map issues, where the disproportionally high percentage of students is striking. Both confirms the assumption that students or presumingly children in general have more difficulties with the utilization of maps and consequently with the navigational tasks. This particularly included trouble with the readability of the map, its accessibility, and the operation of particular map functions (e.g. zooming etc.). It can also be concluded, that the GPS feature made up a fair amount of the technology driven appeal of the system for the children. The findings suggest that a number of improvements need to be made to the navigator service of the current prototype system. With respect to the GPS issues this implies improvement to context capture and processing. Some of the shortcomings are actually related to limitations of GPS sensor device, especially due to outdoor conditions such as signal shading by the tree canopy. Similar issues with positional errors caused by GPS were also reported for the forest education system evaluated by Abe et al. [1]. Still, these obstructions are apparently not sufficiently handled with error control procedures as well as plausibility checks of the data. Moreover, this stresses that alternative positioning strategies such as dead reckoning, user feedback coupled with landmark based navigation need to be implemented in a mobile nature guide system. Finally the problems that especially students tend to experience while using maps need to be accounted for by the implementation of the concepts suggested by Patalavitiute et al. [251] including the personalization of maps for target groups.

Hardware issues Crucial constraints commonly cited for mobile devices, such as small screen real estate and limited input modalities [186, 206, 355] appeared to be less of an issue. But aspects related to outdoor usage of the device led to a number of complaints by participants. Especially additional restrictions to the readability due to lighting conditions (i.e. direct sunlight and reflections) were criticized. Such readability problems due to lighting conditions are a common issue documented also by related studies on mobile guides for EE or environmental tourism such as Abe et al. [1] and Okada et al. [243] as well as other systems designed for outdoor usage [304]. As was described previously, the attempt was made to protect the hardware from adverse weather conditions by placing it in a flexible casing. Although the casing was transparent, it further severed the readability in some cases by adding to the reflection problem. In addition the protective casing, even though designed for mobile devices, caused a slight reduction of the touch screen sensitivity of the device, causing some users to have extra trouble with operating the system with the pen. This makes it clear that for the utilization of a handheld device for a mobile nature guide system a tradeoff between readability, operability, protection and portability has to be achieved, which should be subject to further investigations.

According to observations by the experimenters the casing equipped with a strap did allow the users to still engage in direct exploration activities with both hands-free if needed. As opposed to the evaluation by Abe et al. [1] as well as Okada et al. [243], the weight and handling of the hardware was only perceived as a problem by a small fraction of the participants. Corresponding to the findings of Rogers et al. [275], the device was also passed on to other participants if it was restricting the user in other activities. Still, some social interaction issues have been perceived that also have a certain relevance for the usability of the system. Each mobile guide system had to serve a group of users either family members or students, which actually matches realistic conditions for the deployment in EE institutions.

Due to the small size of the screen and reflections of the LCD-display when observed from an angle, not all subjects were able to view the presentation at a time, making additional cooperation necessary. According to Woodruff et al. [366], such social interactions can also be facilitated by audio presentations. Thus a potential solution to the limitations resulting from this setup may be to enhance multi-modality by offering audio support (i.e. presenting all textual information as alternative audio files). The effectiveness of audio presentations in mobile nature guides needs to be examined further. Since, audio output may also distract the user from directly experiencing nature.

5.6 Conclusions

The study showed that the mobile nature guide system can be successfully employed as an instrument for EE interventions. A significant increase in environmental knowledge was achieved, comparable to the effect of a brochure and a human guide. It can thus be concluded, that a mobile nature guide can function as an effective alternative to traditional interpretive media with respect to the enhancement of environmental knowledge through EE interventions. However, these findings are based on the assessment of immediate effects following a short term EE intervention. Future research should include follow up measurements as proposed in the outcome research literature [32, 38, 195] to determine the persistence of the measured effects over time. Especially since no significant effect of any of the interpretive media on additional environmental literacy components (i.e. attitude, values and concerns, behavior) could be determined, it seems essential that future studies will compare the effects of the different media based on long term EE interventions. This implies a repeated employment of the guide media over several consecutive days (according to Bogner [38] at least 5 days) in a longitudinal study. Interventions designed for such a study should further stress aspects of specific responsible behavior or actual training of such behavior. Also approaches that can enhance emotional affinity, such as perspective taking, should be tested more explicitly.
Since an influence of socio-demographic factors could be observed, at least with reference to environmental knowledge, the impact of such additional factors and their interaction with the different media should also be subject to further investigation. In addition, study design related issues constrain the power of these findings and should subsequently be improved in prospective studies. These include the relatively small sample sizes for each treatment group. Besides, the counterbalancing approach may, in the face of the small sample size, bias the actual intervention and media effects. Also the validity of the scales especially for values and concerns and behavior should be rechecked based on psychometric techniques.

In spite of the lack of differences between the traditional media and the mobile guide concerning the impact on environmental literacy, the examined target groups display differences in judg-

ing the assistance by the guide media. The ”cool factor” or gimmickry, as characterized by Schwarzer [308] and Fleck et al. [106], seems to have the greatest impact on students, which combined with the fascination for technology in general, appears to lead to the result that students in many aspects favored the mobile guide over the traditional interpretive media. Children appear to have no clear preference but tend to favor the brochure and to some extent also the mobile guide. Whereas parents tend to favor the human guide over the other media, especially regarding aspects related to the success of interpretation. Nonetheless, they do acknowledge the interactive qualities of self-guiding media. One of the main drawbacks for parents seemed to be that they felt more distracted from the actual direct experience of nature when using a mobile nature guide.
Regarding the assistance during navigational tasks both families and students generally seemed to be equally satisfied with the mobile nature guide as well as the brochure. Actual performance measures for wayfinding and target recognition did not yield a pronounced benefit of the mobile nature guide over the brochure with the paper-based map. It became clear that the groups using the mobile guide needed a certain period of time to familiarize with the system. Thus some of the disadvantages experienced especially by the families with the navigator service, may be compensated by prior training or will fade out as more people get used to mobile devices. It is striking though that even with the current state of the mobile nature guide prototype the navigator service helps students to achieve slightly better results.
It can be concluded that, in order to offer a true advantage over the paper-based map, the navigator service needs to be further improved including the implementation of actual navigational assistance features such as offering directions and warnings.

An important finding with reference to the effectiveness of a mobile guide for EE activities is, that participants using the mobile nature guide consumed the content more intensively. Not solely the families but especially the student groups using the mobile guide consumed more of the content. Consequently, the mobile nature guide can potentially serve as an important instrument for EE activities, given that it actually leads to a more intensive usage of the environmental information and an increase in motivation to experience nature directly. Future research thus needs to investigate if this effect was caused mainly by interaction modalities of the system or can be attributed to stronger motivation induced by the electronic medium. In addition it will be important to determine if this higher consumption of content leads to a lasting increase in environmental knowledge as well as the other environmental literacy components.

A fair amount of this distraction and frustration that the users experienced with the mobile nature guide, was caused by usability issues. It was, however, also noticed, that while the users became more accustomed to using the device, they started to focus more on the content and the experience of their natural environment. Nonetheless, a number of core usability issues were identified in the study that need to be solved in following mobile nature guide systems. This includes the improvement of performance and stability of the system as well as adaptations to the system architecture. As was stressed by Dunlop et al. [91], user interface design is a significant challenge for any handheld software development. Thus also for the mobile nature guide user interaction needs to be more intuitive also for non computer literate participants of different target groups. This also involves the utilization of standardized interaction mechanisms based on familiar paradigms such as the web browser as well as a more target group specific interaction design. Also the application of wizards or a brief tutorial presented upon starting the system could potentially help to minimize the familiarization period and alleviate some of the usability related difficulties. Yet, in particular readability and interaction modalities need to be further improved for outdoor usage conditions. The provision of acoustic modalities (e.g. audio output) is considered a promising solution but can potentially also distract from experiencing the natural soundscape. As a consequence, additional studies should be conducted that help to shed light on the effect of using audio messages in mobile nature guides.

As part of the improvement of the navigator service, capture and processing of geographical context also needs to be optimized, next to the inclusion of alternative strategies that help to compensate the restriction of positioning technologies in remote areas. Also more intuitive interaction mechanisms such as egocentric maps that help to reduce the cognitive load due to mental rotation, need to be tested in mobile nature guides along with an improved implementation of the approaches to personalized maps and mobile nature maps.

The evaluation also showed, that it remains an essential challenge, especially for the authors of tours, to utilize the mobile nature guide as an actual interpreter that mediates between the visitor and the natural phenomenon. Thus in order to avoid a fixation on the technology and subsequent distraction by the device, it is important to limit the amount of content, especially textual information presented through the device and include elements that promote direct experience.

In conclusion, it should be pointed out, that despite the limitations of the prototype system and the yet restricted realization of the concepts, the mobile nature guide did in many aspects match the performance of traditional interpretive media, especially the brochure. Even though the capabilities of the human guide can, at the time, essentially not be substituted by a mobile guide, it turned out that in particular for students a mobile nature guide can offer an attractive alternative for self-directed EE activities.

Still, even if it were mostly the students who would profit from the employment of mobile nature guides for EE activities, this would already imply a major contribution to the overall goal of the DESD and the long-term promotion of environmental literacy among citizens. For EE institutions especially school classes constitute a critical target group, since they may come from families, which may commonly not visit the institutions during their leisure time. Consequently these students may also have less direct experience of the natural environment. Furthermore, based on the findings of Leeming et al. [196] children engaging in EE programs involving proenvironmental activities throughout a school year, may serve as multipliers influencing their parents to adopt more proenvironmental behavior as well as changes in their awareness of environmental issues.

For EE institutions, supposed to serve as providers for mobile nature guide services, Kindt's [169]

concern about the capability of such institutions to provide and maintain such computer-mediated interpretation tools still remains to be evaluated as the first mobile nature guides are being deployed. Despite the cost of initial investment in the technology, key advantages of mobile nature guides over traditional media are also anticipated for the EE institutions. These include the capacity for easy and dynamic updating of the information on the device and the provision of more target group specific services in one unit. The reusability of the instrument and content components is potentially enhanced. As was illustrated in Chapter 3, a manager service provides new opportunities for visitor monitoring and management. This can also include the evaluation of provided EE activities. The generation of content of course still poses a burden on the educators. Although the approach of a community based environmental interpretation instrument could also facilitate this task.
Yet, with reference to the critique by Payne [254] concerning the potential of rationalizations in the EE domain due to the embracement of computer-mediated instruments, it should be stressed, that a mobile nature guide is not essentially intended as a replacement for traditional interpretation media or services. Instead it should provide EE institutions with an opportunity to extend their programs and attract especially target groups that have so far not been motivated to engage in EE activities.

Still, future mobile nature guides that implement all of the concepts proposed in Chapter 3, can bring a number of advantages over traditional, especially paper-based, media. A number of authors in the mobile guide literature [50, 62, 91] point out that an improved overall acceptance of these systems by all target groups, will require to take even more advantage of the additional capabilities of mobile guides. In accordance with the conceptual design, the assets of a mobile guide are the services, that allow for the context-based adaptation and presentation of content. For the mobile nature guide this implies an improved context capture and processing not only for positioning but also for other components of the environment as well as the user context. On this basis the personalization of the presentation and the adaptation to environmental conditions could be further improved. A similar improvement could be achieved by the implementation of an emotional pedagogical agent. Such a component could potentially, beyond facilitating the interaction with the device, reinforce instruments of perspective taking and emotional binding with the natural environment. Dunlop et al. [91] also underlined benefits resulting from mechanisms of public annotations and recommendations (i.e. social navigation). This could be achieved in the mobile nature guide through the concepts of digital souvenirs and spatial bookmarks which have not been implemented in the prototype system. As was pointed out by Rogers et al. [275], digital augmentation offers a promising way to enhance the process of learning in EE, accordingly the implementation of the proposed concepts for exploration tools would offer crucial advantages to a mobile nature guide over traditional media with respect to visualization of hidden details and also the motivation of direct interaction with and exploration of natural phenomena.

Chapter 6

Summary and Outlook

It was the objective of this thesis to develop a new concept for computer-mediated environmental education. This novel concept should contribute to the resolution of key issues that environmental educators currently face in their efforts to promote environmental literacy. These include especially a lack of interest and motivation of young people to participate in environmental education (EE) activities and in the direct experience of the natural environment in particular. Based on a new model for mobile EE a conceptual design for a mobile nature guide system was developed, which can serve as an effective computer-mediated instrument in EE.

A new model for computer-mediated EE was developed in Chapter 2, incorporating fundamental concepts from the field of environmental education and mobile guide systems, reviewed in Chapter 1. This model for mobile environmental interpretation was devised to specifically apply instruments to influence environmental behavior, which were derived from a comprehensive environmental literacy model.

Based on a requirements analysis, the conceptual design for a mobile nature guide system was generated in Chapter 3. This design encompasses a comprehensive description of the components needed for a new computer-mediated instrument, in correspondence to the proposed model for mobile EE.

Chapter 4 presented the development of IT concepts for the mobile nature guide in accordance with the devised conceptual design. Specifically an overall system architecture was proposed and the design of the different graphical user interface (GUI) modules was illustrated. Finally the implementation of a prototype mobile nature guide system (i.e. the MobiNaG system) was documented and the core implementation issues were discussed.

There is a lack of research on the actual effects of such mobile computer-mediated instruments on the user's environmental literacy, compared to traditional environmental education approaches.

Chapter 5 documented the evaluation of the prototype mobile nature guide system, conducted as a proof-of-concept study. In this field study, an impact evaluation was conducted with families and school classes employing the mobile nature guide prototype in a natural area. The evaluation further comprised an assessment of the key usability issues related to the prototype system.

The main results of this thesis are:

1. Identification of EE instruments, based on a comprehensive environmental literacy model, that can be implemented and enhanced by computer-mediated instruments to further promote environmental literacy. These instruments comprise: transfer of knowledge, guidance and support for direct experience and exploration, visualization of phenomena and behavioral consequences, representation of role models, and presentation of feedback on environmental behavior.

2. Development of a new model for mobile environmental education that facilitates the combination of computer-mediated education and the direct exploration of nature by merging concepts of personal environmental interpretation and context-aware applications.

3. A front-end evaluation revealed the relevant target groups (i.e. students, families, nature lovers) and their respective requirements regarding a mobile guide system. A usage scenario for a mobile nature guide system was further specified by establishing a task model for a general environmental interpretation unit.

4. Development of a conceptual design for mobile nature guide systems in accordance with the proposed model for mobile EE. The conceptual design had to cope in particular with specific environmental requirements unique to mobile environmental guide systems deployed in natural areas (e.g. remoteness, natural dynamics).

5. The specification of a series of context-based services that offer the functionalities derived from the requirements analysis. Next to the core services for navigational assistance and location-based information provision, a mobile nature guide especially needs to offer services that facilitate direct exploration. Additional services need to be included for the purpose of visitor monitor and evaluation.

6. Development of an IT concept for a mobile nature guide, complementary to the conceptual design. A service-oriented software architecture was chosen as foundation for the frame application to which various services can be flexibly added.

7. The guidebook and human guide were determined as suitable interface metaphors. An iconic, intuitive user interface does not suffice, though. Especially the application of an animated pedagogical agent as an "e-Interpreter" constitutes an important interaction paradigm for a mobile nature guide that can serve as a role model.

8. Implementation of a prototype mobile nature guide system as part of the MobiNaG project. A number of issues were documented, especially concerning the application of Scalable Vector Graphics (SVG). It was shown to be a suitable technology for the realization of adaptable, mobile nature maps but demonstrated limitations in the support for other user interface elements.

9. Successful evaluation of the prototype system in the field. The impact evaluation showed that such a mobile nature guide system can be successfully employed for EE interventions, leading to an increase in environmental knowledge similar to that achieved by traditional EE instruments. With respect to navigational performance the evaluation demonstrated, that the navigational assistance through the prototype mobile nature guide needs to be further improved in order to achieve a decisive benefit over traditional EE instruments. The mobile nature guide did improve the utilization of EE content especially among the students owing to a stronger motivational effect of the mobile technology. This indicates that a mobile nature guide can indeed serve as an effective EE instrument that can help educators to motivate especially young people to engage in self-determined exploration of their natural environment.

Next to addressing the encountered implementation issues, future work on mobile nature guides should focus on the realization of further mobile nature guide services, proposed as part of the conceptual design. This encompasses especially the exploration tool services, which are expected to further improve the system's suitability for self-determined exploration. Taking advantage of the full potential of mobile guide technologies, also implies the continuous enhancement of context-awareness, including the capture of extra environmental context features employing additional sensors.

The evaluation also pointed to the issue of distraction from nature by the technology, which should be engaged by further research on interaction modalities for mobile nature guides. This also implies the full integration of an animated pedagogical agent.

These technological enhancements are the groundwork for further empirical research in EE, concerning the impact of mobile nature guides on environmental literacy. It remains to be tested, if an improved system can change environmental literacy components besides knowledge. Furthermore, the effect of an "e-Interpreter" on the perspective taking capabilities and emotional affinity of visitors should be evaluated. Future evaluations should, however, be designed as longitudinal studies including follow-up measurements.

Finally, the applicability of mobile nature guides beyond the field of informal EE needs to be investigated. This should for instance include the application in the environmental conservation community, but it can also be envisioned that such an application can be further integrated into every day life as a "personal environmental literacy advisor".

Appendix A

State of the Art Background Materials

A.1 Background materials on environmental communication

The following collection of recommendations for an effective EC campaign has been identified in the literature [55, 131, 234]:

Reach out and integrate: Effective environmental communication demands partnerships with a number of key stakeholders (e.g. local/regional government, citizens, NGOs, business, and the media). Facilitate the integration of initiatives at different levels.

Create and reinforce awareness: EC should create awareness for environmental issues, educate about the environment and change or reinforce behavior.

Involve and listen: Receivers of the message should be encouraged to participate and contribute based on their own ideas and experiences (e.g. children, members of the community, teachers etc.). Thus it is the challenge to make EC a bidirectional communication process. Greater involvement also creates ownership and long-term commitment.

Build trust and lead by example: Credible information is key. Any communication strategy relating to environment, needs to work on the basis of transparency and accountability. In order to be credible one must lead by example.

Say it right and say it clear: Information regarding the environment and sustainable development needs to be provided to people who need it, when they need it, and in forms they can understand. This means that the information needs to be tailored specifically to the target group and furthermore it needs to be adapted to the context of the individual receiver.

Make it fun and rewarding: Children are not the only ones who respond to fun experiences. If people get enjoyment, relaxation and stimulation from environmental communication, which also requires emotional engagement, they are more likely to invest more time, energy and long-term commitment. Environmental protection is often presented as a battle that cannot be won, this needs to change as doom and gloom de-motivates the general public. Hands-on, practical experiences that are relevant to people's daily lives will send a much clearer message and have a longer-lasting impact.

These recommendations should be taken into consideration for the design of the computer-based EC instrument.

A.2 Background materials on environmental education

A.2.1 A review of environmental education concepts

Literature reviews by Thomson [336] and Archie and McCrea [14] as well as Ramsey and Hungerford [263] show that significant concepts for EE and frameworks for environmental literacy commonly share the following aspects:

Knowledge: The knowledge component usually implies that the learner acquires knowledge of fundamental ecological concepts as well as an understanding for interactions between the social/human systems and natural systems drawing on disciplines in natural sciences, social sciences and humanities. Further more the learner is expected to know about existing environmental problems and strategies for their resolution.

Awareness/Sensitivity: As part of the awareness and sensitivity component the learner should develop positive attitudes and values leading to an empathy for nature and society. This should include a sense of responsibility for personal and group behavior as well as a strong internal locus of control (i.e. a sense that he or she can manifest some influence upon or control over the outcomes of a specific activity).

Skills: Skills that the learner should obtain usually focus on decision making and problem solving. Environmental literate citizens should be able to identify environmental issues and analyze, synthesize and evaluate related information from different sources, considering social, economic, political, technological, cultural, historic and aesthetic aspects of environmental issues. Finally they should also have the skills to develop, implement and evaluate action strategies for the resolution of these environmental problems/issues.

Active Involvement: All concepts stress the aspect of active involvement based on a willingness to apply the obtained skills to help resolve environmental problems and engage in environmental responsible behavior. Such personal or group involvement may include Eco-management (e.g. more energy efficient transportation), Economic or Consumer Action (e.g. purchase environmentally friendly goods with respect to packaging), Persuasion (e.g. encourage others to stop environmentally unfriendly behavior), Political Action (e.g. contact officials/politicians with regard to environmental issues), Legal Action (e.g. reporting violations of environmental laws).

A.2.2 Issues related to the usage of information and communication technologies in environmental education

Table A.1 presents a summary of challenges and promises associated with the application of ICT in EE, which was compiled from the literature [13, 277].

Table A.1: Summary of challenges and promises of the application of ICT in EE based on [13, 277].

ICT Challenges	ICT Promises
Organizational challenges and resource deficits:	**Who wants to be heard must stick with the trend:**
- Lack of computer literacy: Educators more so then learners frequently lack the necessary training and expertise in the use of ICT and preparation of multimedia materials. - Lack of time and money: Multimedia educational materials are generally very resource intensive with respect to time required for their production as well as hardware and software needed for their production and use. Most EE institutions are however under-equipped, due with small budgets. - Inequity of access: In order for computer-aided EE to be effective, it must be affordable and accessible to all, which is at the time not given due to still existing digital divides.	- The "Information and communication era": We life in an age where most messages are transported via multimedia devices. Who wants to be heard especially in order to effectively communicate ecological principles, must make use of multimedia technologies. Low-tech messages will appear out-dated and unprofessional. - New sources of environmental information: To date mass media have already become the primary source of environmental information for European citizens. As will be discussed below this trend continues to increase as more and more environmental institutions provide free access to environmental data via the internet.
Didactic challenges:	**Competition for attention and influence:**
- Lack of didactic methods and multimedia culture: There is still a lack of didactic methods EE as well as best practice examples. Multimedia applications require a multimedial communication culture. - Danger of technological misapplication: Especially among those who readily embrace ICT there is the danger of giving in to the lure of new technology, which results in teaching technology instead of EE, potentially the overriding of traditional approaches. - Making it easy to use and keeping abreast: Developers of ICT for EE need to make access and use of the tools hassle-free for learners and educators. The software should empower students to take a more proactive role in the EE process. Developers as well as educators further have to live up to the rapid change of the technology.	- The media shapes life style: Lifestyle and related role models are presented through the mass media (films, advertisement, etc). This lifestyle also significantly influences the environmental behavior of an individual and hence also his willingness to participate and help to resolve environmental issues. - Perception of the environment and media competence: The perception of the environment is to a large extend shaped by the daily media consumption and the flood of images and other information. Environmentally literate citizens today need to acquire at least a basic media competence, which in turn can be fostered through active use and design of multimedia applications. This may be an opportunity for EE to improve its perceived significance in the society.
Antagonism between experiencing nature and using a computer:	**New tools for educators:**
- Environmental activity substitution: Educators and adult participant fear that ICT will become just an artificial, unproductive substitute for interaction with nature. Thus computer-aided education must be used as a catalyst for field-based exploration. - Loosing touch to reality: ICT in EE should not be just another computer game, which captures all of the users attention. Multimedia should remain a tool and not determine the style of learning or teaching.	- Staying up to date: While the ICT revolution led to an increasingly shorter information life-cycle, it also facilitates the search for up-to-date information and provides tools for production and maintenance. - New roles for educators and learners: ICT offer a particularly supportive environment for constructivist learning. As a consequence the educator's significance as primary source of information is reduced and he becomes motivator and facilitator.
Environmental impacts of ICT:	**New opportunities for learners:**
- Acceleration of globalization: The boom of ICT has been facilitating the globalization of markets, production and habits, leading to new social and environmental issues on a global scale. - Not only "green and clean": There has been a lot of progress made but the production as well as use of devices for ICT still has extensive impacts on the environment and should be analyzed critically and comprehensively.	- New opportunities for participation: ICT can encourage self-directed learning and learners become involved in content production. The production builds on media competence and also requires understanding of environmental concepts. - Experiential learning: ICT in education can motivate learner interaction, experimentation and cooperative learning. ICT offers tools to simulate and model outcomes to scenarios, and can be controlled by the learner at their own pace and style. - Open a window to the world: Open networks enable global exchange on environmental issues. Learners can learn about the impacts, their behavior may have around the world. They can publish and share materials with others, increasing their sense of responsibility and involvement.

A.3 Review of mobile guide systems

A.3.1 Taxonomy and comparison

Table A.2 presents the reviewed mobile guide systems and the categorization based on the proposed taxonomy.

Table A.2: Taxonomy of mobile guide systems.

Name	References	Domain					Type of guide									
		Tourism	Museum & Exhibits	Community events	Mobile Learning	Personal Assistant	City guide	Cultural heritage guide	Environmental tourism guide	Museum guide	Park guide	Conference guide	Fair guide	Campus guide	Nature guide	Personal guide
CRUMPET	Schmidt-Belz/Posland 2003, Schmitdt-Belz et al. 2002	1					1									
Cyberguide	Long et al. 1996, Abowd et al. 1997	1					1							1		
Deep Map	Malaka/Zipf 2000	1					1									
GUIDE	Davies et al. 2001, Cheverst et al. 2000	1					1									
LoL@	Pospischil et al. 2002, Uhrliz/Lechtaler 2001	1					1									
Tellmaris	Coors et al. 2003, Laakso et al. 2003	1					1									
Real	Baus et al. 2002	1					1							1		
AccesSights	Klante et al. 2004	1						1								
ArcheoGuide	Almeida et al. 2000	1						1								
MobiDenk	Krösche et al. 2004	1						1								
PEACH	Kruppa et al. 2004,Goren-Bar et al. 2005,Stock/Zancanaro2002	1						1		1						
PARAMOUNT / VISPA	Löhnert et al 2001	1							1							
WebPark	Burghardt et al. 2003, Dias et al. 2004	1							1							
National Park IS/ReoGeo	Almer et al. 2003, Frech/Koch 2003	1							1							
BUGAbuttler (BLIS)	BLIS 2005		1								1					
eTour	Breitenfeldt 2003		1						1		1					
HIPS	Oppermann/Specht 1999, Broadbent/Marti 1997		1							1						
Imogl	Luyten/Cornix 2004		1							1						
SCALEX	Trummer/Martin 2004		1							1						
Sotto Voce	Aoki et al. 2002, Woodruf et al. 2001, Woodruf et al. 2002		1							1						
Electronic Guidebook	Semper/Spasojevic 2002, Fleck et al. 2002				1					1						
Momuna (companion)	Sauer/Osswald 2003				1					1						
CHIMER	Weiss 2002, Tétard/Patokorpi 2004				1			1								
Ambient Woods (Equator)	Rogers et al 2005, Randell et al. 2003				1										1	
Forest Education Support Sys.	Abe et al. 2005				1										1	
Bird Watching Learning Sys.	Chen et al. 2003				1										1	
DigitelEE 2	Okada et al. 2003				1										1	
ActiveCampus	Griswold et al. 2002			1										1		1
PalmGuide	Sumi/Mase 2002			1								1				
SaiMotion	Hermann/Heidmann 2002			1									1			

Table A.3: Features of mobile guide systems: mobile device, positioning technologies, services. 1: used, c: considered, p: planned, e: envisioned, -/?: no information available.

Name	Hardware				Positioning technologies								Navigation / Map service										Information service				
													Data type		Database			Style									
	Smartphone	PDA	TabletPC / Laptop	Internet PC	Manual / User Interaction	GPS	WLAN / BT	Cell ID	RFID / Barcode	IR	Multimodal positioning	Alternative Strategies	Raster	Vector	On Device (static)	On device (Geo-DB)	Map-server	Sketch-like	Custom 2D	+3D	Other	Interaction	context-independent	context-sensitive	push-service	pull-service	tour-service
CRUMPET		1				1		c									1		1					1	1	1	
Cyberguide		1				1				1			1					1						1		1	
Deep Map		1	1	1	1	1						1					1		1	1				1	?		1
GUIDE			1				1	c				1	1					1						1	1	1	1
LoL@	1	1				1	c	1				1	1	1			1		1			1		1		1	1
Tellmaris	1		1		1	1							1						1	1		1	1			1	
Real		1	1		1	1				1	1	1					1	1		1		1	?	?	?	?	?
AccesSights		1				1											1				1	1		1	1		
ArcheoGuide			1			1														1				1			1
MobiDenk		1				1											1				1	1		1	1		
PEACH		1				1				1			-	-	-	-	-	-	-	-	-	-		1	1	?	
PARAMOUNT / VISPA		1				1							1						1	1				1			
WebPark	1	1				1		c					1	1			1		1					1	1	1	
National Park IS/ReoGeo		1		1		1										1			1	1		1		1		1	1
BUGAbuttler (BLIS)		1				1													1					1	1		1
eTour		1			1	1									1								1	1		1	
HIPS		1		1						1														1	1	1	
Imogl		1				1	1		c	1					1									1	1		
SCALEX		1								1														1		1	1
Sotto Voce		1			1								-	-	-	-	-	-	-	-	-	-		1		1	1
Electronic Guidebook		1	1	1					1	1					1					1			1	1		1	
Momuna (companion)		1					c		1											1				1		1	1
CHIMER	1	1	1			1								1			1							1		1	
Ambient Woods (Equator)		1				1	1		1				-	-	-	-	-	-	-	-	-	-		1	1	1	
Forest Edu. Support Syst.		1				1									1			1						1	1		
Bird Watching Learning Syst.		1																									
DigitelEE 2			1			1												1									
ActiveCampus		1			1		1						?	1					1			1		1	1	1	
PalmGuide		1								1			-	-	-	-	-	-	-	-	-	-		1	?		1
SaiMotion		1				e	1				p			1					1			1	1	1	1	1	1

Table A.4: Features of mobile guide systems: services, interface, evaluation, context, system architecture. 1: used, c: considered, p: planned, e: envisioned, -/?: no information available.

Name	Communication service						Com-mercial		Other services		Interface								Evalutation				Context				Architecture					
											Metaphor					Modality																
	Peers	Other users	Friend-finder	Graffiti / Bookmarks	Emergency service	Diary service	Booking	M-commerce	Search	User-monitoring	Browser	Map	Folder / tabs	Human guide	Other	Visual	Acustic	Haptic	Requirements	Lab-study/Expert evlauation	Field-study	Case study	User context	Geographical context	Environmental context	System context	Stand-alone	Client-server	Interacting Applications	Multi-Agent-Systems	Hybrid	Multi-Blackboard
CRUMPET											1					1			1		1		1	1		1				1		
Cyberguide	1		1	1		1	1		1			1				1					1			1	1				1			
Deep Map							1					1				1	1	1			1		1	1	p	p				1		
GUIDE	1	1				1	1				1					1	1		1	1	1		1	1	1	1		1				
LoL@	1					1	1					1				1	1	1	1		1		1	1	1			1				
Tellmaris			1						1			1				1		1			1			1					1			
Real												1				1					1		1	1	1						1	
AccesSights				1								1				1	1	1					1	1	1			1				
ArcheoGuide															1	1	1	1			1		1	1	?			1				
MobiDenk												1				1	1			1			1	1				1				
PEACH						1		1		1				1		1	1	1			1	1	1	1		1		1				
PARAMOUNT / VISPA		1			1			1		1		1				1				1				1				1				
WebPark	1			1	1	1		1	1	1		1				1	1				1		1	1	1			1				
National Park IS/ReoGeo						1		1		1		1				1	1				p		1	1				1				
BUGAbuttler (BLIS)																1	1							1	1		?					
eTour									1						1	1		1			1			1			?	1				
HIPS		1				1					1					1	1	1		1			1	1	1	1		1				
Imogl													1			1					1		1	1	1		1					
SCALEX															1	1	1	1					1	1		1		1				
Sotto Voce															1	1	1				1		1	1			1					
Electronic Guidebook		1				1					1					1	1				1		1	1				1				
Momuna (companion)	1					1		1		1		1				1	1	1				1	1	1				1				
CHIMER						1										1	1							1				1				
Ambient Woods (Equator)		1								1					?	1	1	?			1		?	1	1							
Forest Edu. Support Syst.		1		1							1					1					1			1				1				
Bird Watching Learning Syst.																																
DigitelEE 2																1	1				1			1				1				
ActiveCampus		1	1	1					1			1				1			1				1	1	1			1				
PalmGuide			1	1		1								1		1					1		1	1	1						1	
SaiMotion			1			1		1			1					1			1				1	1				1				

A.3.2 General characteristics of maps for mobile guides

A variety of different types of maps are used for various services in mobile guide systems [22]. These types of maps vary in a number of aspects, which include:

Data type: Data types used for digital maps are raster or vector [252]. Raster based maps can be easily generated by scanning original 2D maps. However, raster maps (i.e. bitmap, jpeg or tiff format) consume large amounts of memory and loose rapidly in quality during scaling and rotation of the map. Vector based maps on the other hand are more laborious to create requiring digitizing by hand or appropriate geographic information must be available in a database. But the file size is generally smaller than that of similar raster maps and these maps can be zoomed and manipulated without loss of graphical quality (see Abwod et al. [2] for more details).

Database: Either all data is already included in a static view-only map or adaptable maps are used, in which information can be altered, added or removed dynamically. Further all geographic information can be stored on the device or maps and additional geographic information may be hosted on a map-server that may generate maps on demand.

Style and visualization type: Maps are generally considered an abstract geographical representation of the real world [239] the geometric objects, symbols and signatures used to achieve an adequate representation may greatly vary from simple black and white sketches to sophisticated works of art. Traditionally the real world was encoded in a two-dimensional (2D) visualization commonly from the bird's-eye view. Attempting to reduce the degree of abstraction, 2D maps may be overlayed by three-dimensional (3D) symbols or they may be visualized as complete 3D representation commonly from the perspective of the user.

User interface : Aside from the pure representation of geographical information, digital maps are used in the context of different services [22]. In mobile guidance systems, maps frequently serve as means of interaction becoming a part of the user interface that grants access to additional information [22, 218].

Appendix B

Requirements and Concepts

B.1 Front-end evaluation

Front-end evaluations are frequently performed using direct and qualitative methods. Front-end evaluators typically conduct face-to-face interviews asking open-ended questions. This allows visitors to describe their experiences in their own words as opposed to having them fit their experiences into the pre-determined responses that usually appear on standardized questionnaires. Open-ended questions should encourage visitors to think and speak freely about a topic allowing the evaluator to learn more about the visitor [179].

B.1.1 Questionnaires

As has been summarized in chapter 3, the front-end evaluation for the MobiNaG project was conducted based on a semi-standardized face-to-face interview. Questionnaires were worked out in accordance with the recommendations by Savage and James [291] as well as Neubert [233] regarding questionnaire design. Prior to the actual evaluation the questionnaires were pilot-tested and revised. Different questionnaires were designed for students, teachers and adults each including open-ended as well as hybrid (semi-open-ended) questions. The following aspects were addressed by the questions designed for the guided interviews:

- Demographic characteristics
- Parameters of the visit
- Knowledge, interests and experiences especially regarding the specific natural area

- Prior experience with ICT
- Requirements regarding environmental interpretation activities
- Requirements regarding a future mobile nature guide.

The corresponding catalogue of questions is listed below:

Students/Children

- Which grade level do you currently attend?
- Is this your first visit to the center? [Yes; No (prior visits with friends, family, school)]
- Which type of activities did you participate in during your stay? [Guided tour (through exhibition, through natural area); Self-determined visit(exhibition, trails, hike, animal park)]
- What did you like best?
- What do you know about the river Rhine and its floodplain?
- Which part of nature is or particular interest to you?
- What would you like to do and experience during an interpretive tour?
- Where do you use the Internet? [Not at all; At home; At school]
- Have you used a PDA (Personal Digital Assistant) before? [No (No mobile device at all, but I have used a mobile phone); Yes (I own a PDA, I do not own a PDA)]
- Would you like to participate in a tour with a mobile nature guide? [No; Yes (If so, what would you like to do with it?)]
- I would like to use a mobile nature guide for... [Environmental games/scavenger hunt, explore the exhibition, self-guided tour through the natural area with information, animations, audio samples, take photographs and notes that I can download from the web later on]
- What do you think should be part of a good environmental computer game?
- You are... [Male; Female]
- Where are you from? [From this city; from the area; from other regions in the state; from a different state]
- What type of school do you attend? [Elementary school; basic track; middle track; advanced track; others]
- Who accompanies you during this visit? [Friends/classmates; family; no one]

Adults/Teachers

- Is this your first visit to the center? [Yes; No (my second visit, have been here a few times, I am a regular visitor)]
- Have you planned you visit in advance? [Yes; No]
- What is the motivation for your visit to the center and the floodplain conservation area, and what did you expect?
- How long did your visit last? [$<$ 1h ; 1–2 h; 3–4 h; $>$ 4 h]
- What did you know about the river Rhine and its floodplain prior to this visit?
- Which personal experiences have you made with the this river and its floodplain, does it have personal significance for you?
- Which of these topics are most interesting to you? [Development and change of the floodplain; Flood protection; Floodplain ecosystem; Animals and plants; Conservation; Cultural history of the area; Others]
- Which type of activities did you participate in during your stay? [Guided tour (through exhibition, through natural area); Self-determined visit(exhibition, trails, hike, animal park)]
- What do you expect of a good guided tour?
- What do you expect of an interactive interpretive tour?
- How would you rate your computer skills? [Expert; Good; Basic; No skills]
- Do you have access to the Internet? [Not at all; At home; At work/school; At the University; At public institutions]
- Would you use the Internet to prepare for the visit in this or another natural area? [No; Yes]
- If so, which kind of information would you be looking for? [Directions; Opening hours; Infrastructure; Activities of children; Tours, trails, hikes; Other publications, Contact information; Other]
- Have you used a PDA (Personal Digital Assistant) before? [No (No mobile device at all, but I have used a mobile phone); Yes (I own a PDA, I do not own a PDA)]
- Would you like to participate in a tour with a mobile nature guide, if you could check it out at the center by leaving a deposit? [No; Yes]
- Which functionalities should the mobile guide offer? multiple answers [Navigation; Information; Guided tours; Activities and Games; Diary with digital souvenirs (personal photographs and notes)]

- Which media should be used to present the information? [No idea; Text; Audio; Photographs,images; Animations; Maps]
- What type of tour would you prefer? [Guided tour; Self-determined exploration; Mix of both]
- You are... [Male; Female]
- Where are you from? [From this city; from the area; from other regions in the state; from a different state]
- What is the highest degree of education that you are seeking or have achived? [Grammer-school (basic, medium, advanced track); College degree in....; no statement]
- What is your current professional occupation?
- Which year where you born in...?
- Who accompanies you during this visit? [Friends/classmates; family; no one]

B.1.2 Interview methods

The actual interviews were then conducted on-site at the NAZKA institution by two interviewers who were familiarized with the test procedure as well as the project. The majority of participants were recruited from among the regular population of visitors to the center on weekends as well as during weekdays. Some interviews were conducted with members of groups such as school classes or associations who had made appointments to take a guided tour or attended a special event at the center. In order not to disturb the visitors during their actual experience of the center or the natural area, the interview sessions were generally administered as exit-surveys. Due to constraints in time available to conduct the evaluation and in an attempt to survey as many participants as possible, no standard sampling strategy was applied. All of the visitors pertaining to one of the target groups and who were willing to engage in the, on average 10–15 minutes interview, were included in the survey. Korn [179] recommends a minimum sample size of 35 participants for qualitative front-end evaluations. A total of 42 adults (19 females and 23 males) and 48 students (25 females and 23 males) participated in the MobiNaG front-end evaluation.

B.1.3 User profiles

The following tables summarize the user profiles and requirements for the three target groups: Students (see Table B.1), families (see Table B.2), and adults (see Table B.2).

Table B.1: Summary of user profiles for students based on front-end evaluation.

Target group	Primary school	Secondary school
Visit characteristics	- Visit with family - More or less regularly	- Visit with family/friends or school field trip
	- For special events or environmental games	- For special events or exhibitions
Prior knowledge	- Tails - Cultural and historical aspects	- Cultural and historical aspects - Animals, plants and habitats
Experiences	- Positive direct experiences with animals - Enjoy interactive games in visitor center	- Positive recreation related experiences with animals - Enjoy interactive exhibits and models in center
Interest	- Animals + plants	- Aquatic animals - Habitats such as forests, grasslands and wetlands
ICT skills	- Some access internet at home - No experience with mobile devices	- Access internet at home, sometimes at school - Own and use mobile phone, some are familiar with PDA
Expectations: interpretive activities	- Direct observations of animals	- Direct observations of animals - Self-determined exploration including physical activity
Expectations: mobile nature guide	- Learn more about animals and plants by interactive explorat. - Not only fun but also want to learn realistic facts	- Learn more about animals and plants by interactive explorat. - Multimedia information based on animations, photos and graphics

Table B.2: Summary of user profiles for families and adults based on front-end evaluation.

Target group	Families with children	Adults < 50 years	Adults ≥ 50 years
Visit characteristics	- Visit with family (1 to 2 children) - Regularly	- Visit with social group or partner - Frequently first time visit	- Visit with social group or partner, or alone - Generally first time visit
	- Activities for kids - Sometimes also special events or exhibitions	- Frequently spare time naturalists - Recreational activities	- Frequently general interest in nature or spare time naturalists - Recreational activities
	- 0.5-2 h stay	- 1-2 h stay	- 1-2 h stay
Prior knowledge	- Environmental issues - Cultural and historical aspects - Some knowledge about ecosystem	- Local ecosystem - Environmental issues - Cultural and historical aspects	- Cultural and historical aspects - Environmental issues - Local ecosystem
Experiences	- Positive direct experiences with nature (also spiritual and aesthetic) - Some negative experiences	- Positive recreation related experiences with nature (some childhood memories)	- Positive recreation related experiences with nature (some childhood memories also spiritual and aesthetic) - Identification with local natural environment
Interest	- Animals, plants + ecosystem - Some cultural + historic aspects	- Animals, plants + ecosystem	- Local ecosystem + animals, plants - Environmental protection activities
ICT skills	- Basic to expert skills - Daily use of Internet at home and/or work - Mobile phone, some have no experience with mobile device	- Basic to good skills - Daily-weekly use of Internet at home and/or work - Mobile phone, some also use PDA or are familiar with it	- Basic skills - Daily use of Internet or never at home and/or work - Some have no experience, mobile phone, some familiar with PDA
Expectations: interpretive activities	- Personalized, location-based info relating to direct experience - Direct, interactive and hands-on experience of nature - Pointed out special phenomena	- Personalized, location-based, concise + entertaining info from expert - Direct, interactive and self-determined explorat. of nature	- Personalized, location-based, concise + entertaining info from expert - Direct, interactive and self-determined exploration of nature - Background information on plants, animals and ecological interactions
Expectations: mobile nature guide	- Easy to use - Location-based information, navigational assistance, environmental games, instrument to identify animals and plants - Prefer predefined tours or a combination of predefined and self-determined exploration	- Easy to use - Navigational assistance, location-based information, instrument to document visit, environmental games - Prefer self-determined exploration	- Easy to use + Tutorial or assistance for first time users - Navigational assistance, instrument to document visit, location-based information, access to more detail on demand - Prefer self-determined exploration and some combination of predefined and self-determined

B.2 Background to navigational assistance

B.2.1 Map-based means of spatial communication

The cartography literature lists a number of fundamental aspects that can be applied in map design for an effective spatial communication. Chiefly Tversky [341] lists a number mechanisms that help to reduce the user's cognitive load by schematizing the world via maps:

Maps give an overview Traditionally maps have projected the three-dimensional (3D) world as a two-dimensional (2D) representation onto a 2D medium such as paper. Even though today there are also 3D maps readily available, offering various perspectives that may be more realistic [228, 251], the 2D map still has some important advantages. A 2D map is commonly easier to create and, according to spatial cognition research, it is more efficient for navigation and orientation tasks, where the spatial relationship among objects has to be evaluated. The elevation information added by 3D representations is rather needed for determining what objects actually look like during target recognition.

Maps omit information Maps are useful due to the reduction of space. Along with the reduction in size, the amount of information needs to be reduced as well, by omitting information that is not essential. Which information is essential is determined by the user's purpose for using the map and can also vary depending on the type of user and context of use.

Maps regularize information Maps generalize and regularize information in order to reduce complexity and thus facilitate the processing of spatial information. Oversimplification can, however, also easily lead to errors.

Maps use a mix of scales Features of a map can be emphasized by increasing their size (i.e. drawing them at different scale) in relation to other objects (e.g. major roads in a roadmap). This reduces realism and hurts the principle of geometric correctness [367], but it supports the readability and allows to highlight essential information.

Maps use a mix of perspectives Target features or landmarks are frequently represented in frontal view or egocentric perspective on top of an overview. This way maps can support the navigation to targets as well as the recognition of these targets.

Maps exaggerate, distort and visualize Based on the world view or message to be told, maps can also distort or exaggerate certain features. Furthermore they can also visualize features that humans can normally not perceive in the environment (e.g. geological formations below the surface).

B.2.2 Functional requirements of mobile maps

To achieve the proposed adaptive visualization, Reichenbacher [268] lists a number of basic functional requirements that should be met by maps to be displayed in mobile environments. They should be:

- displayed in a fast rendering process
- flexible in content, i.e. the content should be dynamically updateable
- changeable in style
- crisp and easily legible
- linkable to other information
- adaptable to different users, activities, and situations
- compatible to other web services
- derivable from existing information

B.3 Theoretical background to exploration tool service and related work

B.3.1 Mixed reality

The virtuality continuum in-between the two extremes, VR and the real world, is covered by different facets of mixed reality (MR). Milgram and Kishino [224] essentially differentiate between augmented reality (AR) and augmented virtuality (AV). AR refers to any case in which an otherwise real environment is "augmented" by means of virtual (i.e. computer graphic) objects. This applies for instance to head-mounted displays (HMD's) equipped with a see-through capability. This implies that computer generated graphics can be optically superimposed, using half-silvered mirrors, onto directly viewed real-world scenes. Next to displays that grant direct viewing of the real objects AR technologies also include displays that allow only indirect viewing of the object such as "video see-through" systems as well as immersive and non-immersive video displays [224]. AV on the other hand relates to cases, where that which is being augmented is not some direct representation of a real scene, but rather a virtual world, that is generated primarily by computer. This is for instance the case in an completely computer graphics based display, immersive or partially immersive, that is overlayed with video reality.

B.3.2 Mobile augmented reality systems

In the WEM project a "video see-through" HMD, connected to a notebook PC and GPS rover in a backpack, was used. Computer generated as well as photographic information about plants and insects, specific to the test site (i.e. a campus park), was superimposed onto the stereoscopic video signal captured by two charge-coupled device (CCD) cameras mounted on the HMD. Next to video data, audio clips could be added as well. User-interaction with the system was granted via a hand-held 3D input device [105].

In the "Ambient Wood" project an entire framework of different modes and types of digital augmentation, using different devices and media, was established to promote self-determined exploration during outdoor field-trips. As types of augmentation, Rogers et al. [275] differentiated between pre-recorded data that is played (e.g., animations, video, sounds) and live data that is probed (e.g. moisture readings) by physical sensors. Further two main methods of applying digital augmentation were used in the Ambient Wood: student-initiated and environmentally-initiated. The student-initiated mode was designed to enable the students to be in control of the access and interaction with the type of digital augmentation (e.g. they could decide when to take the next sensor reading). The environmentally-initiated mode was designed such that the environment decided when and what to deliver to the students in terms of contextually-relevant information (e.g. visual images or sounds were presented via PDA or wireless speakers at particular locations, previously unknown to the students), adding an element of surprise. In the Ambient Wood project a PDA was used as a hand-held non-immersive display to present prerecorded videoclips, images with voice-over descriptions and simple visualizations of dynamic processes. It was further used to visualize live data collected by a probe tool. In addition a "periscope" was applied as a partially immersive video display at a fixed location. Furthermore a number of auditory displays, presenting prerecorded sounds, were deployed either in the form of wireless loudspeakers installed around the wood or a portable player [275].

Also in the DigitalEE project a hand-held non-immersive display was employed in the form of a tablet computer equipped with a CCD camera. Next to the user of the device on site (i.e. real participant), another user at a distant location shared the experiences of the real participant via the internet through a non-immersive display. The project aimed at integrating AR and AV, and the two participants could exchange nature experiences, opinions, values and expertise etc. by interacting in the "Mobile Cyberspace" via the video- or auditory display. For this purpose the photo-realistic representation of the site is augmented with computer generated information (e.g. avatar representations or map) and live-video data recorded by the real participant [243].

B.4 Background on animated pedagogical agents

In a review of animated pedagogical agents (APAs) Johnson et al. [156] present a collection of functionalities, that this type of agent can offer to support the learning process of interacting students. Key items include:

Interactive demonstrations An APA can teach physical tasks by showing first hand how it is done. He can further explain complex processes by immersing in the procedure and by pointing out particular aspects.

Navigational guidance With the APA as a guide, students or other participants can experience a more personalized form of navigational assistance and the APA can help them to develop a cognitive map or spatial model of the environment.

Attentional guidance The agent can focus the user's attention on certain objects or phenomena by using elements of natural interaction such as gaze and deictic references.

Immediate feedback One of the crucial roles of a tutor as well as an environmental interpreter is to provide the user with feedback to his actions. An APA can provide feedback based on verbal as well as nonverbal communication. Body language indicating to a participant that he has committed an error can have a strong impression on him.

Conveying and eliciting emotion Motivation is strongly influenced by emotions and has itself a paramount effect on learning. Emotions, and with them motivation, may be influenced by an APA in several ways. An agent that appears to care about the learner's progress can encourage him to care more about his own progress and can thus improve self confidence in his knowledge and abilities. An agent exhibiting enthusiasm for the subject, can promote interest and similar enthusiasm in the learner. Further, an agent that is fun to interact with and that may serve as a role model, generates a more positive learning experience and makes more fun. This can help the learner to engage more intensely and longer in the activity. Through displaying emotive behaviors an APA may even convey empathy.

Virtual companion An APA can also act as team mate or peer, helping to complete complex tasks in virtual worlds or supporting social processes related to the learning experience.

Adaptive pedagogical interactions A tutor should have certain pedagogical abilities, such as answering questions, generating explanations, asking probing questions, and tracking the learner's skill level. Further, opportunistic instruction is an important instrument in order to support learning at the right time and to offer the appropriate instruction for the current context.

Appendix C

Impact Evaluation

This appendix section includes materials as well as detailed statistical results of the evaluation along with additional result graphs.

C.1 Instrument

The following section presents the questionnaires for the different target groups, which were employed for the application of the instrument during the impact evaluation.

C.1.1 Questionnaires for adults

B1

Impact evaluation
[Demography+Pretest - Adults]

Test-Team: ____________ Medium: ____________ Date: **.05.05**

Time: ____________ Label: **E6**

To be completed by facilitator

1. Is this your first visit to the visitor center and conservation area?
- ❑ Yes
- ❑ I was here once before
- ❑ I have been here several times
- ❑ I am a regular visitor

2. Have you participated in a guided nature tour before?
- ❑ No
- ❑ Yes, with a human guide
- ❑ Yes, with a brochure
- ❑ Yes, I followed a self-guided trail

3. Which kind of nature experience do you prefer?
- ❑ A guided tour with a human guide (e.g. ranger)
- ❑ Exploring nature on my own; without a guide
- ❑ No preference
- ❑ Other: ____________

4. How often per month are you outdoors, exploring nature with all senses?
- ❑ Never
- ❑ On 1-2 days
- ❑ Every weekend (or 8 days)
- ❑ Daily
- ❑ Other ____________

5. What do you expect of a guided nature tour ? Please give two keywords

1)____________ 2)____________

❑ No idea

6. How did you get to the visitor center today?

Public transportation	By car	By bicycle	On foot
❑	❑	❑	❑

7. How many legs do insects have?

2	4	6	8	No idea
❑	❑	❑	❑	❑

[Demography+Pretest-Adults]

1. How many of your friends are engaged in conservation?

- ❑ The majority
- ❑ Half of them
- ❑ Only a few
- ❑ None
- ❑ No idea

2. What is the name of your hometown (quarter)?

____

3. Which of the following things apply to your neighborhood?
(You're allowed to mark more than one answer)

- ❑ Polluted air / exhaust from cars or factories
- ❑ The area is heavily developed – mostly buildings and paved roads
- ❑ There are a lot of fields close-by
- ❑ There are a lot of gardens and parks in the neighborhood
- ❑ There is a forest close-by

4. Are you familiar with mobile devices (cell phone, PDA)?

- ❑ No
- ❑ Yes, I have a cell phone
- ❑ Yes, I have a PDA / Smartphone
- ❑ Yes, but I do not own a mobile device by myself
- ❑ Other______________________________________

5. You are...

❑ Female ❑ Male

6. What is the highest degree of education that you are seeking or have achieved?

- ❑ Basic track secondary school ("Hauptschule")
- ❑ Medium track secondary school ("Mittlere Reife")
- ❑ Advanced track - grammar school (Hoschulreife)
- ❑ College degree in: natural sciences | engineering | social sciences | humanities (Please underline)
- ❑ No comment

7. What is your current professional occupation?

❑ ____

____

8. Which year where you born in?

____

9. How was this evaluation brought to your attention?

- ❑ newspaper / radio (Please specify):____________________________
- ❑ Leaflet
- ❑ Friends
- ❑ During visit on site ❑ Other

Forschungszentrum Karlsruhe GmbH - 2 - Institut for Applied Computer Science

[Demography+Pretest-Adults]

Please rate the following statements:

		Strongly agree	Agree	Undecided	Disagree	Strongly disagree
1.	I think there currently exist more important issues in this country, than the conservation of plants and animals	❑	❑	❑	❑	❑
2.	We should protect all plants and animals. Not only the once which are useful to us.	❑	❑	❑	❑	❑
3.	I prefer forests where the trees stand in orderly rows	❑	❑	❑	❑	❑
4.	Before I harm a fly, I try to imagine how I would feel if I was in her place	❑	❑	❑	❑	❑
5.	Since my personal impact is very small, I don't feel responsible for the destruction of the floodplain.	❑	❑	❑	❑	❑
6.	If old trees are cut in my neighborhood due to housing development, I don't react emotionally but accept it as a necessity	❑	❑	❑	❑	❑

	I would like to know more about...	Strongly disagree	Disagree	Undecided	Agree	Strongly agree
7.	the consequences of pollution for mankind	❑	❑	❑	❑	❑
8.	the consequences of waldsterben for soils, drinking water, air, animals and plants.	❑	❑	❑	❑	❑
9.	identification of unknown plants	❑	❑	❑	❑	❑

		Never applies	Does not apply	Undecided	Mostly applies	Always applies
10.	I often discuss environmental and conservation issues with my friends.	❑	❑	❑	❑	❑
11.	I intentionally walk short distances, instead of driving to protect the climate and therefore also the forest.	❑	❑	❑	❑	❑
12.	I commonly observing animals	❑	❑	❑	❑	❑

	How much would you like to participate in the following activities in the future:	Not at all	Rather not	Undecided	Probably	Most definitely
13.	Working voluntarily in a conservation organization	❑	❑	❑	❑	❑
14.	Planting trees	❑	❑	❑	❑	❑
15.	Call a persons attention to his irresponsible environmental behavior.	❑	❑	❑	❑	❑

[Demography+Pretest-Adults]

1. How much time would you spent per month to volunteer for the conservation of this floodplain?

- ❑ I'm sure there are enough people in conservation organization, who are already doing this job.
- ❑ I would rather donate €_______ to a conservation organization
- ❑ Half a day
- ❑ One day
- ❑ One weekend
- ❑ More than one weekend
- ❑ None

2. Which functions do trees fulfill for the environment and for the mankind? (You're allowed to mark multiple answer)

- ❑ They are food source for animals and humans.
- ❑ They are a resource for mankind
- ❑ They transform oxygen into carbon dioxide
- ❑ They are a place for social gatherings of people
- ❑ No idea

3. Why is the willow a floodplain specialist?

- ❑ It has no roots and can therefore stand long periods of flooding
- ❑ It has very hard wood that can endure the water for a long time.
- ❑ It can grow additional fine roots during flooding events. So it can get oxygen directly out of the water.
- ❑ It's leaves are an important food source for fishes.
- ❑ No idea

4. Who is responsible for the transformation of leaves into nutrients? (You're allowed to mark multiple answers)

- ❑ The community of soil organisms
- ❑ The dung-beetle
- ❑ The wind
- ❑ Fungi and bacteria
- ❑ No idea

5. Please assign the leaves (a, b or c) to the matching tree.

a) b) c)

Beech__________________

Lime-tree________________

Oak-tree________________

❑ No idea

Thank you!!!

B2

Impact evaluation

[Usability - Adults]

Test-Team: ____________________ Medium: __________________ Date: **.05.05**

Time: __________________________________ Label: **E6** ____________________________

To be completed by facilitator

1. **How do you fell following this guided tour?**
(You're allowed to mark multiple answer)

❑ Frustrated and annoyed
❑ Satisfied and relaxed
❑ Bored
❑ Motivated

❑ Other____

2. **List the two things you liked best about the tour**

1)______________________________ 2)______________________________

❑ **no idea**

3. **List the two thinks you liked least about the tour?**

1)______________________________ 2)______________________________

❑ no idea

		Applies	Partially applies	Undecided	Does partially not apply	Does not apply
4.	I will recommend this tour to my friends	❑	❑	❑	❑	❑
5.	I was not concentrating during the tour	❑	❑	❑	❑	❑
6.	I've learned a lot today	❑	❑	❑	❑	❑
7.	The tour met my expectations	❑	❑	❑	❑	❑
8.	I was moved emotionally by the tour	❑	❑	❑	❑	❑
9.	I got emotionally involved in the tour	❑	❑	❑	❑	❑
10.	The presented content was implausible	❑	❑	❑	❑	❑
11.	The tour was very interactive with lots of opportunities for self-determined exploration	❑	❑	❑	❑	❑
12.	The tour distracted me from actually experience nature	❑	❑	❑	❑	❑
13.	Due to the tour I can better understand why I should protect trees.	❑	❑	❑	❑	❑

Forschungszentrum Karlsruhe GmbH - 1 - Institute for Applied Computer Science

[Usability-Adults]

	Which school grades would you assign to the following aspects of the tour?	**1**	**2**	**3**	**4**	**5**	**6**
1.	Fun	❑	❑	❑	❑	❑	❑
2.	Content	❑	❑	❑	❑	❑	❑
3.	Length/Duration	❑	❑	❑	❑	❑	❑
4.	Presentation (Guide / Brochure)	❑	❑	❑	❑	❑	❑
5.	Usage of different media	❑	❑	❑	❑	❑	❑
6.	Games	❑	❑	❑	❑	❑	❑
7.	Overall impression	❑	❑	❑	❑	❑	❑

8. Could you locate the stations?
❑ Yes, all of them without problems
❑ Some of them
❑ No
❑ No idea

9. Did the map help you to get oriented in this area?
❑ Yes
❑ Only partially
❑ No
❑ No idea

10. Did you read all of the texts completely?
❑ Yes, I've read them all
❑ No, I skipped some of them
❑ No, I glanced over some of them

Please specify the texts you did not read completely____________________________

____

❑ No comment

11. Do you think your children could benefited from this experience?

❑ Yes, because:____

❑ To some extend, because:____

❑ No, because:____
❑ No comment

12. How much would you be willing to pay such a guided tour?

Nothing	2 €	4 €	6 €	8 €	10 €	>10 €	No comment
❑	❑	❑	❑	❑	❑	❑	❑

C.1.2 Questionnaires for children and students

a1

Impact evaluation

[Demography - Children]

Test-Team:____________ Medium:____________ Date: .05.05

Time: ____________ Label: K1

To be completed by facilitator

1. **Which grade do you currently attend?** Grade level:____________

2. **Which type of school do you currently attend?**
- ❑ Elementary school
- ❑ Basic track – secondary school
- ❑ Medium track – secondary school
- ❑ Advanced track - grammar school
- ❑ Another type of school:____________

3. **You are...**
   ❑ Female ❑ Male

4. **How old are you?**

   ____

5. **Is this your first visit to the visitor center and conservation area?**
- ❑ Yes
- ❑ I was here once before
- ❑ I have been here a couple of times
- ❑ I am here regularly

6. **Did you take a guided nature tour before?**
- ❑ No, never
- ❑ Yes, with a human guide
- ❑ Yes, with a brochure
- ❑ Yes, I followed a self-guided trail

7. **What do you expect of a guided nature tour ? Please give two keywords**

   1)____________ 2)____________

8. **How many times a week do you play out in nature, looking for plants and animals?**

Never	1 day	2 days	3 days	4 days	Every day
❑	❑	❑	❑	❑	❑

9. **Which color do the flowers of canola have ?**

White	Yellow	Red	Violet	No idea
❑	❑	❑	❑	❑

[Demography-Children]

1. **How many legs does an ant have?**

2	4	6	8	No idea
❑	❑	❑	❑	❑

2. **Which plant are cornflakes made from?**

____

❑ No idea

3. **Have you ever used a mobile device (Cell phone, pocket computer)?**
(You're allowed to mark multiple answers)

❑ No
❑ Yes, I have got a cell phone
❑ Yes, my parents have got a cell phone/ a PDA (pocket computer) ➔➔➔
❑ Yes, my friends have got a cell phone

❑ Other____

4. **What is the name of your hometown (district)?**

____

5. **Which of the following things apply to your neighborhood?**
(You're allowed to mark multiple answers)

❑ Polluted air / exhaust from cars or factories
❑ The area is heavily developed – mostly buildings and paved roads
❑ There are a lot of fields close-by
❑ There are a lot of gardens and parks in the neighborhood
❑ There is a forest close-by

6. **Do you know somebody who is member of a conservation organization?**
(You're allowod to mark multiple answers)

❑ No
❑ Yes, I'm a member myself
❑ Yes, in my family
❑ Yes, a friend of mine

a2

Impact evaluation

[Usability+Posttest - Children]

Test-Team: ____________ Medium: ____________ Date: .05.05

Time: ____________ Label: K1

To be completed by facilitator

1. How do you feel following this guided tour?
(You're allowed to mark multiple answer)

- ❑ Frustrated and annoyed
- ❑ Satisfied and relaxed
- ❑ Bored
- ❑ Motivated and inspired

Other____________

2. List the two things you liked best about the tour

1)____________ 2)____________

3. List the two thinks you liked least about the tour?

1)____________ 2)____________

		Yes	No idea	No
4.	I will tell my friends about this tour	❑	❑	❑
5.	I would like to take a similar tour again on a different topic	❑	❑	❑
6.	I was distracted during the tour	❑	❑	❑
7.	I've learned a lot today	❑	❑	❑

8. Could you find the stations?

- ❑ Yes, all of them without problems
- ❑ Some of them
- ❑ No
- ❑ No idea

9. Did the map help you to find the right trail and the stations?

- ❑ Yes
- ❑ Partially
- ❑ No
- ❑ No idea

10. For how long during the tour did you handle the mobile guide/ the brochure yourself

Never	Occasionally	For about half of the tour	All the time	No comment
❑	❑	❑	❑	❑

[Usability-Children]

1. Did you read or listen to all texts completely?

- ❑ Yes, all of them
- ❑ No, some of them where skipped
- ❑ No idea

		Applies	Partially applies	Undecided	Does partially not apply	Does not apply
2.	During the tour I was able to explore and discover lots of things by myself	❑	❑	❑	❑	❑
3.	The tour rather distracted me from actually experiencing nature	❑	❑	❑	❑	❑
4.	The presentation on the natural phenomena was easy to understand	❑	❑	❑	❑	❑
5.	The tour has made me understand, that I have to protect trees	❑	❑	❑	❑	❑
6.	The guided tour was exciting like a thrilling movie	❑	❑	❑	❑	❑

	Which grades would you give the guided tour?	1	2	3	4	5	6
7.	Fun	❑	❑	❑	❑	❑	❑
8.	Entertainment	❑	❑	❑	❑	❑	❑
9.	Duration / Length	❑	❑	❑	❑	❑	❑
10.	Guide/brochure	❑	❑	❑	❑	❑	❑
11.	Usage of different media (e.g. Images, texts)	❑	❑	❑	❑	❑	❑
12.	Games	❑	❑	❑	❑	❑	❑
13.	Overall impression	❑	❑	❑	❑	❑	❑

Please rate the following statements by marking the boxes that matches your opinion.

		Strongly agree	Agree	I don't care	Disagree	Strongly disagree
14.	I annoyed with all of the natural landscape being used for roads and buildings.	❑	❑	❑	❑	❑
15.	I like it is better when the trees in a forest stand in orderly rows.	❑	❑	❑	❑	❑
16.	Having to think about protecting nature all the time is annoying	❑	❑	❑	❑	❑

[Usability-Children]

		Strongly agree	Agree	I don't care	Disagree	Strongly disagree
1.	Before I smash a fly, I try to put myself into her situation.	❑	❑	❑	❑	❑
2.	Trees do have a soul	❑	❑	❑	❑	❑
3.	I don't care, if old trees are cut down for new houses in my neighborhood.	❑	❑	❑	❑	❑

	I would like to know more about...	Strongly disagree	Disagr ee	I don't care	Agree	Strongly agree
4.	Which animals are endangered and which are protected.	❑	❑	❑	❑	❑
5.	How do I cut a tree so that I can build furniture?	❑	❑	❑	❑	❑
6.	How can I identify an unknown plant	❑	❑	❑	❑	❑

	This is what I do already....	Never applies	Don't applies	I don't know	Applies	Always applies
7.	I frequently talk to my friends about nature and conservation.	❑	❑	❑	❑	❑
8.	I walk short distances, instead of asking for a ride, because it is better for the climate and also for the forest.	❑	❑	❑	❑	❑
9.	I recycle paper	❑	❑	❑	❑	❑

	I would like to...	Not at all	Rather not	Undecide d	Probabl y	Most definitely
10.	Plant trees	❑	❑	❑	❑	❑
11.	Watch nature documentations in television	❑	❑	❑	❑	❑
12.	Call a persons attention, when it is harming the nature	❑	❑	❑	❑	❑

13. How much time would you spent per month to volunteer for the conservation of this floodplain?

❑ I'm sure there are enough people in conservation organization, who are already doing this job.
❑ I would rather donate my spending money for this month to a conservation organization
❑ Half a day
❑ One day
❑ One weekend
❑ More than one weekend
❑ None

[Usability-Children]

1. **Which is your favorite tree?**

____

2. **Which functions do trees fulfill for the environment and for the mankind?**
(You're allowed to mark multiple answer)
- ❑ Resource for wool
- ❑ Habitat
- ❑ Source of pollution
- ❑ Place for social gatherings
- ❑ No idea

3. **Why is the willow a floodplain specialist?**
- ❑ It has no roots and can therefore stand long periods of flooding
- ❑ It has very hard wood that can endure the water for a long time.
- ❑ It can grow additional fine roots during flooding events. So it can get oxygen directly out of the water.
- ❑ It's leaves are an important food source for fishes.
- ❑ No idea

4. **Who is mostly responsible for turning leaves into nutrients?**
(You're allowed to mark multiple answer)
- ❑ Other trees
- ❑ The community of soil organisms
- ❑ The wind
- ❑ Birds
- ❑ No idea

5. **What can I do for protecting the forest?**
(You're allowed to mark multiple answer)
- ❑ I can save paper and recycle it, to minimize the logging of forests
- ❑ I should use more paper, so that more young trees can be planted
- ❑ Ride my bike or take the train instead of the car
- ❑ Leave the trails and walk cross-country through the forest
- ❑ No idea

6. **Please match the leaves (a, b or c) with the right tree**

a) b) c)

Beech________________

Lime-tree______________

Oak-tree______________

❑ No idea

Thank you!

C.2 Result graphs

C.2.1 Intervention and media effects

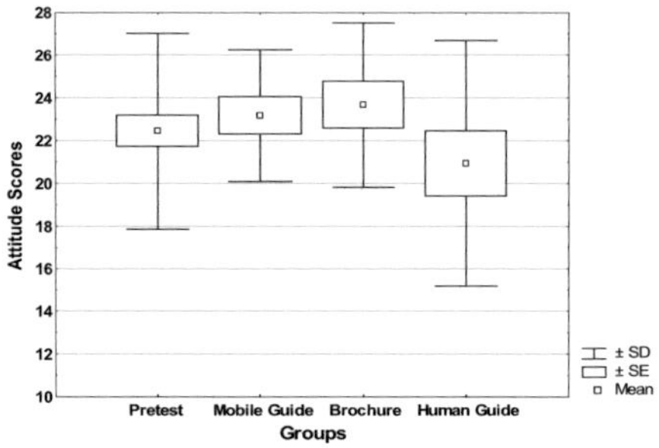

Figure C.1: Mean attitude scale scores of children groups, including: pooled pretest (n=38), mobile nature guide (n=12), brochure (n=12) and human guide (n=14).

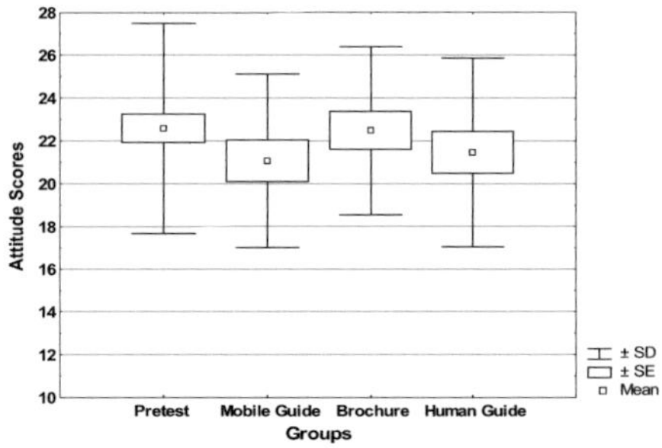

Figure C.2: Mean attitude scale scores of student groups, including: pooled pretest (n=53), mobile nature guide (n=17), brochure (n=19) and human guide (n=20).

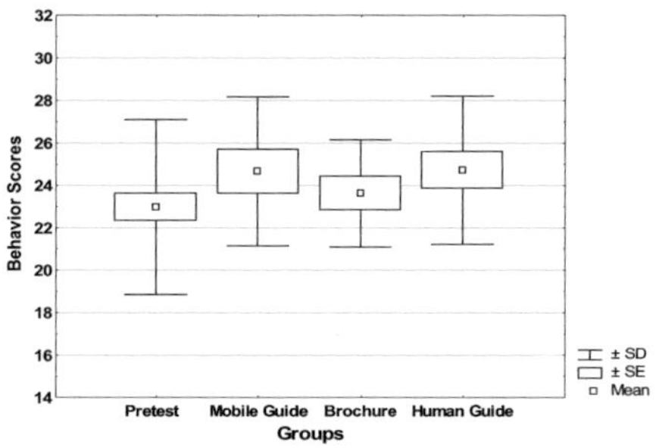

Figure C.3: Mean behavior scale scores of parent groups, including: pooled pretest (n=40), mobile nature guide (n=11), brochure (n=10) and human guide (n=16).

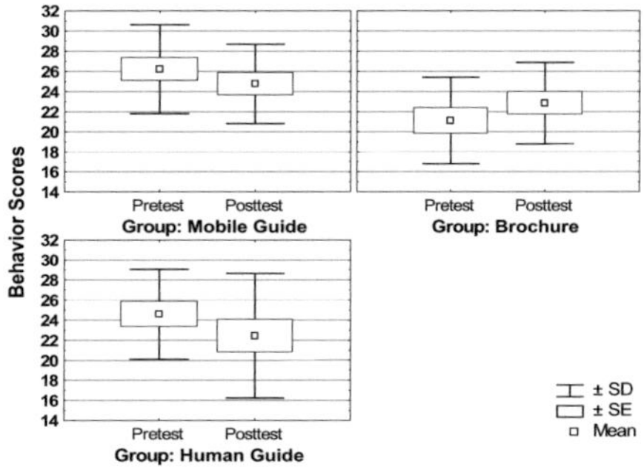

Figure C.4: Mean behavior scale scores of children groups, including: pretests of mobile guide (n=14), brochure (n=11), human guide (n=13) and posttests of mobile guide (n=12), brochure (n=12) human guide (n=14).

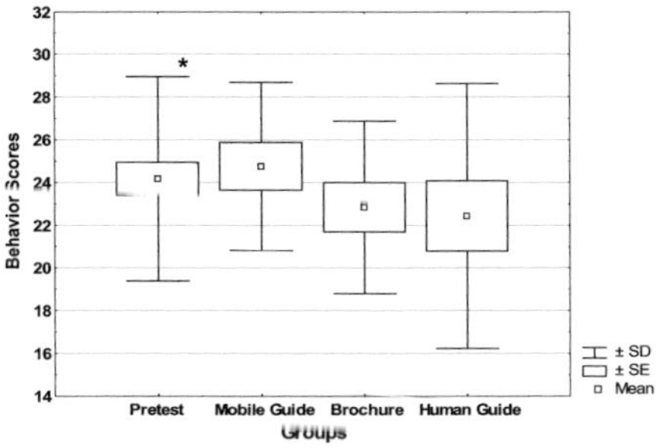

Figure C.5: Mean behavior scale scores of children groups, including: pooled pretest (n=37), mobile nature guide (n=12), brochure (n=12) and human guide (n=14). ∗ = significant difference to pretest (p<0.05).

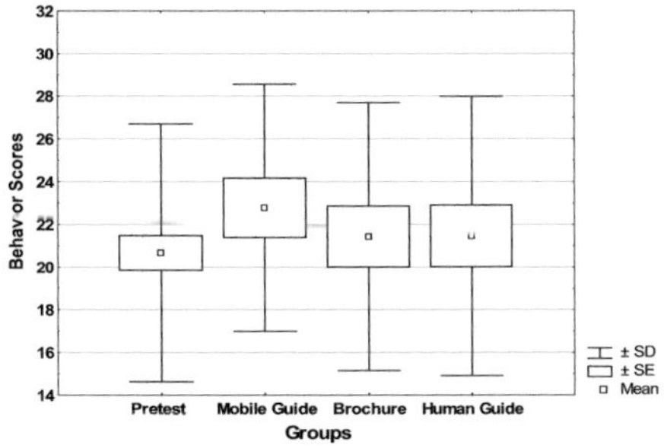

Figure C.6: Mean behavior scale scores of student groups, including: pooled pretest (n=53), mobile nature guide (n=17), brochure (n=19) and human guide (n=20).

C.2.2 Additional factors

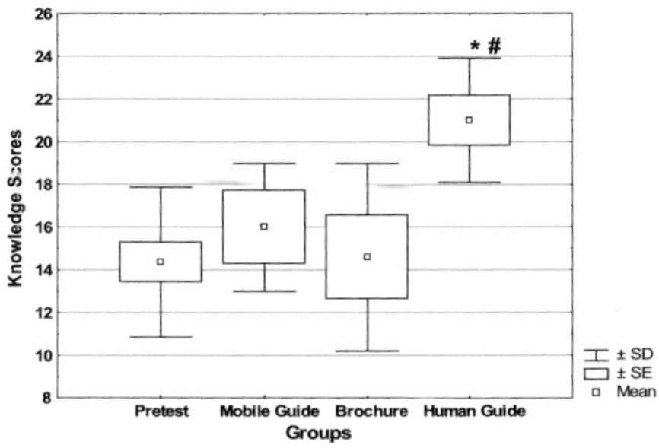

Figure C.7: Mean knowledge scale scores of advanced track students, including: pooled pretest (n=14), mobile nature guide (n=3), brochure (n=5) and human guide (n=6). ∗ = significant difference to pretest (p<0.05), # significant difference between media (p<0.05).

C.2.3 User satisfaction

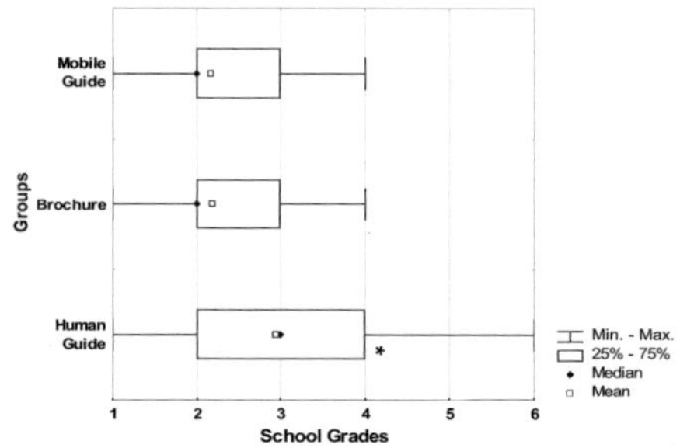

Figure C.8: Average grade for games given by parent groups, including: mobile nature guide (n=23), brochure (n=21) and human guide (n=19). $*$ = significant difference between media (p<0.05).

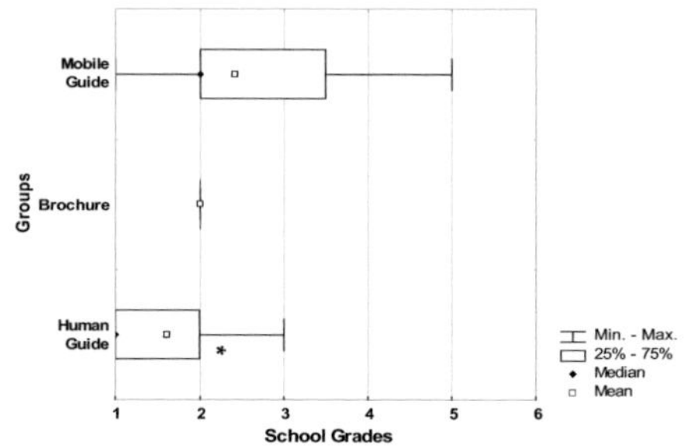

Figure C.9: Average grade for guide given by parent groups, including: mobile nature guide (n=24), brochure (n=21), human guide (n=23) $*$ = significant difference between media (p<0.05).

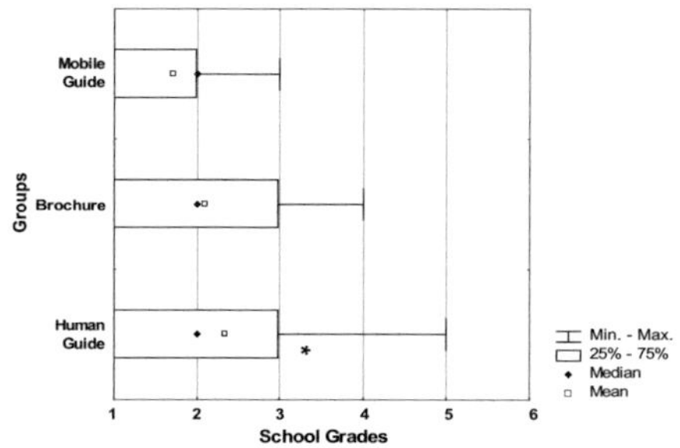

Figure C.10: Average grade for overall tour given by student groups, including: mobile nature guide (n=31), brochure (n=33) and human guide (n=36). $*$ = significant difference between media (p<0.05).

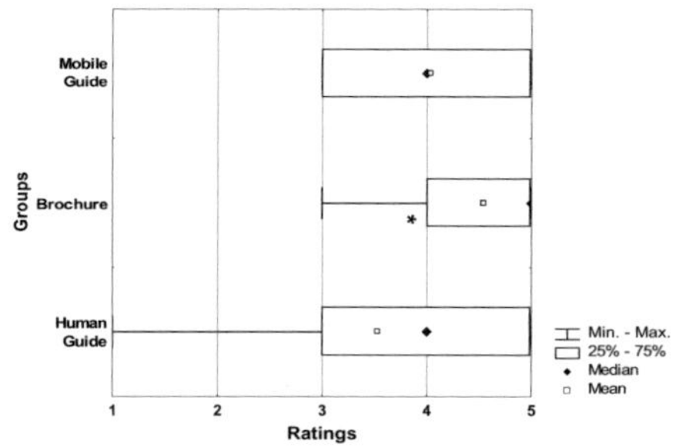

Figure C.11: Average grade for perceived interactivity given by parent groups, including: mobile nature guide (n=26), brochure (n=20), human guide (n=30) $*$ = significant difference between media (p<0.05).

C.2.4 Utilization of content

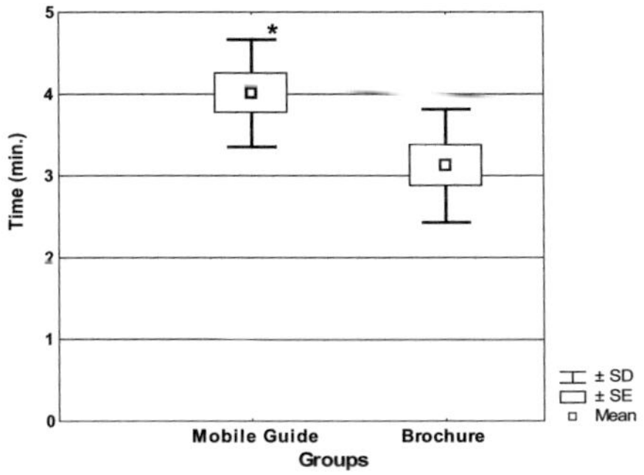

Figure C.12: Average amount of time that student groups spent at stations, including: mobile nature guide (n=6), brochure (n=6). $*$ = significant difference between media ($p<0.05$).

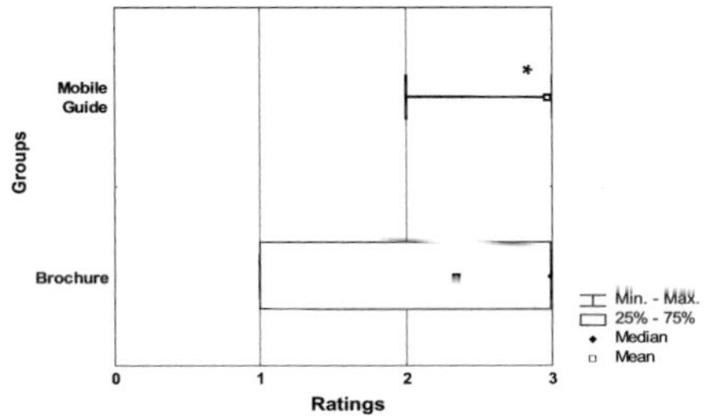

Figure C.13: Self-reported content consumed by parent groups, including: mobile nature guide (n=26), brochure (n=21). $*$ = significant difference between media ($p<0.05$).

C.3 Statistic analysis

C.3.1 Analysis of variance

Table C.1: Results of one-factorial ANOVA testing for pretest differences between media groups and influence of education on knowledge and attitude scores (significant difference (p<0.05)).

One-factorial ANOVA results	df treatment	MSQ treatment	df error	MSQ error	F	p-value
Knowledge scores on pretest						
Parents	2	3.168	37	7.390	0.429	0.655
Children	2	17.790	35	29.380	0.605	0.551
Students	2	6.006	50	15.137	0.397	0.675
Attitude scores on pretest						
Parents	2	1.805	37	3.078	0.587	0.561
Children	2	36.239	35	20.140	1.799	0.180
Students	2	0.450	50	25.002	0.018	0.982
Values/concerns scores on pretest						
Parents	2	15.365	37	7.005	2.193	0.126
Children	2	6.272	35	8.562	0.733	0.488
Students	2	17.727	50	7.311	2.425	0.099
Behavior scores on pretest						
Parents	2	19.392	37	17.003	1.141	0.331
Children	2	82.422	34	19.358	4.258	**0.022**
Students	2	34.257	50	36.507	0.938	0.398
Influence of gender on attitude						
Parents gender groups overall	1	0.488	76	7.467	0.065	0.799
Children gender groups overall	1	6.934	74	20.568	0.337	0.563
Students gender groups overall	1	4.152	107	20.528	0.202	0.654
Influence of gender on knowledge						
Parents gender groups overall	1	1.905	76	9.799	0.194	0.661
Children gender groups overall	1	3.338	74	28.507	0.117	0.733
Students gender groups overall	1	0.477	107	17.630	0.027	0.870
Influence of education on attitude						
Parents education groups overall	1	0.205	76	7.470	0.027	0.869
Influence of education on knowledge						
Parents education groups overall	1	26.531	75	8.581	3.092	0.083
Parents education groups pretest	1	49.965	38	6.048	8.262	**0.007**
Parents education groups posttest	1	0.838	35	4.807	0.174413	0.679

Table C.2: Results of two-factorial ANOVA testing for interaction effects between media and additional factors (significant difference ($p<0.05$)).

Two-factorial ANOVA results	Factors / interaction	df treatment	MSQ treatment	df error	MSQ error	F	p-value
Gender x media interaction effects on attitude							
Parents	Gender	1	2.032	71	3.101	0.655	0.421
Parents	Media	2	0.106	71	3.101	0.034	0.966
Parents	Gender x media	2	2.299	71	3.101	0.742	0.480
Children	Gender	1	0.164	70	20.273	0.008	0.929
Children	Media	2	41.760	70	20.273	2.060	0.135
Children	Gender x media	2	3.867	70	20.273	0.191	0.827
Students	Gender	1	3.702	103	20.929	0.177	0.675
Students	Media	2	6.543	103	20.929	0.313	0.732
Students	Gender x media	2	15.022	103	20.929	0.718	0.490
Gender x media interaction effects on knowledge							
Parents	Gender	1	0.590	71	9.281	0.064	0.802
Parents	Media	2	2.338	71	9.281	0.252	0.778
Parents	Gender x media	2	3.141	71	9.281	0.338	0.714
Children	Gender	1	5.416	70	29.524	0.183	0.670
Children	Media	2	0.615	70	29.524	0.021	0.979
Children	Gender x media	2	20.580	70	29.524	0.697	0.501
Students	Gender	1	0.974	103	17.870	0.054	0.816
Students	Media	2	7.266	103	17.870	0.407	0.667
Students	Gender x media	2	16.138	103	17.870	0.903	0.408
Prior knowledge x media interaction effects on knowledge							
Parents	Prior knowledge	1	0.386	62	9.294	0.042	0.839
Parents	Media	2	18.205	62	9.294	1.959	0.150
Parents	Prior k. x media	2	28.394	62	9.294	3.055	0.054
Students	Prior knowledge	1	54.170	103	16.447	3.294	0.072
Students	Media	2	41.496	103	16.447	2.523	0.085
Students	Prior k. x media	2	51.872	103	16.447	3.154	**0.047**
Education x media interaction effects on knowledge							
Parents	Education	1	30.334	71	8.699	3.487	0.066
Parents	Media	2	4.358	71	8.699	0.501	0.608
Parents	Education x media	2	8.351	71	8.699	0.960	0.388
Students	School from	2	10.890	100	17.882	0.609	0.546
Students	Media	2	12.839	100	17.882	0.718	0.490
Students	School form x media	4	12.114	100	17.882	0.677	0.609

C.3.2 T-test for unpaired sample groups

Table C.3: T-test for unpaired sample groups was applied to determine intervention effects on environmental literacy components ($p<0.05$).

T-test for unpaired samples	Group 1	Group 2	Mean 1	Mean 2	n 1	n 2	StDev 1	StDev 2	df	t-value	p-value
Intervention effect on knowledge											
Parents	Control group	Mobile guide	14.175	17.727	40	11	2.678	2.284	49	-4.009	**0.000**
Parents	Control group	Brochure	14.175	17.600	40	10	2.678	2.675	48	-3.618	**0.001**
Parents	Control group	Human guide	14.175	17.438	40	16	2.678	1.861	54	-4.450	**0.000**
Children	Control group	Mobile guide	16.053	17.917	38	12	5.362	5.017	48	-1.065	0.292
Children	Control group	Brochure	16.053	20.250	38	12	5.362	3.745	48	-2.516	**0.015**
Children	Control group	Human guide	16.053	18.500	38	14	5.362	5.761	50	-1.431	0.159
Students	Control group	Mobile guide	13.415	15.706	53	17	3.845	2.910	68	-2.254	**0.027**
Students	Control group	Brochure	13.415	15.579	53	19	3.845	4.273	70	-2.044	**0.045**
Students	Control group	Human guide	13.415	17.900	53	20	3.845	4.217	71	-4.329	**0.000**
Intervention effect on attitude											
Parents	Control group	Mobile guide	18.250	18.909	40	11	1.736	1.375	49	-1.160	0.252
Parents	Control group	Brochure	18.250	18.500	40	10	1.736	2.321	48	-0.380	0.705
Parents	Control group	Human guide	18.250	19.313	40	16	1.736	1.352	54	-2.192	**0.033**
Children	Control group	Mobile guide	22.447	23.167	38	12	4.584	3.070	48	-0.507	0.614
Children	Control group	Brochure	22.447	23.667	38	12	4.584	3.846	48	-0.832	0.410
Children	Control group	Human guide	22.447	20.929	38	14	4.584	5.744	50	0.989	0.327
Students	Control group	Mobile guide	22.566	21.059	53	17	4.905	4.054	68	1.146	0.256
Students	Control group	Brochure	22.566	22.474	53	19	4.905	3.921	70	0.074	0.941
Students	Control group	Human guide	22.566	21.450	53	20	4.905	4.407	71	0.890	0.376
Intervention effect on values and concerns											
Parents	Control group	Mobile guide	14.950	15.818	40	11	2.726	2.562	49	-0.947	0.348
Parents	Control group	Brochure	14.950	15.400	40	10	2.726	3.062	48	-0.456	0.651
Parents	Control group	Human guide	14.950	15.188	40	16	2.726	3.060	54	-0.284	0.777
Children	Control group	Mobile guide	10.684	11.250	38	12	2.905	2.832	48	-0.592	0.557
Children	Control group	Brochure	10.684	10.750	38	12	2.905	3.519	48	-0.065	0.948
Children	Control group	Human guide	10.684	11.286	38	14	2.905	2.946	50	-0.660	0.512
Students	Control group	Mobile guide	11.434	12.059	53	17	2.777	1.952	68	-0.860	0.393
Students	Control group	Brochure	11.434	11.684	53	19	2.777	2.790	70	-0.337	0.737
Students	Control group	Human guide	11.434	10.800	53	20	2.777	2.587	71	0.886	0.379
Intervention effect on behavior											
Parents	Control group	Mobile guide	22.950	24.636	40	11	4.138	3.529	49	-1.232	0.224
Parents	Control group	Brochure	22.950	23.600	40	10	4.138	2.547	48	-0.473	0.639
Parents	Control group	Human guide	22.950	24.688	40	16	4.138	3.516	54	-1.478	0.145
Children	Control group	Mobile guide	24.162	24.750	37	12	4.781	3.934	47	-0.385	0.702
Children	Control group	Brochure	24.162	22.833	37	12	4.781	4.041	47	0.866	0.391
Children	Control group	Human guide	24.162	22.429	37	14	4.781	6.198	49	1.063	0.293
Students	Control group	Mobile guide	20.660	22.765	53	17	6.035	5.783	68	-1.263	0.211
Students	Control group	Brochure	20.660	21.421	53	19	6.035	6.266	70	-0.467	0.642
Students	Control group	Human guide	20.660	21.450	53	20	6.035	6.533	71	-0.487	0.627

Table C.4: T-test for unpaired sample groups was applied to determine media effects on environmental literacy components (significant difference (p<0.05)).

T-test for unpaired samples	Group 1	Group 2	Mean 1	Mean 2	n 1	n 2	StDev 1	StDev 2	df	t-value	p-value
Media effect on knowledge											
Parents	Mobile guide	Brochure	17.727	17.600	11	10	2.284	2.675	19	0.118	0.908
Parents	Mobile guide	Human guide	17.727	17.438	11	16	2.284	1.861	25	0.363	0.720
Children	Mobile guide	Brochure	17.917	20.250	12	12	5.017	3.745	22	-1.291	0.210
Children	Mobile guide	Human guide	17.917	18.500	12	14	5.017	5.761	24	-0.273	0.787
Students	Mobile guide	Brochure	15.706	15.579	17	19	2.910	4.273	34	0.103	0.919
Students	Mobile guide	Human guide	15.706	17.900	17	20	2.910	4.217	35	-1.809	0.079
Media effect on attitude											
Parents	Mobile guide	Brochure	18.909	18.500	11	10	1.375	2.321	19	0.497	0.625
Parents	Mobile guide	Human guide	18.909	19.313	11	16	1.375	1.352	25	-0.756	0.456
Children	Mobile guide	Brochure	23.167	23.667	12	12	3.070	3.846	22	-0.352	0.728
Children	Mobile guide	Human guide	23.167	20.929	12	14	3.070	5.744	24	1.208	0.239
Students	Mobile guide	Brochure	21.059	22.474	17	19	4.054	3.921	34	-1.064	0.295
Students	Mobile guide	Human guide	21.059	21.450	17	20	4.054	4.407	35	-0.279	0.782
Media effect on values and concerns											
Parents	Mobile guide	Brochure	15.818	15.400	11	10	2.562	3.062	19	0.341	0.737
Parents	Mobile guide	Human guide	15.818	15.188	11	16	2.562	3.060	25	0.561	0.580
Children	Mobile guide	Brochure	11.250	10.750	12	12	2.832	3.519	22	0.383	0.705
Children	Mobile guide	Human guide	11.250	11.286	12	14	2.832	2.946	24	-0.031	0.975
Students	Mobile guide	Brochure	12.059	11.684	17	19	1.952	2.790	34	0.461	0.647
Students	Mobile guide	Human guide	12.059	10.800	17	20	1.952	2.587	35	1.646	0.109
Media effect on behavior											
Parents	Mobile guide	Brochure	24.636	23.600	11	10	3.529	2.547	19	0.764	0.454
Parents	Mobile guide	Human guide	24.636	24.688	11	16	3.529	3.516	25	-0.037	0.971
Children	Mobile guide	Brochure	24.750	22.833	12	12	3.934	4.041	22	1.177	0.252
Children	Mobile guide	Human guide	24.750	22.429	12	14	3.934	6.198	24	1.117	0.275
Students	Mobile guide	Brochure	22.765	21.421	17	19	5.783	6.266	34	0.666	0.510
Students	Mobile guide	Human guide	22.765	21.450	17	20	5.783	6.533	35	0.643	0.525

Table C.5: T-test for unpaired sample groups was applied to determine effects of additional factors on environmental literacy components (significant difference (p<0.05)).

T-test for unpaired samples	Group 1	Group 2	Mean 1	Mean 2	n 1	n 2	StDev 1	StDev 2	df	t-value	p-value
Influence of prior knowledge on knowledge											
Students	Mobile guide	Human guide	13.000	20.125	4	8	4.243	3.044	10	-3.375	**0.007**
Students	Mobile guide	Brochure	13.000	16.222	4	9	4.243	4.024	11	-1.313	0.216
Students	Human guide	Brochure	20.125	16.222	8	9	3.044	4.024	15	2.231	**0.041**
Influence of school type on knowledge											
Basic track	Control	Mobile guide	13.481	15.222	27	9	4.127	2.539	34	-1.186	0.244
Basic track	Control	Brochure	13.481	16.000	27	10	4.127	5.077	35	-1.550	0.130
Basic track	Control	Human guide	13.481	15.800	27	10	4.127	4.290	35	-1.502	0.142
Media x school type effect on knowledge (posttest)											
Basic track	Mobile guide	Brochure	15.222	16.000	9	10	2.539	5.077	17	-0.414	0.684
Basic track	Mobile guide	Human guide	15.222	15.800	9	10	2.539	4.290	17	-0.352	0.729
Basic track	Brochure	Human guide	16.000	15.800	10	10	5.077	4.290	18	0.095	0.925
Influence of school type on knowledge											
Medium track	Control	Mobile guide	12.167	16.400	12	5	3.512	3.912	15	-2.195	**0.044**
Medium track	Control	Brochure	12.167	15.750	12	4	3.512	2.062	14	-1.906	0.077
Medium track	Control	Human guide	12.167	18.500	12	4	3.512	3.000	14	-3.218	**0.006**
Media x school type effect on knowledge (posttest)											
Medium track	Mobile guide	Brochure	16.400	15.750	5	4	3.912	2.062	7	0.298	0.774
Medium track	Mobile guide	Human guide	16.400	18.500	5	4	3.912	3.000	7	-0.882	0.407
Medium track	Brochure	Human guide	15.750	18.500	4	4	2.062	3.000	6	-1.511	0.182
Influence of school type on knowledge											
Advanced track	Control	Mobile guide	14.357	16.000	14	3	3.500	3.000	15	-0.751	0.464
Advanced track	Control	Brochure	14.357	14.600	14	5	3.500	4.393	17	-0.125	0.902
Advanced track	Control	Human guide	14.357	21.000	14	6	3.500	2.898	18	-4.072	**0.001**
Media x school type effect on knowledge (posttest)											
Advanced track	Mobile guide	Brochure	16.000	14.600	3	5	3.000	4.393	6	0.481	0.647
Advanced track	Mobile guide	Human guide	16.000	21.000	3	6	3.000	2.898	7	-2.415	**0.046**
Advanced track	Brochure	Human guide	14.600	21.000	5	6	4.393	2.898	9	-2.904	**0.017**
Influence of school type on knowledge											
Elementary sch.	Control	Mobile guide	13.000	14.667	18	6	4.615	4.803	22	-0.759	0.456
Elementary sch.	Control	Brochure	13.000	18.857	18	7	4.615	4.140	23	-2.925	**0.008**
Elementary sch.	Control	Human guide	13.000	18.444	18	9	4.615	6.386	25	-2.542	**0.018**
Media x school type effect on knowledge (posttest)											
Elementary sch.	Mobile guide	Brochure	14.667	18.857	6	7	4.803	4.140	11	-1.691	0.119
Elementary sch.	Mobile guide	Human guide	14.667	18.444	6	9	4.803	6.386	13	-1.230	0.241
Elementary sch.	Brochure	Human guide	18.857	18.444	7	9	4.140	6.386	14	0.148	0.885

Table C.6: T-test for unpaired sample groups was applied to determine differences in navigation success with guide media (significant difference (p<0.05)).

T-test for unpaired samples	Group 1	Group 2	Mean 1	Mean 2	n 1	n 2	StDev 1	StDev 2	df	t-value	p-value
Time spent between stations - Families											
S_S1	Brochure	Mobile guide	0.004	0.006	7	11	0.001	0.003	16	-1.020	0.323
S1_S2	Brochure	Mobile guide	0.002	0.003	7	11	0.001	0.001	16	-2.627	**0.018**
S2_S3	Brochure	Mobile guide	0.004	0.003	7	11	0.002	0.001	16	1.810	0.089
S3_S4	Brochure	Mobile guide	0.004	0.005	7	11	0.002	0.002	16	-1.302	0.211
S4_S5	Brochure	Mobile guide	0.005	0.005	7	11	0.002	0.002	16	0.152	0.881
S5_S6	Brochure	Mobile guide	0.006	0.006	7	11	0.001	0.002	16	0.145	0.887
S6_S7	Brochure	Mobile guide	0.006	0.005	7	11	0.001	0.002	16	0.940	0.361
S7_E	Brochure	Mobile guide	0.002	0.001	7	11	0.001	0.001	16	0.694	0.498
Time spent between stations - Students											
S_S1	Brochure	Mobile guide	0.005	0.010	9	8	0.002	0.003	15	-3.265	**0.005**
S1_S2	Brochure	Mobile guide	0.003	0.004	9	7	0.002	0.004	14	-0.608	0.553
S2_S3	Brochure	Mobile guide	0.003	0.004	9	7	0.002	0.003	14	-0.727	0.479
S3_S4	Brochure	Mobile guide	0.006	0.004	9	8	0.003	0.002	15	1.543	0.144
S4_S5	Brochure	Mobile guide	0.007	0.006	9	8	0.003	0.003	15	0.467	0.647
S5_S6	Brochure	Mobile guide	0.006	0.005	8	8	0.002	0.002	14	0.270	0.791
S6_S7	Brochure	Mobile guide	0.005	0.005	9	7	0.001	0.003	14	-0.241	0.813
S7_E	Brochure	Mobile guide	0.001	0.001	9	7	0.001	0.001	14	0.587	0.567
Time spent at stations											
Families	Brochure	Mobile guide	0.003	0.004	7	7	0.001	0.001	12	-1.822	0.093
Students	Brochure	Mobile guide	0.003	0.004	7	7	0.001	0.001	12	-2.458	**0.030**

C.3.3 U-test

Table C.7: Part 1 of U-test results. U-test was applied to determine differences in user satisfaction with guide media (significant difference ($p<0.05$)).

U-test	Group 1	Group 2	Rank sum 1	Rank sum 2	n 1	n 2	U-value	Z-value	p-value
Differences in emotional state									
Parents	Mobile guide	Brochure	618.500	509.500	26	21	267.500	-0.118	0.906
Parents	Mobile guide	Human guide	691.000	849.000	26	29	340.000	-0.624	0.533
Children	Mobile guide	Brochure	538.000	590.000	25	22	213.000	-1.322	0.186
Children	Mobile guide	Human guide	604.000	722.000	25	26	279.000	-0.867	0.386
Students	Mobile guide	Brochure	1243.500	1102.500	33	35	472.500	1.288	0.198
Students	Mobile guide	Human guide	1327.000	1301.000	33	39	521.000	1.384	0.166
Differences in perceived knowledge gain									
Parents	Mobile guide	Brochure	546.500	581.500	26	21	195.500	1.658	0.097
Parents	Mobile guide	Human guide	569.000	1084.000	26	31	218.000	2.964	**0.003**
Children	Mobile guide	Brochure	658.000	518.000	25	22	265.000	0.435	0.664
Children	Mobile guide	Human guide	745.000	686.000	25	26	308.000	0.765	0.444
Students	Mobile guide	Brochure	1183.000	1232.000	33	35	566.000	0.336	0.737
Students	Mobile guide	Human guide	1291.000	1410.000	33	39	590.000	0.776	0.438
Differences in perceived entertainment qualities									
Children	Mobile guide	Brochure	657.500	518.500	25	23	242.500	0.929	0.353
Children	Mobile guide	Human guide	741.500	533.500	25	25	208.500	2.018	**0.044**
Students	Mobile guide	Brochure	964.000	1181.000	31	34	468.000	-0.775	0.438
Students	Mobile guide	Human guide	916.500	1568.500	31	39	420.500	-2.176	**0.030**
Differences in perception of games									
Parents	Human guide	Brochure	469.000	351.000	19	21	120.000	2.153	**0.031**
Parents	Mobile guide	Human guide	404.000	499.000	23	19	128.000	2.287	**0.022**
Children	Mobile guide	Brochure	640.000	488.000	25	22	235.000	0.853	0.394
Children	Mobile guide	Human guide	611.000	614.000	25	24	286.000	-0.280	0.779
Students	Mobile guide	Brochure	960.500	1055.500	29	34	460.500	0.448	0.654
Students	Mobile guide	Human guide	854.000	1357.000	29	37	419.000	-1.518	0.129
Differences in satisfaction with guide media									
Parents	Mobile guide	Human guide	687.000	441.000	24	23	165.000	-2.362	**0.018**
Children	Mobile guide	Brochure	586.500	494.500	24	22	241.500	0.495	0.621
Children	Mobile guide	Human guide	627.500	597.500	24	25	272.500	0.550	0.582
Students	Mobile guide	Brochure	990.500	1287.500	31	36	494.500	-0.799	0.425
Students	Mobile guide	Human guide	1004.000	1411.000	31	38	508.000	-0.977	0.329
Differences in satisfaction with overall tour									
Children	Mobile guide	Brochure	491.500	498.500	22	22	238.500	-0.082	0.935
Children	Mobile guide	Human guide	514.000	567.000	22	24	261.000	-0.066	0.947
Students	Mobile guide	Brochure	892.000	1188.000	31	33	396.000	-1.552	0.121
Students	Mobile guide	Human guide	885.500	1392.500	31	36	389.500	-2.119	**0.034**

Table C.8: Part 2 of U-test results. U-test was applied to determine differences in user satisfaction with guide media (significant difference (p<0.05)).

U-test	Group 1	Group 2	Rank sum 1	Rank sum 2	n 1	n 2	U-value	Z-value	p-value
Differences in perception of interactivity									
Parents	Mobile guide	Brochure	516.000	565.000	26	20	165.000	2.105	**0.035**
Parents	Mobile guide	Human guide	812.500	783.500	26	30	318.500	1.175	0.240
Children	Mobile guide	Brochure	596.000	629.000	26	23	245.000	-1.082	0.279
Children	Mobile guide	Human guide	737.500	640.500	26	26	289.500	0.888	0.375
Students	Mobile guide	Brochure	1131.500	1214.500	32	36	548.500	0.338	0.735
Students	Mobile guide	Human guide	1316.500	1239.500	32	39	459.500	1.901	0.057
Differences in perception of distraction									
Parents	Mobile guide	Brochure	495.000	633.000	26	21	144.000	2.760	**0.006**
Parents	Mobile guide	Human guide	503.500	1036.500	26	29	152.500	3.785	**0.000**
Children	Mobile guide	Brochure	606.500	521.500	25	22	268.500	0.139	0.890
Children	Mobile guide	Human guide	648.000	678.000	25	26	323.000	-0.038	0.970
Students	Mobile guide	Brochure	1127.000	1288.000	33	36	566.000	-0.336	0.737
Students	Mobile guide	Human guide	1193.000	1363.000	33	38	622.000	0.058	0.954
Differences in perceived excitement									
Children	Mobile guide	Brochure	610.500	614.500	26	23	259.500	-0.791	0.429
Children	Mobile guide	Human guide	587.000	844.000	26	27	236.000	-2.046	**0.041**
Students	Mobile guide	Brochure	1226.500	1188.500	33	36	522.500	0.859	0.390
Students	Mobile guide	Human guide	1337.000	1364.000	33	40	544.000	1.286	0.199
Differences in perceived involvement									
Children	Mobile guide	Brochure	730.500	397.500	24	23	121.500	3.288	**0.001**
Students	Mobile guide	Brochure	1031.000	1180.000	31	35	535.000	-0.096	0.923
Differences in perceived content consumption									
Parents	Mobile guide	Brochure	720.500	407.500	26	21	176.500	-2.065	**0.039**
Children	Mobile guide	Brochure	570.000	465.000	24	21	234.000	0.410	0.682
Students	Mobile guide	Brochure	1119.000	1296.000	33	36	558.000	-0.432	0.665
Differences in perceived navigational assistance									
Parents	Mobile guide	Brochure	504.500	576.500	25	21	179.500	1.830	0.067
Children	Mobile guide	Brochure	567.500	608.500	25	23	242.500	-0.929	0.353
Students	Mobile guide	Brochure	1152.000	1194.000	32	36	528.000	0.590	0.555

C.3.4 Fisher's exact test

Table C.9: 2x2 Tables and Fisher's exact test statistics for gender effects in family groups.

2 x 2 Contingency Table: Parents-Pretest (Gender distribution)

	Human	Mobile G	Total
Female	8	5	13
% of Total	27.59%	17.24%	44.83%
Male	7	9	16
% of Total	24.14%	31.03%	55.17%
Total	15	14	29
% of Total	51.72%	48.28%	
Chi-square (df=1)		0.91	p= .3404
V-square (df=1)		0.88	p= .3489
Yates-corr. Chi-squ.		0.34	p= .5621
Phi-square	0.03134		
Fisher's exact p, 1-sided			p= .2817
Fisher's exact p, 2-sided			p= .4621

	Mobile G	Brochure	Total
Female	5	6	11
% of Total	20.00%	24.00%	44.00%
Male	9	5	14
% of Total	36.00%	20.00%	56.00%
Total	14	11	25
% of Total	56.00%	44.00%	
Chi-square (df=1)		0.89	p= .3464
V-square (df=1)		0.85	p= .3563
Yates-corr. Chi-squ.		0.29	p= .5922
Phi-square		0.03546	
Fisher's exact p, 1-sided			p= .2962
Fisher's exact p, 2-sided			p= .4347

	Human	Brochure	Total
Female	8	6	14
% of Total	30.77%	23.08%	53.85%
Male	7	5	12
% of Total	26.92%	19.23%	46.15%
Total	15	11	26
% of Total	57.69%	42.31%	
Chi-square (df=1)		0	p= .9512
V-square (df=1)		0	p= .9521
Yates-corr. Chi-squ		0.11	p= .7362
Phi-square		0.00014	
Fisher's exact p, 1-sided			p= .6319
Fisher's exact p, 2-sided			p=1.0000

2 x 2 Contingency Table: Parents-Posttest (Gender distribution)

	Human	Mobile G	Total
Female	10	8	18
% of Total	35.71%	28.57%	64.29%
Male	6	4	10
% of Total	21.43%	14.29%	35.71%
Total	16	12	28
% of Total	57.14%	42.86%	
Chi-square (df=1)		0.05	p= .8199
V-square (df=1)		0.05	p= .8231
Yates-corr. Chi-squ.		0.03	p= .8644
Phi-square	0.00185		
Fisher's exact p, 1-sided			p= .5696
Fisher's exact p, 2-sided			p=1.0000

	Mobile G	Brochure	Total
Female	8	5	13
% of Total	36.36%	22.73%	59.09%
Male	4	5	9
% of Total	18.18%	22.73%	40.91%
Total	12	10	22
% of Total	54.55%	45.46%	
Chi-square (df=1)		0.63	p= .4285
V-square (df=1)		0.6	p= .4392
Yates-corr. Chi-squ.		0.13	p= .7216
Phi-square		0.02849	
Fisher's exact p, 1-sided			p= .3607
Fisher's exact p, 2-sided			p= .6656

	Human	Brochure	Total
Female	10	5	15
% of Total	38.46%	19.23%	57.69%
Male	6	5	11
% of Total	23.08%	19.23%	42.31%
Total	16	10	26
% of Total	61.54%	38.46%	
Chi-square (df=1)		0.39	p= .5302
V-square (df=1)		0.38	p= .5383
Yates-corr. Chi-squ		0.05	p= .8261
Phi-square		0.02	
Fisher's exact p, 1-sided			p= .4116
Fisher's exact p, 2-sided			p= .6891

2 x 2 Contingency Table: Children-Pretest (Gender distribution)

	Human	Mobile G	Total
Female	8	5	13
% of Total	29.63%	18.52%	48.15%
Male	5	9	14
% of Total	18.52%	33.33%	51.85%
Total	13	14	27
% of Total	48.15%	51.85%	
Chi-square (df=1)		1.8	p= .1796
V-square (df=1)		1.73	p= .1879
Yates-corr. Chi-squ.		0.91	p= .3389
Phi-square	0.06669		
Fisher's exact p, 1-sided			p= .1697
Fisher's exact p, 2-sided			p= .2568

	Mobile G	Brochure	Total
Female	5	4	9
% of Total	20.00%	16.00%	36.00%
Male	9	7	16
% of Total	36.00%	28.00%	64.00%
Total	14	11	25
% of Total	56.00%	44.00%	
Chi-square (df=1)		0	p= .9732
V-square (df=1)		0	p= .9738
Yates-corr. Chi-squ.		0.15	p= .6994
Phi-square		0.00005	
Fisher's exact p, 1-sided			p= .6479
Fisher's exact p, 2-sided			p=1.000

	Human	Brochure	Total
Female	8	4	12
% of Total	33.33%	16.67%	50.00%
Male	5	7	12
% of Total	20.83%	29.17%	50.00%
Total	13	11	24
% of Total	54.17%	45.83%	
Chi-square (df=1)		1.51	p= .2191
V-square (df=1)		1.45	p= .2289
Yates-corr. Chi-squ		0.67	p= .4126
Phi-square		0.06294	
Fisher's exact p, 1-sided			p= .2068
Fisher's exact p, 2-sided			p= .4136

2 x 2 Contingency Table: Children-Posttest (Gender distribution)

	Human	Mobile G	Total
Female	10	6	16
% of Total	38.46%	23.08%	61.54%
Male	4	6	10
% of Total	15.39%	23.08%	38.46%
Total	14	12	26
% of Total	53.85%	46.15%	
Chi-square (df=1)		1.25	p= .2629
V-square (df=1)		1.21	p= .2723
Yates-corr. Chi-squ.		0.51	p= .4744
Phi-square	0.04821		
Fisher's exact p, 1-sided			p= .2375
Fisher's exact p, 2-sided			p= .4216

	Mobile G	Brochure	Total
Female	6	6	12
% of Total	25.00%	25.00%	50.00%
Male	6	6	12
% of Total	25.00%	25.00%	50.00%
Total	12	12	24
% of Total	50.00%	50.00%	
Chi-square (df=1)		0	p=1.000
V-square (df=1)		0	p=1.000
Yates-corr. Chi-squ.		0.17	p= .6831
Phi-square		0	
Fisher's exact p, 1-sided			p= .6579
Fisher's exact p, 2-sided			p=1.000

	Human	Brochure	Total
Female	10	6	16
% of Total	38.46%	23.08%	61.54%
Male	4	6	10
% of Total	15.39%	23.08%	38.46%
Total	14	12	26
% of Total	53.85%	46.15%	
Chi-square (df=1)		1.25	p= .2629
V-square (df=1)		1.21	p= .2723
Yates-corr. Chi-squ		0.51	p= .4744
Phi-square		0.04821	
Fisher's exact p, 1-sided			p= .2375
Fisher's exact p, 2-sided			p= .4216

Table C.10: Contingency Tables and Fisher's exact test statistics for gender effects in student groups and feedback on guide medium by parents.

2 x 2 Contingency Table: Students-Pretest (Gender distribution)

	Human	Mobile Gu	Total
Female	9	8	17
% of Total	25.00%	22.22%	47.22%
Male	11	8	19
% of Total	30.56%	22.22%	52.78%
Total	20	16	36
% of Total	55.56%	44.44%	
Chi-square (df=1)		0.09	p= .7652
V-square (df=1)		0.09	p= .7684
Yates-corr. Chi-squ		0	p= .9702
Phi-square	0.00248		
Fisher's exact p, 1-sided			p= .6110
Fisher's exact p, 2-sided			p=1.0000

	Mobile Gu	Brochure	Total
Female	8	7	15
% of Total	24.24%	21.21%	45.46%
Male	8	10	18
% of Total	24.24%	30.30%	54.55%
Total	16	17	33
% of Total	48.49%	51.52%	
Chi-square (df=1)		0.26	p= .6109
V-square (df=1)		0.25	p= .6164
Yates-corr. Chi-squ.		0.03	p= .8737
Phi-square	0.00784		
Fisher's exact p, 1-sided			p= .4369
Fisher's exact p, 2-sided			p= .7319

	Human	Brochure	Total
Female	9	7	16
% of Total	24.32%	18.92%	43.24%
Male	11	10	21
% of Total	29.73%	27.03%	56.76%
Total	20	17	37
% of Total	54.05%	45.95%	
Chi-square (df=1)		0.05	p= .8150
V-square (df=1)		0.05	p= .8175
Yates-corr. Chi-squ		0.01	p= .9212
Phi-square	0.00148		
Fisher's exact p, 1-sided			p= .5398
Fisher's exact p, 2-sided			p=1.0000

2 x 2 Contingency Table: Students-Posttest (Gender distribution)

	Human	Mobile Gu	Total
Female	9	8	17
% of Total	24.32%	21.62%	45.95%
Male	11	9	20
% of Total	29.73%	24.32%	54.05%
Total	20	17	37
% of Total	54.05%	45.95%	
Chi-square (df=1)		0.02	p= .9003
V-square (df=1)		0.02	p= .9017
Yates-corr. Chi-squ		0.04	p= .8370
Phi-square	0.00042		
Fisher's exact p, 1-sided			p= .5810
Fisher's exact p, 2-sided			p=1.0000

	Mobile Gu	Brochure	Total
Female	8	10	18
% of Total	22.22%	27.78%	50.00%
Male	9	9	18
% of Total	25.00%	25.00%	50.00%
Total	17	19	36
% of Total	47.22%	52.78%	
Chi-square (df=1)		0.11	p= .7385
V-square (df=1)		0.11	p= .7420
Yates-corr. Chi-squ.		0	p=1.000
Phi-square	0.0031		
Fisher's exact p, 1-sided			p= .5000
Fisher's exact p, 2-sided			p=1.000

	Human	Brochure	Total
Female	9	10	19
% of Total	23.08%	25.64%	48.72%
Male	11	9	20
% of Total	28.21%	23.08%	51.28%
Total	20	19	39
% of Total	51.28%	48.72%	
Chi-square (df=1)		0.23	p= .6337
V-square (df=1)		0.22	p= .6380
Yates-corr. Chi-squ		0.02	p= .8759
Phi-square	0.00582		
Fisher's exact p, 1-sided			p= .4381
Fisher's exact p, 2-sided			p= .7524

2 x 2 Contingency Table: Parents (Positive feedback)

	Human	Mobile Gu	Total
Other	20	18	38
% of Total	36.36%	32.73%	69.09%
Guide	9	8	17
% of Total	16.36%	14.55%	30.91%
Total	29	26	55
% of Total	52.73%	47.27%	
Chi-square (df=1)		0	p= .9830
V-square (df=1)		0	p= .9832
Yates-corr. Chi-squ		0.07	p= .7864
Phi-square	0.00001		
Fisher's exact p, 1-sided			p= .6073
Fisher's exact p, 2-sided			p=1.0000

	Mobile Gu	Brochure	Total
Other	18	21	39
% of Total	38.30%	44.68%	82.98%
Guide	8	0	8
% of Total	17.02%	0.00%	17.02%
Total	26	21	47
% of Total	55.32%	44.68%	
Chi-square (df=1)		7.79	p= .0053
V-square (df=1)		7.62	p= .0058
Yates-corr. Chi-squ.		5.76	p= .0164
Phi-square	0.16568		
Fisher's exact p, 1-sided			**p= .005**
Fisher's exact p, 2-sided			p= .0056

	Human	Brochure	Total
Other	20	21	41
% of Total	40.00%	42.00%	82.00%
Guide	9	0	9
% of Total	18.00%	0.00%	18.00%
Total	29	21	50
% of Total	58.00%	42.00%	
Chi-square (df=1)		7.95	p= .0048
V-square (df=1)		7.79	p= .0053
Yates-corr. Chi-squ		5.98	p= .0144
Phi-square	0.15896		
Fisher's exact p, 1-sided			**p= .0040**
Fisher's exact p, 2-sided			p= .0065

2 x 2 Contingency Table: Parents (Negativ feedback)

	Human	Mobile Gu	Total
Other	24	7	31
% of Total	48.98%	14.29%	63.27%
Guide	1	17	18
% of Total	2.04%	34.69%	36.74%
Total	25	24	49
% of Total	51.02%	48.98%	
Chi-square (df=1)		23.53	p= .0000
V-square (df=1)		23.05	p= .0000
Yates-corr. Chi-squ		20.75	p= .0000
Phi-square	0.48029		
Fisher's exact p, 1-sided			**p= .0000**
Fisher's exact p, 2-sided			p= .0000

	Mobile Gu	Brochure	Total
Other	7	16	23
% of Total	16.67%	38.10%	54.76%
Guide	17	2	19
% of Total	40.48%	4.76%	45.24%
Total	24	18	42
% of Total	57.14%	42.86%	
Chi-square (df=1)		14.81	p= .0001
V-square (df=1)		14.46	p= .0001
Yates-corr. Chi-squ.		12.5	p= .0004
Phi-square	0.35259		
Fisher's exact p, 1-sided			**p= .0001**
Fisher's exact p, 2-sided			p= .0001

	Human	Brochure	Total
Other	24	16	40
% of Total	55.81%	37.21%	93.02%
Guide	1	2	3
% of Total	2.33%	4.65%	6.98%
Total	25	18	43
% of Total	58.14%	41.86%	
Chi-square (df=1)		0.82	p= .3665
V-square (df=1)		0.8	p= .3722
Yates-corr. Chi-squ		0.09	p= .7670
Phi-square	0.01896		
Fisher's exact p, 1-sided			p= .3761
Fisher's exact p, 2-sided			p= .5624

Table C.11: 2x2 Tables and Fisher's exact test statistics for feedback on guide medium by children and students.

2 x 2 Contingency Table: Children (Positive feedback)

	Human	Mobile Gu	Total
Other	25	15	40
% of Total	48.08%	28.85%	76.92%
Guide	1	11	12
% of Total	1.92%	21.15%	23.08%
Total	26	26	52
% of Total	50.00%	50.00%	
Chi-square (df=1)		10.83	p= .0010
V-square (df=1)		10.63	p= .0011
Yates-corr. Chi-squ		8.77	p= .0031
Phi-square	0.20833		
Fisher's exact p, 1-sided			**p= .001**
Fisher's exact p, 2-sided			p= .0020

	Mobile Gu	Brochure	Total
Other	15	22	37
% of Total	30.61%	44.90%	75.51%
Guide	11	1	12
% of Total	22.45%	2.04%	24.49%
Total	26	23	49
% of Total	53.06%	46.94%	
Chi-square (df=1)		9.51	p= .0020
V-square (df=1)		9.32	p= .0023
Yates-corr. Chi-squ.		7.57	p= .0059
Phi-square	0.19407		
Fisher's exact p, 1-sided			**p= .002**
Fisher's exact p, 2-sided			p= .0024

	Human	Brochure	Total
Other	25	22	47
% of Total	51.02%	44.90%	95.92%
Guide	1	1	2
% of Total	2.04%	2.04%	4.08%
Total	26	23	49
% of Total	53.06%	46.94%	
Chi-square (df=1)		0.01	p= .9294
V-square (df=1)		0.01	p= .9301
Yates-corr. Chi-squ		0.4	p= .5256
Phi-square	0.00016		
Fisher's exact p, 1-sided			p= .7236
Fisher's exact p, 2-sided			p=1.0000

2 x 2 Contingency Table: Children (Negative feedback)

	Human	Mobile Gu	Total
Other	25	19	44
% of Total	48.08%	36.54%	84.62%
Guide	1	7	8
% of Total	1.92%	13.46%	15.39%
Total	26	26	52
% of Total	50.00%	50.00%	
Chi-square (df=1)		5.32	p= .0211
V-square (df=1)		5.22	p= .0224
Yates-corr. Chi-squ		3.69	p= .0546
Phi-square	0.10227		
Fisher's exact p, 1-sided			**p= .0248**
Fisher's exact p, 2-sided			p= .0496

	Mobile Gu	Brochure	Total
Other	19	22	41
% of Total	38.78%	44.90%	83.67%
Guide	7	1	8
% of Total	14.29%	2.04%	16.33%
Total	26	23	49
% of Total	53.06%	46.94%	
Chi-square (df=1)		4.55	p= .0329
V-square (df=1)		4.46	p= .0347
Yates-corr. Chi-squ.		3.05	p= .0807
Phi-square	0.09292		
Fisher's exact p, 1-sided			**p= .037**
Fisher's exact p, 2-sided			p= .0522

	Human	Brochure	Total
Other	25	22	47
% of Total	51.02%	44.90%	95.92%
Guide	1	1	2
% of Total	2.04%	2.04%	4.08%
Total	26	23	49
% of Total	53.06%	46.94%	
Chi-square (df=1)		0.01	p= .9294
V-square (df=1)		0.01	p= .9301
Yates-corr. Chi-squ		0.4	p= .5256
Phi-square	0.00016		
Fisher's exact p, 1-sided			p= .7236
Fisher's exact p, 2-sided			p=1.0000

2 x 2 Contingency Table: Students (Positive feedback)

	Human	Mobile Gu	Total
Other	37	23	60
% of Total	50.69%	31.51%	82.19%
Guide	3	10	13
% of Total	4.11%	13.70%	17.81%
Total	40	33	73
% of Total	54.80%	45.21%	
Chi-square (df=1)		6.42	p= .0113
V-square (df=1)		6.34	p= .0118
Yates-corr. Chi-squ		4.96	p= .0259
Phi-square	0.088		
Fisher's exact p, 1-sided			**p= .0125**
Fisher's exact p, 2-sided			p= .0148

	Mobile Gu	Brochure	Total
Other	23	30	53
% of Total	33.33%	43.48%	76.81%
Guide	10	6	16
% of Total	14.49%	8.70%	23.19%
Total	33	36	69
% of Total	47.83%	52.17%	
Chi-square (df=1)		1.8	p= .1800
V-square (df=1)		1.77	p= .1832
Yates-corr. Chi-squ.		1.11	p= .2913
Phi-square	0.02605		
Fisher's exact p, 1-sided			p= .1457
Fisher's exact p, 2-sided			p= .2548

	Human	Brochure	Total
Other	37	30	67
% of Total	48.68%	39.47%	88.16%
Guide	3	6	9
% of Total	3.95%	7.90%	11.84%
Total	40	36	76
% of Total	52.63%	47.37%	
Chi-square (df=1)		1.53	p= .2169
V-square (df=1)		1.5	p= .2199
Yates-corr. Chi-squ		0.77	p= .3792
Phi-square	0.02007		
Fisher's exact p, 1-sided			p= .1899
Fisher's exact p, 2-sided			p= .2937

2 x 2 Contingency Table: Students (Negative feedback)

	Human	Mobile Gu	Total
Other	33	24	57
% of Total	45.21%	32.88%	78.08%
Guide	7	9	16
% of Total	9.59%	12.33%	21.92%
Total	40	33	73
% of Total	54.80%	45.21%	
Chi-square (df=1)		1.01	p= .3151
V-square (df=1)		1	p= .3185
Yates-corr. Chi-squ		0.52	p= .4713
Phi-square	0.01382		
Fisher's exact p, 1-sided			p= .2353
Fisher's exact p, 2-sided			p= .3978

	Mobile Gu	Brochure	Total
Other	24	24	48
% of Total	35.82%	35.82%	71.64%
Guide	9	10	19
% of Total	13.43%	14.93%	28.36%
Total	33	34	67
% of Total	49.25%	50.75%	
Chi-square (df=1)		0.04	p= .8460
V-square (df=1)		0.04	p= .8472
Yates-corr. Chi-squ.		0.01	p= .9387
Phi-square	0.00056		
Fisher's exact p, 1-sided			p= .5308
Fisher's exact p, 2-sided			p=1.000

	Human	Brochure	Total
Other	33	24	57
% of Total	44.60%	32.43%	77.03%
Guide	7	10	17
% of Total	9.46%	13.51%	22.97%
Total	40	34	74
% of Total	54.05%	45.95%	
Chi-square (df=1)		1.47	p= .2248
V-square (df=1)		1.45	p= .2279
Yates-corr. Chi-squ		0.88	p= .3489
Phi-square	0.01991		
Fisher's exact p, 1-sided			p= .1745
Fisher's exact p, 2-sided			p= .2737

Table C.12: 2x2 Tables and Fisher's exact test statistics for distribution of education levels of parents.

2 x 2 Contingency Table: Parents (Distributiong of education - pretest)

	Human	Mobile G	Total
Academic	9	8	17
% of Total	31.03%	27.59%	58.62%
Non Acad.	6	6	12
% of Total	20.69%	20.69%	41.38%
Total	15	14	29
% of Total	51.72%	48.28%	
Chi-square (df=1)		0.02	p= .8759
V-square (df=1)		0.02	p= .8781
Yates-corr. Chi-squ		0.05	p= .8250
Phi-square	0.00084		
Fisher's exact p, 1-sided			p= .5869
Fisher's exact p, 2-sided			p=1.0000

	Mobile G	Brochure	Total
Other	8	4	12
% of Total	32.00%	16.00%	48.00%
Guide	6	7	13
% of Total	24.00%	28.00%	52.00%
Total	14	11	25
% of Total	56.00%	44.00%	
Chi-square (df=1)		1.07	p= .3019
V-square (df=1)		1.02	p= .3118
Yates-corr. Chi-squ.		0.4	p= .5293
Phi-square		0.04262	
Fisher's exact p, 1-sided			p= .2655
Fisher's exact p, 2-sided			p= .4283

	Human	Brochure	Total
Other	9	4	13
% of Total	34.62%	15.39%	50.00%
Guide	6	7	13
% of Total	23.08%	26.92%	50.00%
Total	15	11	26
% of Total	57.69%	42.31%	
Chi-square (df=1)		1.42	p= .2337
V-square (df=1)		1.36	p= .2429
Yates-corr. Chi-squ		0.63	p= .4273
Phi-square		0.05455	
Fisher's exact p, 1-sided			p= .2142
Fisher's exact p, 2-sided			p= .4283

2 x 2 Contingency Table: Parents (Distributiong of education - posttest)

	Human	Mobile G	Total
Academic	9	6	15
% of Total	32.14%	21.43%	53.57%
Non Acad.	7	6	13
% of Total	25.00%	21.43%	46.43%
Total	16	12	28
% of Total	57.14%	42.86%	
Chi-square (df=1)		0.11	p= .7428
V-square (df=1)		0.1	p= .7473
Yates-corr. Chi-squ		0	p= .9564
Phi-square	0.00385		
Fisher's exact p, 1-sided			p= .5212
Fisher's exact p, 2-sided			p=1.0000

	Mobile G	Brochure	Total
Other	6	3	9
% of Total	27.27%	13.64%	40.91%
Guide	6	7	13
% of Total	27.27%	31.82%	59.09%
Total	12	10	22
% of Total	54.55%	45.46%	
Chi-square (df=1)		0.9	p= .3421
V-square (df=1)		0.86	p= .3533
Yates-corr. Chi-squ.		0.26	p= .6068
Phi-square		0.04103	
Fisher's exact p, 1-sided			p= .3050
Fisher's exact p, 2-sided			p= .4149

	Human	Brochure	Total
Other	9	3	12
% of Total	34.62%	11.54%	46.15%
Guide	7	7	14
% of Total	26.92%	26.92%	53.85%
Total	16	10	26
% of Total	61.54%	38.46%	
Chi-square (df=1)		1.71	p= .1915
V-square (df=1)		1.64	p= .2002
Yates-corr. Chi-squ		0.81	p= .3671
Phi-square		0.06562	
Fisher's exact p, 1-sided			p= .1842
Fisher's exact p, 2-sided			p= .2475

2 x 2 Contingency Table: Parents (Distributiong of education - within media)

Human	Pre	Post	Total
Academic	9	9	18
% of Total	29.03%	29.03%	58.07%
Non Acad.	7	6	13
% of Total	22.58%	19.36%	41.94%
Total	16	15	31
% of Total	51.61%	48.39%	
Chi-square (df=1)		0.040	p= .8325
V-square (df=1)		0.040	p= .8352
Yates-corr. Chi-squ		0.020	p= .8786
Phi-square		0.001	
Fisher's exact p, 1-sided			p= .5607
Fisher's exact p, 2-sided			p=1.0000

Brochure	Pre	Post	Total
Other	4	3	7
% of Total	19.05%	14.29%	33.33%
Guide	7	7	14
% of Total	33.33%	33.33%	66.67%
Total	11	10	21
% of Total	52.38%	47.62%	
Chi-square (df=1)		0.100	p= .7574
V-square (df=1)		0.090	p= .7630
Yates-corr. Chi-squ.		0.020	p= .8772
Phi-square		0.005	
Fisher's exact p, 1-sided			p= .5619
Fisher's exact p, 2-sided			p=1.000

Mobile G.	Pre	Post	Total
Other	8	6	14
% of Total	30.77%	23.08%	53.85%
Guide	6	6	12
% of Total	23.08%	23.08%	46.15%
Total	14	12	26
% of Total	53.85%	46.15%	
Chi-square (df=1)		0.130	p= .7157
V-square (df=1)		0.130	p= .7210
Yates-corr. Chi-squ		0.000	p= .9758
Phi-square		0.005	
Fisher's exact p, 1-sided			p= .5117
Fisher's exact p, 2-sided			p=1.0000

Table C.13: 2x2 Tables and Fisher's exact test statistics for navigation success and content consumption of families & students.

2 x 2 Contingency Table: Families (Navigation success - Junctions)			
	Mobile G	Brochure	Total
Not found	11	1	12
% of Total	8.09%	0.74%	8.82%
Found	77	47	124
% of Total	56.62%	34.56%	91.18%
Total	88	48	136
% of Total	64.71%	35.29%	
Chi-square (df=1)		4.19	p= .0407
V-square (df=1)		4.16	p= .0414
Yates-corr. Chi-squ		2.99	p= .0836
Phi-square	0.0308		
Fisher's exact p, 1-sided			**p= .0343**
Fisher's exact p, 2-sided			p= .0559

2 x 2 Contingency Table: Families (Navigation success - Stations)			
	Mobile G.	Brochure	Total
Not found	13	8	21
% of Total	10.92%	6.72%	17.65%
Found	64	34	98
% of Total	53.78%	28.57%	82.35%
Total	77	42	119
% of Total	64.71%	35.29%	
Chi-square (df=1)		0.09	p= .7672
V-square (df=1)		0.09	p= .7682
Yates-corr. Chi-squ.		0	p= .9646
Phi-square	0.00074		
Fisher's exact p, 1-sided			p= .4758
Fisher's exact p, 2-sided			p= .8042

2 x 2 Contingency Table: Families (Content consumption)			
	Mobile G	Brochure	Total
Not read	64	37	101
% of Total	10.18%	5.88%	16.06%
Read	343	185	528
% of Total	54.53%	29.41%	83.94%
Total	407	222	629
% of Total	64.71%	35.29%	
Chi-square (df=1)		0.09	p= .7585
V-square (df=1)		0.09	p= .7587
Yates-corr. Chi-squ		0.04	p= .8463
Phi-square	0.00015		
Fisher's exact p, 1-sided			p= .0857
Fisher's exact p, 2-sided			p= .7346

2 x 2 Contingency Table: Students (Navigation success - Junctions)			
	Mobile G	Brochure	Total
Not found	17	20	37
% of Total	13.28%	15.63%	28.91%
Found	47	44	91
% of Total	36.72%	34.38%	71.09%
Total	64	64	128
% of Total	50.00%	50.00%	
Chi-square (df=1)		0.34	p= .5586
V-square (df=1)		0.34	p= .5601
Yates-corr. Chi-squ		0.15	p= .6966
Phi-square	0.00267		
Fisher's exact p, 1-sided			p= .3484
Fisher's exact p, 2-sided			p= .6969

2 x 2 Contingency Table: Students (Navigation success - Stations)			
	Mobile G.	Brochure	Total
Not found	28	27	55
% of Total	25.00%	24.11%	49.11%
Found	28	29	57
% of Total	25.00%	25.89%	50.89%
Total	56	56	112
% of Total	50.00%	50.00%	
Chi-square (df=1)		0.04	p= .8501
V-square (df=1)		0.04	p= .8507
Yates-corr. Chi-squ.		0	p=1.000
Phi-square	0.00032		
Fisher's exact p, 1-sided			p= .5000
Fisher's exact p, 2-sided			p=1.000

2 x 2 Contingency Table: Students (Content consumption)			
	Mobile G	Brochure	Total
Not read	84	122	206
% of Total	14.17%	20.61%	34.80%
Read	212	174	386
% of Total	35.81%	29.39%	65.20%
Total	296	296	592
% of Total	50.00%	50.00%	
Chi-square (df=1)		10.75	p= .0010
V-square (df=1)		10.73	p= .0011
Yates-corr. Chi-squ		10.19	p= .0014
Phi-square	0.01816		
Fisher's exact p, 1-sided			**p= .0003**
Fisher's exact p, 2-sided			p= .0013

Bibliography

[1] ABE, M.; YOSHIMURA, T.; YASUKAWA, N.; KOBA, K.; MORIYA, K.; SAKAI, T.: Development and evaluation of a support system for forest education. *Journal of Forest Research* 10 (2005), pp. 43–50.

[2] ABOWD, G.; ATKESON, C.; HONG, J.; LONG, S.; KOOPER, R.; PINKERTON, M.: Cyberguide: A Mobile Context-Aware Tour Guide. *Wireless Networks* 3 (1997), pp. 421–433.

[3] ABOWD, G.; MYNATT, E.: Charting Past, Present, and Future Research in Ubiquitous Computing. *ACM Transactions on Computer-Human Interaction* 7 (2000) 1, pp. 29–58.

[4] AJZEN, I.: The Theory of Planned Behavior. *Organizational Behavior and Human Decision Processes* 50 (1991), pp. 179–211.

[5] AJZEN, I.: Perceived Behavioral Control, Self-Efficacy, Locus of Control, and the Theory of Planned Behavior. *Journal of Applied Social Psychology* 32 (2002), pp. 1–20.

[6] AJZEN, I.; FISHBEIN, M.: Attitude-behavior Relations: A Theoretical Analysis and Review of Empirical Research. *Psychological Bulletin* 84 (1977), pp. 888–918.

[7] AJZEN, I.; MADDEN, T.: Prediction of goal-directed behavior: Attitudes, intentions, and perceived behavioral control. *Journal of Experimental Social Psychology* 22 (1986), pp. 453–473.

[8] ALMER, A.; ZEINER, H.; DERLER, C.; STELZL, H.; LULEY, P.: National Park Information System. In: *Proc. of 17th Int. Conf. Informatics for Environmental Protection (EnviroInfo 2003), Cottbus, Germany*, pp. 25–27, Marburg: Metropolis-Verlag, 2003.

[9] ALTNER, G.: Umweltethik und Umweltbildung. In: *Umweltbildung: Theoretische Konzepte – empirische Erkenntnisse – praktische Erfahrungen* (BEYERSDORF, M.; MICHELSEN, G.; SIEBERT, H., Eds.), pp. 20–26, Neuwied: Luchterhand, 1998.

[10] ANTIKAINEN, H.; BENDAS, D.; MARJONIEMI, K.; MYLLYAHO, M.; OIVO, M.; COLPAERT, A.; JAAKO, N.; KUVAJA, P.; LAINE, K.; RUSANEN, J.; SAARI, E.; SIMILÄ, J.: Mobile Environmental Information Systems. *Cybernetics & Systems* 35 (2004), pp. 737–751.

[11] Aoki, P.; Woodruff, A.: Improving Electronic Guidebook Interfaces Using a Task-Oriented Design Approach. In: *Proc. of 3rd ACM Conf. On Design Interactive Sys. (DIS 2000), New York, NY, USA*, pp. 319–325, New York: ACM, 2000.

[12] Apel, H.: Herausforderungen elektronischer Umweltkommunikation. In: *Umweltbildung: Theoretische Konzepte - empirische Erkenntnisse - praktische Erfahrungen* (Beyersdorf, M.; Michelsen, G.; Siebert, H., Eds.), pp. 334–343, Neuwied: Luchterhand, 1998.

[13] Apel, H.: Multimedia in der Umweltkommunikation. *DGU Nachrichten* 21 (2000), pp. 4–7.

[14] Archie, M.; McCrea, E.: From Environmental Education in the United States: Past, Present and Future. NAAEE publication, 1998.

[15] Attewell, J.; Savill-Smith, C.: Learning with Mobile Devices: Research and Development. In: *Proc. of MLEARN 2003 conference, London, UK*, Learning and Skills Development Agency, 2004.

[16] Avots, D.; Lim, E.; Thibaux, R.; Thrun, S.: A probabilistic technique for simultaneous localization and door state estimation with mobile robots in dynamic environments. In: *Proc. of Conf. on Intelligent Robots and Systems (IROS 2002), Lausanne, Switzerland*, 2002.

[17] Baldauf, M.; Dustdar, S.: A survey on context-aware systems. Technical report tuv-1841-2004-24, Technical University of Vienna, Vienna, Austria, 2004.

[18] Bandelloni, R.; Mori, G.; Paternò, F.: Dynamic Generation of Web Migratory Interfaces. In: *Proc. of 7th Int. Conf. on Human Computer Interaction with Mobile Devices and Services (Mobile HCI 2005), Salzburg, Austria*, New York: ACM, 2005.

[19] Bandura, A.: Self-efficacy. In: *Encyclopedia of human behavior* (Ramachaudran, V. S., Ed.), pp. 71–81, Burlington: Academic Press Inc., 1994.

[20] Barrows, H.: *The Tutorial Process.* Springfield: Southern Illinois University School of Medicine, 1992.

[21] Bartneck, C.; Forlizzi, J.: A Design-Centered Framework for Social Human-Robot Interaction. In: *Proc. of 13th IEEE Int. Workshop on Robot and Human Interactive Communication (RO-MAN2004), Kurashiki, Japan*, 2004.

[22] Baus, J.; Cheverst, K.; Kray, C.: A Survey of Map-based Mobile Guides. In: *Map-based Mobile Services* (Meng, L.; Zipf, A.; Reichenbacher, T., Eds.), pp. 197–213, Berlin: Springer-Verlag, 2005.

[23] BAUS, J.; KRÜGER, A.; WAHLSTER, W.: A Resource Adaptive Mobile Navigation System. In: *Proc. of Int. Conf. on Intelligent User Interface (IUI 2002), San Francisco, CA, USA*, New York: ACM, 2002.

[24] BEINAT, E.: Geographically relevant information for mobile users in protected areas. Final report d1.1.6 public deliverable, IST Information Desk, Brussels, 2004.

[25] BERINGER, A.: The Healing Power of Trees: On organized camping's role in human moral-spiritual development. In: *Proc. of 6th Int. Camping Congress Research Forum, Bendigo, Victoria, Austral*, Daylesford: The St. Francis Institute for Earth Healing, 2003.

[26] BETGÉ-BREZETZ, S.; HÉBERT, P.; CHATILA, R.; DEVY, M.: Uncertain Map Making in Natural Environments. In: *Proc. of IEEE Int. Conf. on Robotics and Automation (ICRA 1996), Minneapolis, MN ,USA*, Washington: IEEE, 1996.

[27] BEYERSDORF, M.: Didaktik und Methoden der Umweltbildung. In: *Umweltbildung: Theoretische Konzepte - empirische Erkenntnisse - praktische Erfahrungen* (BEYERSDORF, M.; MICHELSEN, G.; SIEBERT, H., Eds.), pp. 289–312, Neuwied: Luchterhand, 1998.

[28] B'FAR, R.: *Mobile Computing Principles : Designing and Developing Mobile Applications with UML and XML*. Cambridge, UK.: Cambridge University Press, 2005.

[29] BICKMORE, T.: *Relational Agents: Effecting Change through Human-Computer Relationships*. Ph.D. thesis, Massachusetts Institute of Technology, Cambridge, MA, USA, 2003.

[30] BICKMORE, T.: Unspoken Rules of Spoken Interaction. *Communications of the ACM* 47 (2004) 4, pp. 38–44.

[31] BIEBER, G.; LECK, V.: Scalable Vector Graphics for SaiMotion. *Computer Graphik topics* 15 (2003) 1, pp. 23–24.

[32] BITTNER, A.: *Außerschulische Umweltbildung in der Evaluation - Wirkungen kurzzeitpädagogischer Maßnahmen auf Umwelt- und Naturschutzinteressen von Schülerinnen und Schülern der Sekundarstufe I*. Ph.D. thesis, Universität Göttingen, Verlag Dr. Kovac, 2003.

[33] BIXLER, R.; FLOYD, M.; HAMMITT, W.: Environmental Socialization - Quantitative Tests of the Childhood Play Hypothesis. *Environment and Behavior* 34 (2002) 6, pp. 795–818.

[34] BLAKE, E.; STEVENTON, L.; EDGE, J.; FOSTER, A.: A Field Computer for Animal Trackers. In: *Proc. of 2nd South African Conf. on Human-Computer Interaction (CHI-SA 2001), Pretoria, South Africa*, New York: ACM, 2001.

[35] BLIS: Der BUGAbutler begleitet Sie durch den schönsten Sommer aller Zeiten. Das BLIS_Magazin, 2005.

[36] BÖGEHOLZ, S.: Nature experience and its importance for environmental knowledge, values and action: recent German empirical contributions. *Environmental Education Research* 12 (2006) 1, pp. 65–84.

[37] BOGNER, F.: Environmental perceptions of Irish and Bavarian pupils: an empirical study. *The Environmentalist* 18 (1998), pp. 27–38.

[38] BOGNER, F.: The Influence of Short-Term Outdoor Ecology Education on Long-Term Variables of Environmental Perspective. *Journal of Environmental Education* 29 (1998) 4, pp. 17–29.

[39] BOGNER, F.; BRENGELMANN, J.; WISEMAN, M.: Risk-taking and environmental perception. *The Environmentalist* 20 (2000), pp. 49–62.

[40] BOGNER, F.; WILHELM, M.: Environmental Perspectives of Pupils: The Development of an Attitude and Behavior Scale. *The Environmentalist* 16 (1996), pp. 95–110.

[41] BONDI, L.; DAVIDSON, J.; SMITH, M.: Introduction: Geographys Emotional Turn. In: *Emotional Geographies* (BONDI, L.; DAVIDSON, J.; SMITH, M., Eds.), pp. 1–16, Abingdon: Ashgate, 2005.

[42] BORNTRÄGER, C.; CHEVERST, K.: Social and Technical Pitfalls Designing a Tourist Guide System. In: *Proc. of 3rd Workshop HCI in Mobile Guides in conjunction with Mobile HCI 2003, Udine, Italy*, 2003.

[43] BOYCE, T.; GELLER, E.: Encouraging College Students to Support Pro-Environment Behavior: Effects of Direct Versus Indirect Rewards. *Environment and Behavior* 33 (2001) 1, pp. 107–125.

[44] BRÄMER, R.: Das Bambi-Syndrom. Vorläufige Befunde zur jugendlichen Naturentfremdung. *Natur und Landschaft* 73 (1998) 5, pp. 218–222.

[45] BRÄMER, R.: Zurück zur Natur? Die Wald-und-Wiesen-Therapie. *Psychologie Heute* April (2003), pp. 22–28.

[46] BREIT, H.; ECKENSBERGER, L.: Moral, Alltag und Umwelt. In: *Umweltbildung und Umweltbewußtsein. Forschungsperspektiven im Kontext nachhaltiger Entwicklung* (DE HAAN, G.; KUCKARTZ, U., Eds.), pp. 69–89, Opladen: Leske u. Budrich, 1998.

[47] BREITENFELDT, A.: Pflanzenkunde via Bluetooth. *Mobile Computer & Kommunikation* 1 (2003), pp. 30–31.

[48] BROADBENT, J.; MARTI, P.: Location aware mobile interactive guides: usability issues. In: *Proc. of Int. Cultural Heritage Informatics Meeting (ICHIM 1997), Paris, France*, pp. 1–30, Toronto: Archives & Museum Informatics, 1997.

[49] BROSSARD, D.; LEWENSTEIN, B.; BONNEY, R.: Scientific knowledge and attitude change: The impact of a citizen science project. *International Journal of Science Education* 27 (2005) 9, pp. 1099–1121.

[50] BROWN, B.; CHALMERS, M.: Tourism and mobile technology. In: *Proc. of the 8th European Conf. on Computer Supported Cooperative Work CSCW 2003, Helsinki, Finland* (KUUTTI, K.; KARSTEN, E. H., Eds.), Dordrecht: Kluwer Academic Publishers, 2003.

[51] BROWN, F.: Characterizing effective environmental education and its impact on students' environmental attitudes. In: *Proc. of Seventieth Annual Meeting of the National Association for Research in Science Teaching, Oak Brook, IL, USA*, NARST, 1997.

[52] BURGHARDT, D.; EDWARDES, A.; WEIBEL, R.: WebPark- Entwicklung eines mobilen Informationssystems für Erholungs- und Naturschutzgebiete. *Kartographische Nachrichten* 53 (2003) 2, pp. 58–64.

[53] BUTLIN, C.: Equipment: EasySenseLink and EasySense Flash Logger - Review. *Physics Education* 40 (2005) 3, pp. 289–291.

[54] BYUN, H.; CHEVERST, K.: Exploiting User Models and Context-Awareness to Support Personal Daily Activities. In: *Proc. of Workshop on User Modelling for Context-Aware Applications, 8th Int. Conf. on User Modelling (UM 2001), Sonthofen, Germany*, 2001.

[55] CAMERON, E.: Local Innovations in the Field of Environmental Communication - A study on environmental communication practices throughout Europe. Report, European Commision DG Environment, Communication & Civil Society Unit, Brussels, 2002.

[56] CAMPBELL, N.; REECE, J.: *Biology*. New York: Benjamin Cummings, 2004.

[57] CARTER, J.: *A Sense of Place - An Interpretive Planning Handbook*. Edinburgh, 2001.

[58] CHARVAT, K.; FRYML, J.; SIDA, A.; GNIP, P.: Wireless Supporting of Agricultural and Forestry Information Systems. In: *Proc. of 4th AGILE Conf. on Geographic Information Science 2001, Brno, Czech Republic*, 2001.

[59] CHEN, G.; KOTZ, D.: A Survey of Context-Aware Mobile Computing Research. Dartmouth computer science technical report tr2000-381, Dartmouth College, Hanover, 2000.

[60] CHEN, J.; STANNEY, K.: A Theoretical Model of Wayfinding in Virtual Environments: Proposed Strategies for Navigational Aiding. *Presence* 8 (1999) 6, pp. 671–685.

[61] CHEVERST, K.; DAVIES, N.; MITCHELL, K.: A Reflective Study of the GUIDE System. In: *Proc. of 1st Workshop on HCI in Mobile Tourism Support, Sankt Augustin, Germany* (SCHMIDT-BELZ, B.; CHEVERST, K., Eds.), 2002.

[62] CHEVERST, K.; DAVIES, N.; MITCHELL, K.; FRIDAY, A.; EFSTRATIOU, C.: Developing a Context-aware Electronic Tourist Guide: Some Issues and Experiences. In: *Proc. of ACM Conf. on Human Factors in Computing Systems (CHI 2000), The Hague, Netherlands*, New York: ACM, 2000.

[63] CHITTARO, L.; BURIGAT, S.: Location-aware visualization of a 3D world to select tourist information on a mobile device. In: *Proc. of 3rd Workshop HCI in Mobile Guides at Mobile HCI 04, Glasgow, GB*, Lancaster: http://www.mguides.info, 2004.

[64] COBCROFT, R.; TOWERS, S.; SMITH, J.; BRUNS, A.: Mobile learning in review: Opportunities and challenges for learners, teachers, and institutions. In: *Proc. of the Online Learning and Teaching Conference (OLT2006), Brisbane, Australia*, pp. 21–30, Brisbane, Australia: Queensland University of Technology, 2006.

[65] CONROY, R.: *Spatial Navigation in Immersive Virtual Environments*. Ph.D. thesis, Department of Architecture, University College London, 2001.

[66] COORS, V.; GJESDAL, O.; SULEBAK, J.; LAAKSO, K.: 3D Maps for Boat Tourists. In: *Proc. of Conf. for the IT and Travel & Tourism Industries (ENTER 2003), Helsinki, Finland*, 2003.

[67] COPPOLA, N.: Greening the Technological Curriculum: A Model for Environmental Literacy. *Journal of technology Studies* VXXVI(1), Summer-Fall (1999), pp. 39–46.

[68] COPUS: To Know Science is to Love it? Observations on Public understanding of Science Research. Committee on the public understanding of science of the royal society, British Association for the Advancement of Science, London, 1995.

[69] CORNELL, J. B.: *Sharing Nature with Children II*. Nevada City: Dawn Publications, 1989.

[70] COWELL, A.; TANASSE, T. E.; STANNEY, K.: Using Anthropomorphic Embodied Conversational Agents in Mobile Guides and Information Appliances. In: *Proc. of 3rd Workshop HCI in Mobile Guides in conjunction with Mobile HCI 2003, Udine, Italy*, 2003.

[71] COX, S.; DAISEY, P.; LAKE, R.; PORTELE, C.; WHITESIDE, A.: OpenGIS - Geography Markup Language (GML) Encoding Specification. http://www.opengeospatial.org/standards/gml, Open Geospatial Consortium, Inc (OGC), 2002.

[72] CREWS, T.; BISWAS, G.; GOLDMAN, S.; BRANSFORD, J.: Anchored Interactive Learning Environments. *International Journal of Artificial Intelligence in Education* 8 (1997), pp. 142–178.

[73] DALLWITZ, M.; PAINE, T.; ZURCHER, E.: Interactive identification using the Internet. Technical report 70, EEA, Copenhagen, 2002.

[74] DANIELS, M.; MARION, J.: Communicating Leave No Trace Ethics and Practices: Efficacy of Two-Day Trainer Courses. *Journal of Park and Recreation Administration* 23 (2005) 4, pp. 1–19.

[75] DARKEN, R. P.; ALLARD, T.; ACHILLE, L.: Spatial Orientation and Wayfinding in Large-Scale Virtual Spaces: An Introduction. *Presence* 7 (1998) 2, pp. 101–107.

[76] DAUDI, S.; HEIMLICH, J.: Environmental Education: can it be defined? EETAP Resource Library 14, Environmental Education and Training Partnership, Stevens Point, 1997.

[77] DAVIES, N.; CHEVERST, K.; MITCHELL, K.; EFRAT, A.: Using and Determining Location in a Context-Sensitive Tour Guide. *IEEE Computer: Special Issues on Location Aware Computing* 34 (2001) 8, pp. 35–41.

[78] DAVIS, M.: A multidimensional approach to individual differences in empathy. *JSAS Catalog of Selected Documents in Psychology* 10 (1980) 85.

[79] DELAY, R.: Forming knowledge: Constructivist learning and experiential education. *Journal of Experiential Education* 19 (1996) 2, pp. 76–81.

[80] DEVLIN, A.; BERNSTEIN, J.: Interactive Wayfinding: Use of Cues by Men and Women. *Journal of Environmental Psychology* 18 (1995), pp. 23–38.

[81] DEY, A.; ABOWD, G.: Towards a Better Understanding of Context and Context-Awareness. In: *Proc. of Workshop on The What, Who, Where, When, and How of Context-Awareness in conjunction with CHI 2000, The Hague, The Netherlands*, New York: ACM, 1999.

[82] DEY, A.; ABOWD, G.; SALBER, D.: A Context-Based Infrastructure for Smart Environments. In: *Proc. of 1st Int. Workshop on Managing Interactions in Smart Environments (MANSE 99), Dublin, Ireland*, pp. 114–128, 1999.

[83] DIAMANTIS, D.: The Concept of Ecotourism: Evolution and Trends. *Current Issues in Tourism* 2 (1999) 2&3, pp. 93–122.

[84] DIAS, E.; BEINAT, E.; RHIN, C.; HALLER, R.; SCHOLTEN, H.: Adding Value and Improving Processes Using Location-Based Services in Protected Areas. In: *Series Research on Computing Science (e-Environment)* (PRASTACO, P.; CORTÉS, U.; DÍAZ DE LEÓN, J.-L.; MURILLO, M., Eds.), Mexico: Instituto Politécnico Nacional, 2004.

[85] DIAS, E.; BEINAT, E.; RHIN, C.; SCHOLTEN, H.: Location Aware ICT in Addressing Protected Areas Goals. In: *Series Research on Computing Science (e-Environment)* (PRASTACO, P.; CORTÉS, U.; DÍAZ DE LEÓN, J.-L.; MURILLO, M., Eds.), Mexico: Instituto Politécnico Nacional, 2004.

[86] DIAS, E.; BEINAT, E.; SCHOLTEN, H.: Effects of Mobile Information Sharing in Natural Parks. In: *Proc. of 18th Int. Conf. Informatics for Environmental Protection - Sh@ring (EnviroInfo 2004), Geneva, Switzerland*, Geneva: Tricorne Editions, 2004.

[87] DICKINGER, A.; HEINZMANN, P.; MURPHY, J.: Mobile Environmental Applications. In: *Proc. of 38th Hawaii Int. Conf. on System Sciences, Maui, Hawaii*, Washington: IEEE, 2005.

[88] DILLEMUTH, J.: Map Design for Mobile Display. In: *Proc. of American Congress on Surveying and Mapping (AutoCarto 2005), Las Vegas, USA*, 2005.

[89] DIX, A.; RODDEN, T.; DAVIES, N.; TREVOR, J.; FRIDAY, A.; PALFREYMAN, K.: Exploiting space and location as a design framework for interactive mobile systems. *ACM Transactions on Computer-Human Interaction (TOCHI)* 7 (2000) 3, pp. 285–321.

[90] DONLON, J.; FORBUS, K.: Using a Geographic Information System for Qualitative Spatial Reasoning Trafficability. In: *Proc. of 13th Int. workshop on qualitative reasoning (QR 1999), Loch Awe, Scotland*, pp. 6–9, 1999.

[91] DUNLOP, M.; PTASINSKI, P.; MORRISON, A.; MCCALLUM, S.; RISBEY, C.; STEWART, F.: Design and development of Taeneb City Guide - From Paper Maps and Guidebooks to Electronic Guides. In: *Proc. of Int. Conf. Information and Communication Technologies in Tourism - (ENTER 2004), Cairo, Egypt* (FREW, A. J., Ed.), Berlin: Springer-Verlag, 2004.

[92] DÜPMEIER, C.; GEIGER, W.; EBEL, R.; SCHMID, E.; LINNENBACH, M.: Theme park soil-geology-nature: a case study for the evaluation of new concepts of EIS for the general public. In: *Proc.of 19th Int.Conf. Informatics for Environmental Protection (EnviroInfo 2005), Brno, Czech Republic* (HREBICEK, J., Ed.), Brno: Masaryk University, 2005.

[93] DÜPMEIER, C.; GEIGER, W.; GRECEANU, C.; WEIDEMANN, R.; RUCHTER, M.; EBEL, R.; LEHLE, M.; LINNENBACH, M.: Themenpark Umwelt. Integration von anderen Informationssystemen in das Internetportal Themenpark Umwelt. Wissenschaftliche Berichte, FZKA-7250 Phase I 2005/06, Forschungszentrum Karlsruhe GmbH, Karlsruhe, 2006.

[94] DÜPMEIER, C.; RUCHTER, M.: Mobile Guide Interface Design for exploring the Wild. In: *Proc. of 3rd Workshop HCI in Mobile Guides at Mobile HCI 04, Glasgow, GB*, 2004.

[95] EDWARDES, A.; BURGHARDT, D.; DIAS, E.; PURVES, R.; WEIBEL, R.: Geo-Enabling Spatially Relevant Data for Mobile Information Use and Visualization. *Lecture Notes in Computer Science* 3833 (2005), pp. 78–92.

[96] EDWARDES, A.; BURGHARDT, D.; WEIBEL, R.: WebPark - Location Based Services for Species Search in Recreation Area. In: *Proc. of 21st Int. Cartographic Conf. (ICC 2003), Durban, South Africa*, pp. 10–16, Durban: International Cartographic Association (ICA), 2003.

[97] EEA: A New Model of Environmental Communication for Europe from Consumption to Use of Information. Expert corner report, European Environment Agency, Copenhagen, 1998.

[98] EISENHAUER, M.; KLEMKE, R.: Contextualization in Nomadic Computing. European Research Consortium for Informatics and Mathematics News 47, ERCIM, Bern, 2001.

[99] EISENHAUER, M.; LORENZ, A.; ZIMMERMANN, A.; DUONG, T.; JAMES, F.: Interaction by Movement - One Giant Leap for Natural Interaction in Mobile Guides. In: *Proc. of 4th Workshop HCI in Mobile Guides at Mobile HCI 05, Salzburg, Austria*, 2005.

[100] EISENHAUER, M.; OPPERMANN, R.; SCHMIDT-BELZ, B.: Mobile information systems for all. In: *Proc. of 10th Int. Conf. on Human - Computer Interaction, Heraklion 2003, Crete, Greece*, 2003.

[101] ERICKSON, T.: Working with interface metaphors. In: *The art of human-computer interface design* (LAUREL, B., Ed.), pp. 65–73, Reading, MA: Addison-Wesley, 1990.

[102] FENNELL, D.: A Content Analysis of Ecotourism Definitions. *Current Issues in Tourism* 4 (2001) 5.

[103] FERRAIOLO, J.; FUJISAWA, J.; JACKSON, D.: Scalable Vector Graphics (SVG) 1.1 Specification. http://www.w3.org/tr/svg11/, W3C, Sophia-Antipolis Cedex, Fujisawa, Cambridge, 2003.

[104] FISCHER-STABEL, P.: *Umweltinformationssysteme*, Heidelberg: Herbert Wichmann, 2005.

[105] FISHER, S.: Environmental Media: Accessing Virtual Representations of Real-time Sensor Data and Site-specific Annotations Embedded in Physical Environments. In: *Proc. of the 7th Int. Conf. on Virtual Systems and Multimedia 2001, Berkeley, USA*, Washington: IEEE, 2001.

[106] FLECK, M.; FRID, M.; KINDBERG, T.; O'BRIEN-STRAIN, E.; RAJANI, R.; SPASOJEVIC, M.: From Informing to Remembering: Ubiquitous Systems in Interactive Museums. *IEEE Pervasive Computing Magazine* 1 (2002) 2, pp. 13–21.

[107] FRANSSON, N.; GÄRLING, T.: Environmental Concern: Conceptual Definitions, Measurement Methods, and Research Findings. *Journal of Environmental Psychology* 19 (1999), pp. 369–382.

[108] FRECH, I.; KOCH, B.: Multimedia Geoinformation in Rural Areas with Eco-Tourism: the ReGeo-System. In: *Proc. of Int. Conf. Information and Communication Technologies in Tourism (ENTER 2003), Helsinki, Finland* (FREW, A.; O'CONNOR, P.; HITZ, M., Eds.), Wien: Springer-Verlag, 2003.

[109] FRICK, J.: *Umweltbezogenes Wissen: Struktur, Einstellungsrelevanz und Verhaltenswirksamkeit.* Ph.D. thesis, Faculty of Philosophy, University of Zürich, 2003.

[110] FRICK, J.; KAISER, F.; WILSON, M.: Environmental knowledge and conservation behavior: exploring prevalence and structure in a representative sample. *Personality and Individual Differences* 37 (2004), pp. 1597–1613.

[111] FRITZ, R.: Mobil in Wald und Flur. *GeoBIT* 7 (2003), pp. 13–15.

[112] FUCHS, M.: Children and Juveniles as Respondents: Experiments on Question Order, Response Order and Scale Effects. In: *Proc. of International Conference on Improving Surveys, Copenhagen, Denmark*, Copenhagen: SFI-SURVEY, 2002.

[113] GARTNER, G.; UHLIRZ, S.: Cartographic Concepts for Realizing a location-based UMTS Service: Vienna City Guide Lol@. In: *Proc. of 20th Int. Cartographic Conf. (ICC 2001), Beijing, China*, 2001.

[114] GASTON, K.; O'NEILL, M.: Automated species identification: why not? *Philosophical Transactions of the Royal Society B: Biological Sciences* 359 (2004), pp. 655–667.

[115] GEIGER, W.; DÜPMEIER, C.; RUCHTER, M.; SCHLACHTER, T.; WEIDEMANN, R.; EBEL, R.; SCHMID, E.: Examples of Public Environmental Information Systems and Portals in Baden-Württemberg. In: *Proc. of 6th Int. Symp. on Environmental Software Systems (ISESS 2005), Sesimbra, Portugal*, Laxenburg: IFIP, 2005.

[116] GEIGER, W.; DÜPMEIER, C.; RUCHTER, M.; SCHLACHTER, T.; WEIDEMANN, R.; MAYER-FÖLL, R.; EBEL, R.; LINNENBACH, M.: Experience gained from the use of environmental information systems for the public in the state of Baden-Württemberg. In: *Proc.of the iEMS 3rd Biennial Meeting (Summit on Environmental Modelling and Software 2006), Burlington, Vt., USA* (VOINOV, A., Ed.), Burlington: Environmental Modelling and Software Society, 2006.

[117] GOODMAN, J.; GRAY, P.; KHAMMAMPAD, K.; BREWSTER, S.: Using Landmarks to Support Older People in Navigation. In: *Proc. of 6th Int. Symposium on Human-Computer Interaction with Mobile Devices (Mobile HCI 2004), Glasgow, UK* (BREWSTER, S.; DUNLOP, M., Eds.), Berlin: Springer-Verlag, 2004.

[118] GOTCH, C.; HALL, T.: Understanding nature-related behaviors among children through a theory of reasoned action approach. *Environmental Education Research* 10 (2004) 2, pp. 157–177.

[119] GOUGH, N.: Keynote: Effective sustainability education: What works? Why? Where Next? In: *Proc. of Linking Research and Practice Conf., Sydney, Australia*, NSW EPA, 2004.

[120] GRAW, M.; BORCHARDT, D.: *Ein Bach ist mehr als Wasser: Materialien für einen fächerverbindenden, projektorientierten Unterricht zum Thema Ökologie und Schutz von Fließgewässern*. Hessisches Ministerium für Umwelt, Landwirtschaft und Forsten, 1999.

[121] GREENWALD, A.: Within-Subjects Designs: To Use or Not To Use? *Psychological Bulletin* 83 (1976) 2, pp. 314–320.

[122] GRISWOLD, W.; BOYER, R.; BROWN, S.; TRUONG, T.; BHASKER, E.; JAY, G.; SHAPIRO, R.: ActiveCampus - Sustaining Educational Communities through Mobile Technology. *Computer Science and Engineering* Jul. (2002).

[123] GUNDERSON, K.; BARNS, C. V.; HENDRICKS, W.; MCAVOY, L.: Wilderness Education: An Updated Review of the Literature and New Directions for Research and Practice. In: *Proc. of Wilderness Science in a Time of Change Conf. 2000, Missoula, MT, USA* (COLE, D. N.; MCCOOL, S. F.; BORRIE, W. T.; OLOUGHLIN, J., Eds.), Ogden: USDA Forest Service, 2000.

[124] GÜNTHER, O.: *Environmental Information Systems*. Berlin: Springer-Verlag, 1998.

[125] HAID, E.; KIECHLE, G.; LEITINGER, S.: Multimediale Beschreibung geo-referenzierter touristischer Objects of Interest. In: *Proc. of CORP 2005 & Geomultimedia05, Vienna, Austria* (SCHRENK, M., Ed.), Wien: Inst. EDV Methoden in Architektur/Raumplanung, 2005.

[126] HAKLAY, M.: From Environmental Information Systems to Environmental Informatics - Evolution and Meaning. CASA working paper 7, Department of Geography, University College London, London, 1999.

[127] HAKLAY, M.: Conceptual Models of Urban Environmental Information Systems - Toward Improved Information Provision. CASA working paper 38, Department of Geography, University College London, London, 2001.

[128] HAKLAY, M.: *Public Environmental Information Systems: Challenges and Perspectives*. Ph.D. thesis, Department of Geography, University College London, 2002.

[129] HAM, S.: *Environmental Interpretation -A Practical for People with Big Ideas and Small Budgets*. Colorado: Fulcrum Publishing, 1992.

[130] HAM, S.; WEILER, B.: 100,000 Beating Bird Hearts: Tourism, Wildlife and Interpretation. In: *Proc. of Sustainable Wildlife Tourism Convention, Hobart, Tasmania*, p. 36, Gold Coast: CRC For Sustainable Tourism, 2001.

[131] HAMACHER, W.; HOLLENBACH, W.; HORSTMANN, K.; SCHOENMAKERS, J.; SCHREIBER, R.; WEILENMANN, M.: Umweltkommunikation in der TZ - Dokumentation eines Fachgesprächs. Techrep., Deutsche Gesellschaft für Technische Zusammenarbeit (GTZ) GmbH (Ed.), Universum Verlagsanstalt, Wiesbaden, 1996.

[132] HAMPE, M.; ELIAS, B.: Integrating topographic information and landmarks for mobile applications. In: *Proc. of 2nd Symposium on Location Based Services and TeleCartography 2004, Vienna, Austria, Geowissenschaftliche Mitteilungen No.66* (GARTNER, G., Ed.), Wien: TU Wien, 2003.

[133] HARRIE, L.; JOHANSSON, M.: Real-time data generalization and integration using Java. *Geoforum Perspective* Febr. (2003), pp. 29–34.

[134] HASSAN, R.; SCHOLES, R.; ASH, N.: Ecosystems and Human Well-being: Current State and Trends. In: *Millennium Ecosystem Assessment*, pp. 4–47, Washington: IslandPress, 2005.

[135] HEDE, A.: An integrated model of multimedia effects on learning. *Journal of Educational Multimedia and Hypermedia* 11 (2002), pp. 177–191.

[136] HEIDMANN, F.; HERMANN, F.; PEISSNER, M.: Interactive Maps On Mobile, Location-Based Systems: Design Solutions and Usability Testing. In: *Proc. of 21st Int. Cartographic Conf. (ICC 2003), Durban, South Africa*, Durban, Coruna: Interantional Cartographic Association, 2003.

[137] HEIMLICH, J.; DAUDI, S.; CUMMONS, H.: Thesaurus of Environmental Education Terms. EETAP Resource Library, 1999.

[138] HELLBRÜCK, J.; FISCHER, M.: *Umweltpsychologie: Ein Lehrbuch*. Göttingen: Hogrefe, 1999.

[139] HENRICKSEN, K.; INDULSKA, J.; RAKOTONIRAINY, A.: Modeling Context Information in Pervasive Computing Systems. In: *Proc. of 1st Int. Conf. on Pervasive Computing (Pervasive 2002), Zürich, Switzerland, LNCS 2414* (MATTERN, F.; NAGHSHINEH, M., Eds.), Berlin: Springer-Verlag, 2002.

[140] HERMANN, F.; HEIDMANN, F.: User Requirement Analysis and Interface Conception for a Mobile, Location-Based Fair Guide. In: *Proc. of 4th Int. Symposium on Human-Computer Interaction with Mobile Devices (Mobile HCI 2002), Pisa, Italy* (PATERNÒ, F., Ed.), Berlin: Springer-Verlag, 2002.

[141] HERRMANN, T.; SCHWEIZER, K.; JANZEN, G.; KATZ, S.: Routen- und Überblickswissen - konzeptuelle Überlegungen. *Kognitionswissenschaft* 7 (1998), pp. 145–159.

[142] HIGHTOWER, J.; BORRIELLO, G.: Location Systems for Ubiquitous Computing. *IEEE Computer: Special Issues on Location Aware Computing* 34 (2001) 8, pp. 57–66.

[143] HÖÖK, K.: Social Navigation: From The Web to the Mobile. In: *Proc. of Mensch & Computer 2003, Stuttgart, Germany* (SZWILLUS, G.; ZIEGLER, J., Eds.), Stuttgart: B.G. Teubner, 2003.

[144] HUNECKE, M.; BLÖBAUM, A.; MATTHIES, E.; HÖGER, R.: Responsibility and environment: Ecological norm orientation and external factors in the domain of travel mode choice behavior. *Environment and Behavior* 33 (2001) 6, pp. 830–852.

[145] HUNTER, K.: *Affective Empathy with Children: Measurement and empathy*. Ph.D. thesis, School of Applied Psychology, Griffith University, 2003.

[146] HUNTER, M.: *Wildlife, forests, and forestry: Principles of managing forests for biological diversity*. Toronto: Prentice-Hall Canada Inc., 1990.

[147] HURRELMANN, K.; ALBERT, M.: *Jugend 2002, 14. Shell-Jugendstudie*. Berlin: Verlag Ullstein, 2002.

[148] HWANG, Y.-H.; KIM, S.-I.; JENG, J.: Examining the Causal Relationships among Selected Antecedents of Responsible Environmental Behavior. *Journal of Environmental Education* 31 (2000) 4, pp. 19–25.

[149] HYDE, J.: *Multi-Modal Usability Evaluation*. Ph.D. thesis, School of Computing Science, Middlesex University, London, UK, 2002.

[150] ISBISTER, K.; DOYLE, P.: Touring Machines: Guide Agents for Sharing Stories about Digital Places. In: *Working Notes of the 1999 AAAI Fall Symposium on Narrative Intelligence*, Menlo Park, CA: AAAI, 1999.

[151] ISBISTER, K.; DOYLE, P.: Design and Evaluation of Embodied Conversational Agents: A Proposed Taxonomy. In: *Proc. of 1st Int. Joint Conf. on Autonomous Agents and Multi-Agent Systems, Bologna, Italy*, New York: ACM, 2002.

[152] ISO/IEC: Information technology – User system interfaces and symbols – Icon symbols and functions. ISO/IEC 11581-1:2000 - Part 1: Icons – General Part 1, International Standards Organization, Geneva , Switzerland, 2005.

[153] JELITTO, M.: Einsatzmöglichkeiten digitaler Medien in der Umweltbildung – Ein Überblick. *DGU Nachrichten* 21 (2000) Mai, pp. 16–20.

[154] JOHNSON, D.: Student-Student Interaction: The Neglected Variable in Education. *Educational Researcher* 10 (1981) 1, pp. 5–10.

[155] JOHNSON, W.; LABORE, C.; CHIU, Y.-C.: A Pedagogical Agent for Psychosocial Intervention on a Handheld. In: *Proc. of AAAI Fall Symposium on Dialogue Systems for Health Communication 2004, Washington, DC, USA*, Menlo Park, CA: AAAI, 2004.

[156] JOHNSON, W.; RICKEL, J.; LESTER, J.: Animated Pedagogical Agents: Face-to-Face Interaction in Interactive Learning Environments. *International Journal of Artificial Intelligence in Education* 11 (2000), pp. 47–78.

[157] JONES, M.; GAY, G.; ARMS, W.; LAZOGE, C.; MELIN, E.; WENDERLICH, R.: Integrating User Requirements into Mobile Information Systems Design: Findings and Challenges. Hci group, Cornell University, Ithaca, NY, 2000.

[158] KAHN, P.: Developmental Psychology and the Biophilia Hypothesis: Children's Affiliation with Nature. *Developmental Review* 17 (1997), pp. 1–61.

[159] KAISER, F.; SHIMODA, T.: Responsibility as Predictor of Ecological Behavior. *Journal of Environmental Psychology* 19 (1999), pp. 243–253.

[160] KAISER, F.; WÖLFING, S.; FUHRER, U.: Environmental Attitude and Ecological Behavior. *Journal of Environmental Psychology* 19 (1999), pp. 1–19.

[161] KALIA, A.; STANKIEWICZ, B.: Acquisition of Object and Structural Landmark Knowledge in Virtual Environment Navigation. In: *Proc. of 10th Annual Workshop on Object Perception, and Memory, Kansas City, MO, USA*, p. 8, http://www.opam.net, 2002.

[162] KALMAN, R.: On the general theory of control systems. *Automatic Control, IRE Transactions on* 4 (1959) 3, pp. 110–110.

[163] KALS, E.; SCHUMACHER, D.; MONTADA, L.: Emotional Affinity toward Nature as a motivational Basis to protect Nature. *Environment and Behavior* 31 (1999) 2, pp. 178–202.

[164] KANEMITSU, H.; KAMADA, T.: POIX: Point Of Interest eXchange Language Specification. http://www.w3.org/tr/poix/, W3C, Sophia-Antipolis Cedex, Fujisawa, Cambridge, 1999.

[165] KARATZAS, K.; MASOURAS, A.; KAPRARA, A.; BASSOUKOS, A.; PAPAIOANNOU, I.; SLINI, T.; MOUSSIOPOULOS, N.: Environmental Information Systems and the Concept of Environmental Informatics. In: *Environmental Online Communication* (SCHARL, A., Ed.), pp. 3–10, Tokyo: Springer, 2004.

[166] KASAPOGLU, M.; ECEVIT, M.: Attitudes and Behavior Toward the Environment: The Case of Lake Burdur in Turkey. *Environment and Behavior* 34 (2002) 3, pp. 363–377.

[167] KELLY, L.; SULLIVAN, T.: Front-end evaluation - beyond the Field of Dreams. *Museum National* 5 (1997) 4, pp. 7–8.

[168] KIM, T. J.; YOU, J.; LEE, J.; JANG, S.-G.; OH, S.: Enabling Technologies for Location-Based Services and Development of Interface Standards for Multi-modal LBS. EPIL Discussion Paper Series 02-1030 WD-1, Organization for Standards, Geneva, 2002.

[169] KINDT, S.: Die Zukunft ist digital, unsere Umwelt auch? Kritische Fragen zur Nutzung von Multimedia in der Umweltbildung. *DGU Nachrichten* 21 (2000), pp. 32–36.

[170] KITCHIN, R.: Cognitive Maps: What Are They and Why Study Them? *Journal of Environmental Psychology* 14 (1994), pp. 1–19.

[171] KJELDSKOV, J.; GRAHAM, C.: A Review of MobileHCI Research Methods. In: *Proc. of 5th Int. Symposium on Human-Computer Interaction with Mobile Devices (Mobile HCI 2003), Udine, Italy, LNCS*, Berlin: Springer-Verlag, 2003.

[172] KJELDSKOV, J.; GRAHAM, C.; PEDELL, S.; VETERE, F.; HOWARD, S.; BALBO, S.; DAVIES, N.: Evaluating the Usability of a Mobile Guide: The Influence of Location, Participants and Resources. In: *Proc. of 7th Int. Conf. on Human Computer Interaction with Mobile Devices and Services (Mobile HCI 2005), Salzburg, Austria*, 2005.

[173] KLIPPEL, A.; WINTER, S.: Structural salience of landmarks for route directions. In: *Proc. of Int. Conf. on Spatial Information Theory (COSIT 2005), Ellicottville, NY, USA, LNCS* (COHN, A. G.; MARK, D. M., Eds.), pp. 347–362, Berlin: Springer-Verlag, 2005.

[174] KLOPFER, E.; SQUIRE, K.; JENKINS, H.: Environmental Detectives: PDAs as a window into a virtual simulated world. In: *Didaktik der Notebook-Universität (Reihe Medien in der Wissenschaft)* (KERRES, M.; KALZ, M.; STRATMANN, J.; DE WITT, C., Eds.), pp. 259–274, Münster: Waxmann Verlag, 2004.

[175] KNAPP, D.; POFF, R.: A Qualitative Analysis of the Immediate and Short-term Impact of an Environmental Interpretive Program. *Environmental Education Research* 7 (2001) 1, pp. 55–65.

[176] KOCH, F.; RAHWAN, I.: Classification of Agents-based Mobile Assistants. In: *Proc. of 3rd Int. Joint Conf. On Autonomous Agents & Multi Agent Systems (AAMAS 2004) - Workshop on Agents for Ubiquitous Computing (UbiAgents), New York, USA*, IEEE Computer Society, 2004.

[177] KOKKONEN, P.; PELTONEN, A.: Mapping Lakelands: Challenges of Map Design for Tourism. In: *Proc. of 19th Int. Cartographic Conf., Ottawa, Canada*, Int. Cartographic Association, 1999.

[178] KOLLMUSS, A.; AGYEMAN, J.: Mind the Gap: why do people act environmentally and what are the barriers to pro-environmental behavior? *Environmental Education Research* 8 (2002) 3, pp. 239–260.

[179] KORN, R.: Studying your Visitors: Where to begin. *History News* 49 (1994) 2.

[180] KRASNER, G.; POPE, S.: A Cookbook for Using the Model-View Controller User Interface Paradigm in Smalltalk-80. *Journal of Object-Oriented Programming* 1 (1988) 3, pp. 26–49.

[181] KRAY, C.: *Situated Interaction on Spatial Topics*. Ph.D. thesis, DISKI series 274, Akademische Verlagsgesellschaft Aka GmbH, 2003.

[182] KRAY, C.; BAUS, J.: A Survey of Mobile Guides. In: *Proc. of 3rd Workshop HCI in Mobile Guides in conjunction with Mobile HCI 2003, Udine, Italy*, 2003.

[183] KRAY, C.; BAUS, J.; KRÜGER, A.: Positionsinformation und Navigationsaufgaben. In: *Geoinformation mobil* (STROBL, J.; ZIPF, A., Eds.), pp. 98–108, Heidelberg: Hüthig Verlag, 2002.

[184] KREFT-BURMAN, K.: Raising environmental awareness in the Baltic Sea area: results and experience gained from the SPA Project. *Int. Journal of Environment and Sustainable Development* 1 (2002) 1, pp. 88–96.

[185] KRENN, B.; NEUMAYR, B.; GSTREIN, E.; GRICE, M.: Life-like agents for the internet: a cross-cultural case study. In: *Agent Culture: Designing Human-Agent Interaction in a Multicultural World* (PAYR, S.; TRAPPL, R., Eds.), London: Lawrence Erlbaum Associates, 2004.

[186] KRISTOFFERSEN, S.; LJUNGBERG, F.: Designing Interaction Styles for a Mobile Use Context. In: *Proc. of Int. Symposium on Handheld and Ubiquitous Computing (HUC 99), Karlsruhe, Germany*, New York: ACM, 1999.

[187] KRÖSCHE, J.; BALDZER, J.; BOLL, S.: MobiDENK Mobile Multimedia in Monument Conservation. *IEEE Computer and Society* April/June (2004), pp. 72–77.

[188] KRUG, K.; MOUNTAIN, D.; PHAN, D.: WebPark - Location-Based Services For Mobile Users in Protected Areas. *GeoInformatics* 6 (2003) Mar.

[189] KRÜGER, A.; BUTZ, A.; MÜLLER, B.; STAHL, C.; WASINGER, R.; STEINBERG, K.-E.; DIRSCHL, A.: The Connected User Interface: Realizing a Personal Situated Navigation Service. In: *Proc. of Int. Conf. on Intelligent User Interface (IUI 2004), Funchal, Madeira, Portugal*, New York: ACM, 2004.

[190] KRUPPA, M.; KRÜGER, A.; ROCCHI, C.; STOCK, O.; ZANCANARO, M.: Seamless Personalized TV-Like Presentations on Mobile and Stationary Devices in a Museum. In: *Proc. of Int. Cultural Heritage Informatics Meeting (ICHIM 2003), Paris, France*, 2003.

[191] KÜHN, P.; ROTHERMEL, K.: Editorial: Orts- und kontextbezogene Dienste. *Informationstechnik und Technische Informatik* 5 (2002), pp. 239–241.

[192] KULL, K.; TOROP, P.: Biotranslation: Translation between umwelten. In: *Translation Translation* (PETRILLI, S., Ed.), pp. 313–328, Amsterdam: Rodopi, 2003.

[193] KUTTER, S.; SPÄTH, V.: *Rheinauen - Bedrohtes Paradies am Oberrhein*. Karlsruhe: G.Braunsche Hofbuchdruckerei und Verlag, 1993.

[194] LANE, M.: Environmentally Responsible Behavior - Does it Really Matter What We Believe? *Planning Forum* 6 (2000), pp. 33–39.

[195] LEEMING, F.; DWYER, W.; PORTER, B.; COBERN, M.: Outcome Research in Environmental Education: A Critical Review. *Journal of Environmental Education* 24 (1993) 4, pp. 8–21.

[196] LEEMING, F.; PORTER, B.; DWYER, W.; COBERN, M.; OLIVER, D.: Effects of Participation in Class Activities on Childrens Environmental Attitudes and Knowledge. *Journal of Environmental Education* 28 (1997) 2, pp. 33–42.

[197] LEHIKOINEN, J.: *Interacting with Wearable Computers: Techniques and their Application in Wayfinding Using Digital Maps*. Ph.D. thesis, Department of Computer and Information Science, University of Tampere, 2002.

[198] LÖHNERT, E.; WITTMANN, E.; PIELMEIER, J.; SAYDA, F.: PARAMOUNT-Public Safety & Commercial Info-Mobility Applications & Services in the Mountains. In: *Proc. of 14th Int. Technical Meeting of the Satellite Division of the Institute of Navigation (ION GPS 2001), Salt Lake City, UT,*, Fairfax: ION Publications department, 2001.

[199] LUDWIG, T.: Einführung in die Naturinterpretation. *Mitteilungen der NNA* 1 (2003), pp. 20–27.

[200] LUYTEN, K.; CONINX, K.: ImogI: Take Control over a Context Aware Electronic Mobile Guide for Museums. In: *Proc. of 3rd Workshop HCI in Mobile Guides at Mobile HCI 04, Glasgow, GB*, 2004.

[201] MAGON, A.; SHUKLA, R.: LBS, the ingredients and the alternatives. Techrep., GISdevelopment.net, Noida, 2001.

[202] MAINWARING, A.; POLASTRE, J.; SZEWCZYK, R.; CULLER, D.; ANDERSON, J.: Wireless Sensor Networks for Habitat Monitoring. In: *Proc. of 1st ACM Int. Workshop on Wireless Sensor Networks and Applications (WSNA 2002), Atlanta, Georgia, USA*, New York: ACM, 2002.

[203] MALAKA, R.; ZIPF, A.: DEEP MAP Challenging IT research in the framework of a tourist info.sys. In: *Proc. of 7th. Int. Conf. Information & communication technologies in tourism (ENTER 2000), Barcelona, Spain* (FESENMAIER, D. R.; KLEIN, S.; BUHALIS, D., Eds.), Berlin: Springer-Verlag, 2000.

[204] MALINOWSKI, J.; GILLESPIE, W.: Individual differences in performance on a large-scale, real-world wayfinding task. *Journal of Environmental Psychology* 21 (2001), pp. 73–82.

[205] MANNING, R.: Emerging Principles for Using Information/Education in Wilderness Management. *International Journal of Wilderness* 1 (2003) 9.

[206] MARCUS, A.: Babyface Design for Mobile Devices and the Web. In: *Proc. of Human-Computer Interface Int. Conf. (HCII 2001), New Orleans, LA, USA*, Mahwah: Lawrence Eribaum Associates, 2001.

[207] MARENTAKIS, G.; BREWSTER, S.: A Comparison of Feedback Cues for Enhancing Pointing Efficiency in Interaction with Spatial Audio Displays. In: *Proc. of 7th Int. Conf. on Human Computer Interaction with Mobile Devices and Services (Mobile HCI 2005), Salzburg, Austria*, New York: ACM, 2005.

[208] MARK, D.; FREKSA, C.; HIRTLE, S.; LLOYD, R.; TVERSKY, B.: Cognitive models of geographical space. *Int. Journal of Geographical Information Science* 13 (1999) 8, pp. 747–774.

[209] MARMASSE, N.; SCHMANDT, C.: Location-aware information delivery with comMotion. In: *Proc. of 2nd Int. Symposium on Handheld and Ubiquitous Computing (HUC 2000), Bristol, UK*, Berlin: Springer-Verlag, 2000.

[210] MARTENS, T.: *Kognitive und Affektive Bedingungen von Umwelthandeln*. Ph.D. thesis, Philospohische Fakultät, Christian-Albrechts-Universität Kiel, 1999.

[211] MARTIN, J.: Mobile Computing: Palm vs. Pocket PC: The products are different, but also the same. How to choose? PC World, 2003.

[212] MARYNOWSKI, S.; JACOBSON, S.: Ecosystem management education for public lands. *Wildlife Society Bulletin* 27 (1999) 1, pp. 134–145.

[213] MAYER-FÖLL, R.; KEITEL, A.; GEIGER, W.: Projekt AJA - Anwendung JAVA-basierter und anderer leistungsfähiger Lösungen in den Bereichen Umwelt, Verkehr und Verwaltung - Phase V 2004. Wissenschaftliche berichte, fzka-7077, Forschungszentrum Karlsruhe GmbH, Karlsruhe, 2004.

[214] MAYRING, P.: Qualitative Inhaltsanalyse. *Forum: Qualitative Social Research* 2 (2000), p. 10.

[215] MCDUFF, M.: Needs Assessment for Participatory Evaluation of Environmental Education Programs. *Applied Environmental Education and Communication* 1 (2002), pp. 25–36.

[216] MENDELSON, Y.; FLOROFF, V.; JOHNSTON, W.; BRANCHE, P.: A Mobile PDA-based Wireless Pulse Oximeter. In: *Proc. of TELEHEALTH 2005, Banff, Canada* (JENNETT, P.; RANGAYYAN, R. M., Eds.), Calgary: ACTA Press, 2005.

[217] MENG, L.: About egocentric geovisualization. In: *Proc. of 12th Int. Conf. on Geoinformatics, University of Gävle, Sweden*, Sweden: University of Gavle, 2004.

[218] MENG, L.; REICHENBACHER, T.: Map-based mobile services. In: *Map-based mobile services - Theories, Methods and Implementations* (ZIPF, A.; MENG, L.; REICHENBACHER, T., Eds.), Berlin Heidelberg New York: Springer Geosciences. Springer-Verlag, 2005.

[219] MEYERS, R.: *A Heuristic for Environmental Values and Ethics, and a Psychometric Instrument to Measure Adult Environmental Ethics and Willingness to Protect the Environment.* Ph.D. thesis, Department of Natural Resources, Ohio State University, 2002.

[220] MICHELSEN, G.: Theoretische Diskussionsstränge der Umweltbildung. In: *Umweltbildung: Theoretische Konzepte - empirische Erkenntnisse - praktische Erfahrungen* (BEYERSDORF, M.; MICHELSEN, G.; SIEBERT, H., Eds.), pp. 61–65, Neuwied: Luchterhand, 1998.

[221] MICHELSEN, G.: Umweltbildung im internationalen Kontext. In: *Umweltbildung: Theoretische Konzepte - empirische Erkenntnisse - praktische Erfahrungen* (BEYERSDORF, M.; MICHELSEN, G.; SIEBERT, H., Eds.), pp. 27–41, Neuwied: Luchterhand, 1998.

[222] MICROSOFT-CORP.: Designing Web Sites for the Internet Explorer for Pocket PC. White paper, 2001.

[223] MILFONT, T.; DUCKITT, J.; CAMERON, L.: A Cross-Cultural Study of Environmental Motive Concerns and Their Implications for Proenvironmental Behavior. *Environment and Behavior* 38 (2006) 6, pp. 745–767.

[224] MILGRAM, P.; KISHINO, A.: A Taxonomy of Mixed Reality Visual Displays. *IEICE Transactions on Information and Systems* E77-D (1994) 12, pp. 1321–1329

[225] MOLNAR, L.; EBBY, D.; HOPP, M.: Developing Information Systems for the Driving Tourist: A Literature Review. Technical report umtri-96-11, University of Michigan, Ann Arbor, 1996.

[226] MOSCARDO, G.: Mindful visitors - Heritage and tourism. *Annals of Tourism Research* 23 (1996) 2, pp. 376–397.

[227] MOUNTAIN, D.: Information Anywhere: User-friendly Location Based Services. *GI News LBS supplement* April/May (2001), pp. 10–22.

[228] MOUNTAIN, D.; LIAROKAPIS, F.; RAPER, J.; BRUJIC-OKRETIC, V.: Interacting with Virtual Reality models on mobile devices. In: *Proc. of 4th Workshop HCI in Mobile Guides at Mobile HCI 05, Salzburg, Austria*, 2005.

[229] MOUNTAIN, D.; RAPER, J.: Location-based services in remote areas. In: *Proc. of AGI Conf. 2002, London, GB*, 2002.

[230] NAAEE: *Nonformal Environmental Education Programs: Guidelines for Excellence.* Washington: NAAEE, 2004.

[231] NADIN, M.: Interface design: A semiotic paradigm. *Semiotica* 69 (1988) 3/4, pp. 269–302.

[232] NAISMITH, L.; LONSDALE, P.; VAVOULA, G.; SHARPLES, M.: Literature Review in Mobile Technologies and Learning. Nesta Futurelab Series Report 11, Nesta Futurelabs, Bristol, 2004.

[233] NEUBERT, S.: Methodische Orientierung für kurze und praxisnahe Forschungsprojekte in Entwicklungsländern - Ein Leitfaden für Länderarbeitsgruppen und Gutachter. Techrep., Deutsches Institut für Entwicklungspolitik, Bonn, 2001.

[234] NEUERBURG, H.-J.; WILKEN, T.: Umweltkommunikation im Sport. In: *Schriftenreihe Sport und Umwelt*, Frankfurt: Deutscher Sportbund, 2002.

[235] NEUVILLE, Y.; LAFARGUE, F.: Icons in the eye of the beholder. *ISO BULLETIN* Nov. (2000), pp. 13–17.

[236] NIELSEN, C.; SØNDERGAARD, A.: Designing for mobility- an integration approach supporting multiple technologies. In: *Proc. of NordiCHI 2000, Stockholm, Sweden*, Stockholm: Royal Institute of Technology, 2000.

[237] NIELSEN, J.; YSSING, C.: What kind of information does an HCI expert want? - On concurrent usability testing. In: *Proc. of 4th Danish HCI Research Symposium 2004, Aalborg University, Denmark*, 2004.

[238] NIVALA, A.-M.; SARJAKOSKI, L.: An Approach to Intelligent Maps: Context Awareness. In: *Proc. of 3rd Workshop HCI in Mobile Guides in conjunction with Mobile HCI 2003, Udine, Italy*, 2003.

[239] NIVALA, A.-M.; SARJAKOSKI, L.: Preventing Interruptions in Mobile Map Reading Process by Personalization. In: *Proc. of 3rd Workshop HCI in Mobile Guides at Mobile HCI 04, Glasgow, GB*, 2004.

[240] NOTHEGGER, C.; WINTER, S.; RAUBAL, M.: Computation of the Salience of Features. *Spatial Cognition and Computation* 4 (2003) 2, pp. 113–136.

[241] NPS: *Planning For Interpretation and Visitor Experience*. Division of Interpretive Planning, Harpers Ferry Center, National Park Service, Washington D.C., 1998.

[242] OECD: Environmental Communication: Applying Communication Tools Towards Sustainable Development. Working paper of the working party on development cooperation and environment, OECD Publications, Paris, 1999.

[243] OKADA, M.; YAMADA, A.; YOSHIDA, M.; MORIYA, K.: DIGITAL-EE II: RV-Augmented Interface Design for networked collaborative Environmental Learning. In: *Proc. of the Int. Conf. on Computer Support for Collaborative Learning (CSCL 2003),*

Bergen, Norway (WASSON, B.; LUDVIGSEN, S.; HOPPE, U., Eds.), Dordrecht, Norwell: Kluwer Academic Publishers, 2003.

[244] OLLI, E.; GRENDSTAD, G.; WOLLEBAEK, D.: CORRELATES OF ENVIRONMENTAL BEHAVIORS Bringing Back Social Context. *Environment and Behavior* 33 (2001) 3, pp. 181–208.

[245] OPPERMANN, R.; SPECHT, M.: A Nomadic Information System for Adaptive Exhibition Guidance. In: *Proc. of Int. Cultural Heritage Informatics Meeting (ICHIM 1999), Washington, DC, USA*, Toronto: Archives & Museum Informatics, 1999.

[246] OSBORNE, J.; HENNESSY, S.: Literature Review in Science Education and the Role of ICT: Promise, Problems and Future Directions. Nesta Futurelab Series Report 6, Nesta Futurelabs, Bristol, 2003.

[247] PANKHURST, R.: *Practical taxonomic computing*. Cambridge, UK.: Cambridge University Press, 1991.

[248] PARKER, L.: *Environmental Communication: Messages, Media, and Methods*. Dubuque: Kendall / Hunt Publishing Company, 1997.

[249] PASCOE, J.: Adding Generic Contextual Capabilities to Wearable Computers. In: *Proc. of 2nd Int. Symposium on Wearable Computers (ISWC 1998), Pittsburgh, PA, USA*, Washington: IEEE, 1998.

[250] PASCOE, J.: *Context-Aware Software*. Ph.D. thesis, Department of Computer Science, University of Kent, 2001.

[251] PATALAVICIUTE, V.; DÜPMEIER, C.; FRECKMANN, P.; RUCHTER, M.: Using SVG-based Maps for Mobile Guide Systems - A Case Study for the Design, Adaptation, and Usage of SVG-based Maps for Mobile Nature Guide Applications. In: *Proc. of 4th Annual Conf. on Scalable Vector Graphics (SVGOpen 2005),Enschede, The Netherlands*, Sophia-Antipolis Cedex, Fujisawa, Cambridge: W3C, 2005.

[252] PATALAVICIUTE, V.; FRECKMANN, P.; DÜPMEIER, C.: SVG basierte Karten für mobile Guidesysteme: Erstellung, Gestaltung und Interaktionsmechanismen unter besonderer Berücksichtigung von Karten für den Naturtourismus. Wissenschaftliche berichte, fzka-7063, Forschungszentrum Karlsruhe GmbH, Karlsruhe, 2005.

[253] PAVLICKO, P.; PETERSON, M.: Topographic Maps with SVG. In: *Proc. of 3rd Annual Conf. on Scalable Vector Graphics (SVGOpen 2004), Tokyo, Japan*, Tokyo: Keio University, 2004.

[254] PAYNE, P.: The technics of environmental education. *Environment and Behavior* 9 (2003) 4, pp. 525–541.

[255] PETRIDES, G.; WEHR, J.: *A Field Guide to Eastern Trees (Peterson Field Guides)*. Boston: Houghton Mifflin, 1998.

[256] PICCIANO, A.: Beyond student perceptions: Issues of interaction, presence, and performance in an online course. *Journal of Asynchronous Learning Networks* 6 (2002) 1, pp. 21–40.

[257] PILLMANN, W.: Environmental Communication - Systems Analysis of Environmentally Related Information Flows as a Basis for the Popularization of the Framework for Sustainable Development. In: *Proc. of 14th Symposium Computer Science for Environmental Protection, Bonn, Germany*, Marburg: Metropolis-Verlag, 2000.

[258] POMBERGER, G.; BISCHOFBERGER, W.; KOLB, D.; PREE, W.; SCHLEMM, H.: Prototyping-Oriented Software Development - Concepts and Tools. *Structured Programming* 12 (1991), pp. 43–60.

[259] POOLEY, J.; O'CONNOR, M.: Environmental Education and Attitudes: Emotions and Beliefs Are What Is Needed. *Environment and Behavior* 32 (2000) 5, pp. 711–723.

[260] POSPISCHIL, G.; UMLAUFT, M.; MICHLMAYR, E.: Designing LoL@, a Mobile Tourist Guide for UMTS. In: *Proc. of 4th Int. Symposium on Human-Computer Interaction with Mobile Devices (Mobile HCI 2002), Pisa, Italy* (PATERNÒ, F., Ed.), Berlin: Springer-Verlag, 2002.

[261] PREECE, J.; ROGERS, Y.; SHARP, H.: *Interaction design; beyond human-computer interaction*. West Sussex: John Wiley & Sons, Inc., 2002.

[262] PYLE, R.: Finding the Wild in a Pavement Crack: Commentary on Peter Kahns Encountering the Other. *Children, Youth and Environments* 15 (2005) 2, pp. 398–400.

[263] RAMSEY, J.; HUNGERFORD, H.: Perspectives on Environmental Education in the United States. In: *New Tools for Environmental Protection: Education, Information and Volutary Measures* (DIETZ, T.; STERN, P. C., Eds.), pp. 147–160, Washington: National Academy of Sciences, 2002.

[264] RANDELL, C.; PHELPS, T.; ROGERS, Y.: Ambient Wood: Demonstration of a Digitally Enhanced Field Trip for Schoolchildren. In: *Proc. of 5th Int. Conf. on Ubiquitous Computing (Ubicomp 2003), Seattle, WA, USA*, Washington: IEEE, 2003.

[265] RAUBAL, M.; WINTER, S.: Enriching Wayfinding Instructions with Local Landmarks. In: *Proc. 2nd of Geographic Information Science (GIScience 2002), Boulder, USA, LNCS 2478* (EGENHOFER, M.; MARK, D., Eds.), Berlin: Springer-Verlag, 2002.

[266] REAL, P.; DÜPMEIER, C.; RUCHTER, M.: Konzeption eines Emotionalen Pädagogischen Agenten und Realisierung von ausgewählten Komponenten. Wissenschaftliche Berichte FZKA 7176, Forschungszentrum Karlsruhe GmbH, Karlsruhe, 2005.

[267] REICHENBACHER, T.: Adaptive Concepts for a Mobile Cartography. *Supplement Journal of Geographical Sciences* 11 (2001), pp. 43–53.

[268] REICHENBACHER, T.: SVG for adaptive visualizations in mobile situations. In: *Proc. of 1st Annual Conf. on Scalable Vector Graphics (SVGOpen 2002), Zürich, Switzerland*, 2002.

[269] REICHENBACHER, T.: *Mobile Cartography - Adaptive Visualization of Geographic Information on Mobile Devices*. Ph.D. thesis, Department of Cartography, TU München, Verlag Dr. Hut, München, 2004.

[270] REICHENBACHER, T.; ANGSÜSSER, S.; MENG, L.: Mobile Kartographie - eine offene Diskussion. *Kartographische Nachrichten* 52 (2002) 4, pp. 164–166.

[271] REID, N.: Thoughts on attitude measurement. *Research in Science & Technological Education* 24 (2006) 1, pp. 3–27.

[272] REID, S.; MARION, J.: The efficacy of visitor education programs. Leave no trace web publications, Leave No Trace Center for Outdoor Ethics, Boulder, 2003.

[273] REILLY, D.; RODGERS, M.; ARGUE, R.; NUNES, M.; INKPEN, K.: Marked-up Maps: Combining Paper Maps and Electronic Information Resources. *Journal of Personal and Ubiquitous Computing* 10 (2006) 4, pp. 215–226

[274] REIMANN, C.: Kick-Real - a Mobile Mixed Reality Game using Computer Vision Based Interaction. In: *Proc. of 7th Int. Conf. on Human Computer Interaction with Mobile Devices and Services (Mobile HCI 2005), Salzburg, Austria*, New York: ACM, 2005.

[275] ROGERS, Y.; PRICE, S.; FITZPATRICK, G.; FLECK, R.; HARRIS, E.; SMITH, H.; RANDELL, C.; MULLER, H.; O'MALLEY, C.; STANTON, D.; THOMPSON, M.; WEAL, M.: Ambient Wood: Designing New Forms of Digital Augmentation for Learning Outdoors. In: *Proc. of Interaction Design and Children (IDC 2004), College Park, MD, USA*, New York: ACM, 2004.

[276] ROGERS, Y.; PRICE, S.; RANDELL, C.; STANTON FRASER, D.; WEAL, M.; FITZPATRICK, G.: Ubi-learning Integrates Indoor and Outdoor Experiences. *Communications of the ACM* 48 (2005) 1, pp. 55–59.

[277] ROHWEDDER, W. J.; ALM, A.: *Using Computers in Environmental Education: Interactive Multimedia and on-Line Learning*. Dubuque: Kendall/Hunt Publishing Company, 1994.

[278] ROSENBERG, M.; HOVLAND, C.: *Cognitive, affective, and behavioral components of attitudes*. New Haven: Yale University Press, 1960.

[279] ROSS, T.; MAY, A.; THOMPSON, S.: The use of Landmarks in Pedestrian Navigation. Instructions and the Effects of Context. In: *Proc. of 6th Int. Symposium on Human-Computer Interaction with Mobile Devices (Mobile HCI 2004), Glasgow, UK* (BREWSTER, S.; DUNLOP, M., Eds.), Berlin: Springer-Verlag, 2004.

[280] ROTH, C.: Environmental Literacy: Its Roots, Evolution, and Directions in the 1990s. U.S. Eric, National Energy Education Department, Washington, 1992.

[281] RUCHTER, M.: Web-basierte Umweltinformationssysteme für die Öffentlichkeit: Konzepte, Potentiale und Realisierbarkeit. Wissenschaftliche Berichte FZKA 6760, Forschungszentrum Karlsruhe, Karlsruhe, 2002.

[282] RUCHTER, M.; DÜPMEIER, C.: Web-based environmental information systems for the public: concepts, potentials, and applicability based on a case study. In: *Proc.of 5th Int.Symp. on Environmental Software Systems (ISESS 2003), Semmering, Austria* (SCHIMAK, G., Ed.), Laxenburg: IFIP, 2003.

[283] RUCHTER, M.; DÜPMEIER, C.; GEIGER, W.; SOBEK, R.; WEIDEMANN, R.; DANNENMAYER, H.; HOFMANN, K.; KRAX, J.: MobiNaf. Erster Prototyp eines Mobilen Naturführers für das Naturschutzzentrum Karlsruhe-Rappenwört. Wissenschaftliche Berichte FZKA 7077, Forschungszentrum Karlsruhe GmbH, Karlsruhe, 2004.

[284] RUCHTER, M.; DÜPMEIER, C.; GEIGER, W.; WEIDEMANN, R.; DANNENMAYER, H.; HOFMANN, K.; EBEL, R.; LINNENBACH, M.; SCHELKLE, E.: MobiNaf - Ergebnisse der Evaluation und Studie zu weiteren Einsatzmöglichkeiten mobiler Naturführer. Wissenschaftliche Berichte FZKA 7250, Forschungszentrum Karlsruhe, Karlsruhe, 2005.

[285] RUCHTER, M.; DÜPMEIER, C.; MAYER-FÖLL, R.; SCHELKLE, E.: Studie zum Einsatz von Mobilen Naturführern in Naturerlebnisgebieten in Baden-Württemberg. Wissenschaftliche Berichte FZKA 7278, Forschungszentrum Karlsruhe, Karlsruhe, 2007.

[286] RUCHTER, M.; REAL, P.; DÜPMEIER, C.: Comparing a Mobile Nature Guide and a Paper Guidebook in the Field. In: *Proc. of 4th Workshop HCI in Mobile Guides at Mobile HCI 05, Salzburg, Austria*, 2005.

[287] SALBER, D.; DEY, A.; ABOWD, G.: The Context Toolkit: Aiding the Development of Context-Enabled Applications. In: *Proc. of ACM Conf. on Human Factors in Computing Systems (CHI 1999), Pittsburgh, PA, USA*, New York: ACM, 1999.

[288] SAUCIER, D.; GREEN, S.; LEASON, J.; MACFADDEN, A.; BELL, S.; ELIAS, L.: Are Sex Differences in Navigation Caused by Sexually Dimorphic Strategies or by Differences in the Ability to Use the Strategies? *Behavioral Neuroscience* 116 (2002) 3, pp. 403–410.

[289] SAUER, S.; GÖBEL, S.: Focus on your young visitors: Kids Innovation, Fundamental changes in digital edutainment. In: *Proc. of Museums and the Web 2003, Toronto, Canada* (BEARMAN, D.; TRANT, J., Eds.), Toronto: Archives & Museum Informatics, 2003.

[290] SAUER, S.; OSSWALD, K.: Momuna - Interactive Integrated Information and Knowledge Environment for Museums. In: *Proc. of 2nd Int. Workshop on Ubiquitous Sys. for Supporting Social Interaction & Face-to-Face Com. in Pub. Space at Ubicomp04, Nottingham, GB*, New York: ACM, 2004.

[291] SAVAGE, G.; JAMES, J.: *A Practical Guide to Evaluating Natural and Cultural Heritage Interpretation.* Australia, 2001.

[292] SAVERY, J.; DUFFY, T.: Problem-based learning: An instructional model and its constructivist framework. In: *Constructivist learning environments: Case studies in instructionaldesign* (WILSON, B., Ed.), pp. 135–148, Englewood Cliffs: Educational Technology Publications, 1995.

[293] SCHAHN, J.; BOHNER, G.: Methodische Aspekte sozialwissenschaftlicher Evaluations forschung im Umweltbereich. *Umweltsoziologie, Sonderheft der Kölner Zeitschrift für Soziologie und Sozialpsychologie* 36 (1996), pp. 548–570.

[294] SCHAHN, J.; BOHNER, G.: Evaluation Research in Environmental Sociology and in Environmental Psychology: Methodological Aspects. Diskussionspapier - Bericht aus dem Psychologischen Institut 85, Universität Heidelberg, Heidelberg, 2002.

[295] SCHANK, R.; FANO, A.; BELL, B.; JONA, M.: The Design of Goal-Based Scenarios. *The Journal of the Learning Sciences* 3 (1993) 4, pp. 305–345.

[296] SCHARL, A.: *Environmental Online Communication (Edited Volume).* Berlin: Springer-Verlag, 2004.

[297] SCHILIT, B.; ADAMS, N.; WANT, R.: Context-Aware Computing Applications. In: *Proc. of Workshop on Mobile Computing Systems and Applications, Santa Cruz, CA, USA*, Washington: IEEE, 1994.

[298] SCHILLING, A.; COORS, V.; LAAKSO, K.: Dynamic 3D Maps for Mobile Tourism Applications. In: *Map-based Mobile Services* (MENG, L.; ZIPF, A.; REICHENBACHER, T., Eds.), pp. 233–244, Berlin: Springer-Verlag, 2005.

[299] SCHMIDT, A.; BEIGL, M.; GELLERSEN, H.-W.: There is more to Context than Location. *Computers & Graphics Journal* 23 (1999) 6, pp. 893–902.

[300] SCHMIDT-BELZ, B.; HERMANN, F.: User Validation of a Nomadic Exhibition Guide. In: *Proc. of 6th Int. Symposium on Human-Computer Interaction with Mobile Devices (Mobile*

HCI 2004), Glasgow, UK (BREWSTER, S.; DUNLOP, M., Eds.), Berlin: Springer-Verlag, 2004.

[301] SCHMIDT-BELZ, B.; LAAMANEN, H.; POSLAD, S.; ZIPF, A.: Location-based Mobile Tourist Services - First User Experiences. In: *Proc. of Conf. for the IT and Travel & Tourism Industries (ENTER 2003), Helsinki, Finland*, Berlin: Springer-Verlag, 2003.

[302] SCHMIDT-BELZ, B.; POSLAD, S.: User Validation of a mobile Tourism Service. In: *Proc. of 3rd Workshop HCI in Mobile Guides in conjunction with Mobile HCI 2003, Udine, Italy*, 2003.

[303] SCHMIDT-BELZ, B.; POSLAD, S.; ZIPF, A.: Creation of User-friendly Mobile Tourism Services. In: *Proc. of Workshop Mobile Tourism Support in adjunction to 4th Int. Symposium on HCI with Mobile Devices 2002, Pisa, Italy*, 2002.

[304] SCHNÄDELBACH, H.; KOLEVA, B.; FLINTHAM, M.; FRASER, M.; IZADI, S.; CHANDLER, P.; FOSTER, M.; BENFORD, S.; GREENHALGH, C.; RODDEN, T.: The Augurscope: A Mixed Reality Interface for Outdoors. In: *Proc. of ACM Conf. on Human Factors in Computing Systems (CHI 2002), Minneapolis, MN, USA*, New York: ACM, 2002.

[305] SCHULTZ, P.-W.: Empathizing With Nature: The Effects of Perspective Taking on Concern for Environmental Issues. *Journal of Social Issues* 56 (2000) 3, pp. 391–406.

[306] SCHULTZ, P.-W.; GOUVEIA, V.; CAMERON, E.; TANKHA, G.; SCHMUCK, P.; FRANEK, M.: Values and Their Relationship to Environmental Concern and Conservation Behavior. *Journal of Cross-Cultural Psychology* 36 (2005) 4, pp. 457–475.

[307] SCHULTZ, P.-W.; SHRIVER, C.; TABANICO, J.; KHAZIAN, A.: Implicit connections with nature. *Journal of Environmental Psychology* 24 (2004), pp. 31–42.

[308] SCHWARZER, M.: Art & Gadgetry. *Museum News* Jul./Aug. (2001).

[309] SCOULLOS, M.; MALOTIDI, V.: *Handbook on methods used in Environmental Education and Education for Sustainable Development*. Athens: MIO-ECSDE, 2004.

[310] SEMPER, R.; SPASOJEVIC, M.: The Electronic Guidebook: Using Portable Devices And A Wireless Web-Based Network To Extend The Museum Experience. In: *Proc. of 6th Int. Conf. on Museums and the Web (MW 2002), Boston, MA, USA*, Toronto: Archives & Museum Informatics, 2002.

[311] SHARPLES, M.; BEALE, R.: A technical review of mobile computational devices. *Journal of Computer Assisted Learning* 19 (2003) 3, pp. 392–395.

[312] SHAW, E.; LABORE, C.; CHIU, Y.-C.; JOHNSON, W.: Animating 2D Digital Puppets with Limited Autonomy. In: *Proc.of 4th Int. Symposium on Smart Graphics, SG 2004, Banff, Canada, LNCS* (BUTZ, A.; KRÜGER, A.; OLIVER, P., Eds.), Berlin: Springer-Verlag, 2004.

[313] SHULTIS, J.: Consuming nature: the uneasy relationship between technology, outdoor recreation and protected areas. *The George Wright Forum* 18 (2001) 1, pp. 56–66.

[314] SIEBER, R.; SCHMID, C.; WIESMANN, S.: Smart Legend - Smart Atlas! In: *Proc. of XXII Int. Cartographic Conf. (ICC2005), Coruna, Spain*, Durban, Coruna: Interantional Cartographic Association, 2005.

[315] SIEBERT, H.: Self-directed learning. In: *Umweltbildung: Theoretische Konzepte - empirische Erkenntnisse - praktische Erfahrungen* (BEYERSDORF, M.; MICHELSEN, G.; SIEBERT, H., Eds.), pp. 94–98, Neuwied: Luchterhand, 1998.

[316] SOBEK, R.: *Design und Implementierung eines mobilen, multimedialen Guides für Pocket PC basierte Systeme*. Master's thesis, Hochschule Karlsruhe - Technik und Wirtschaft, Karlsruhe, 2004.

[317] SOH, B.: *Developing Outdoor Map Design Guidelines Using a Real-world Wayfinding Task*. Master's thesis, Virginia Polytechnic Institute, Virginia State University, 2002.

[318] SONG, D.; GOLDBERG, K.: Networked Robotic Cameras for Collaborative Observation of Natural Environments. In: *Proc. of 12th International Symposium of Robotics Research (ISRR), San Francisco, CA, USA* (THRUN, S.; BROOKS, R.; DURRANT-WHYTE, H., Eds.), Durban, Coruna: Springer, 2005.

[319] SORCINELLI, D.: Which Windows Mobile Device Should I Buy? Pocket PC magazine, 2006.

[320] SPIERLING, U.; GRASBON, D.; BRAUN, N.; IURGEL, I.: Setting the Scene: Playing Digital Director in Interactive Storytelling and Creation. *Computers & Graphics* 26 (2002) 1, pp. 31–44.

[321] SQUIRE, K.; JOHNSON, K.; HOLLAND, W.; ZACHARY, N.; KLOPFER, E.: A Platform For Augmented Reality Gaming; Environmental Detectives. Techrep. 2002, Comparative Media Studies Department, MIT/Microsoft Research, Boston, 2002.

[322] STANKIEWICZ, B.; KALIA, A.: Acquisition and Retention of Structural versus Object Landmark Knowledge When Navigating through a Large-Scale Space. Manuscript - http://stankiewiczlab.psy.utexas.edu/manuscripts/landmarks_s_k.pdf (accessed may'07), Shape & Space Perception Lab, University of Texas, Austin, 2004.

[323] STANKIEWICZ, K.; WIŚNIEWSKA, E.; ZAWILA-NIEDŹWIECKI, T.: Multimedia Geoinformation System ReGeo - Basic Concepts and Current Status. In: *Proc. of 10th EC GI & GIS Workshop, ESDI State of the Art, Warsaw, Poland*, http://www.ec-gis.org, 2004.

[324] STAPP, W.; COX, D.: Environmental Education Model. Environmental education activities manual, Thomson-Shore, Inc., Dexter, 1974.

[325] STATSOFT, I.: *Electronic Statistics Textbook*. Tulsa: StatSoft, 2006.

[326] STECK, S.; MALLOT, H.: The Role of Global and Local Landmarks in Virtual Environment Navigation. *Presence* 9 (2000) 1, pp. 69–83.

[327] STEVENSON, R.: Electronic Field Guides and Citizen Science: Steering Society in a More Earth-Friendly Direction. Conservation Perspectives, 2004.

[328] STEVENSON, R.; HABER, W.; MORRIS, R.: Electronic field guides and user communities in the eco-informatics revolution. *Conservation Ecology* 7(1): 3 (2003).

[329] STOCK, O.; ZANCANARO, M.: Intelligent Interactive Information Presentation for Cultural Tourism. In: *Proc. of Int. CLASS Workshop on Natural, Intelligent and Effective Interaction in Multimodal Dialogue Systems 2002, Copenhagen, Denmark* (VAN KUPPEVELT, J.; DYBKJAER, L.; BERNSEN, L. O., Eds.), The Class Project, 2002.

[330] STORCH, H.: Mobiler Zugriff auf Umweltinformationen - Konzeptionelle Anforderungen an integrative Umweltbewertungsverfahren auf der Basis von individuellen Präferenzen in zeit-räumlichen Kontexten. In: *Geoinformation Mobil - Grundlagen und Perspektiven von Location Based Services* (ZIPF, A. UND STROBL, J., Ed.), pp. 157–166, Heidelberg: H. Wichmann Verlag, 2002.

[331] STORCH, H.: Mobile Portals for Community-Based Collaborative Environmental Monitoring. In: *Proc. of 17th Int. Conf. Informatics for Environmental Protection (EnviroInfo 2003), Cottbus, Germany* (GNAUCK, A.; HEINRICH, R., Eds.), Marburg: Metropolis-Verlag, 2003.

[332] SUMI, Y.; MASE, K.: Supporting the awareness of shared interests and experiences in communities. *International Journal of Human-Computer Studies* 56 (2002), pp. 127–146.

[333] SZEWCZYK, R.; MAINWARING, A.; POLASTRE, J.; CULLER, D.: An Analysis of a Large Scale Habitat Monitoring Application. In: *Proc. of 2nd ACM Conf. on Embedded Networked Sensor Systems (SenSys 2004), Baltimore, MD, USA*, New York: ACM, 2004.

[334] TÉTARD, F.; PATOKORPI, E.: Design of a Mobile Guide for Educational Purposes. In: *Proc. of 3rd Workshop HCI in Mobile Guides at Mobile HCI 04, Glasgow, GB*, 2004.

[335] THOEGERSEN, J.: A cognitive dissonance interpretation of consistencies and inconsistencies in environmentally responsible behavior. *Journal of Environmental Psychology* 24 (2004), pp. 93–103.

[336] THOMSON, G.: What is good environmental education? A draft backgrounder for practitioners, Canadian Parks and Wilderness Society, Ottawa, 2002.

[337] TILDEN, F.: *Interpreting Our Heritage*. Chapel Hill: The University of South Carolina Press, 1957.

[338] TOL: Studie: Mehr als eine Million Windows-Mobilgeräte in EMEA. Heise online news, 2005.

[339] TRUMBULL, D.; BONNEY, R.; BASCOM, D.; CABRAL, A.: Thinking scientifically during participation in a citizen-science project. *Science Education* 84 (2000) 2, pp. 265–275.

[340] TUBB, K.: An Evaluation of the Effectiveness of Interpretation within Dartmoor National Park in Reaching the Goals of Sustainable Tourism Development. *Journal of Sustainable Tourism* 11 (2003) 6, pp. 476–498.

[341] TVERSKY, B.: Some Ways That Maps and Diagrams Communicate. In: *Spatial cognition II: Integrating abstract theories, empirical studies, formal methods, and practical applications.* (TVERSKY, B., Ed.), Berlin: Springer-Verlag, 2000.

[342] UMLAUFT, M.; POSPISCHIL, G.; NIKLFELD, G.; MICHLMAYR, E.: LoL@, a Mobile Tourist Guide for UMTS. *Journal of Information Technology and Tourism* 5 (2003) 3, pp. 151–164.

[343] UNEP: *Global Environment Outlook 3 (GEO-3)*. London: Earthscan Publications Ltd, 2002.

[344] UNESCO: Convention Concerning the Protection of the World Cultural and Natural Heritage. Unesco world heritage, UNESCO, Paris, 1972.

[345] UNESCO: Draft International implementation scheme for the United Nations Decade of Education for Sustainable Development (2005-2014). 172 ex/11, United Nations publication, New York, 2005.

[346] UNESCO; UNEP: The Belgrade Charter. Connect: UNESCO-UNEP Environmental Education Newsletter, 1976.

[347] UNTERBRUNER, U.; UNTERBRUNER, G.: Wirkung verarbeitungsfördernder multimedialer Programmgestaltung auf den Lernprozess von 10- bis 12-jährigen. In: *Lehr- und Lernforschung in der Biologiedidaktik* (KLEE, R.; SANDMANN, A.; VOGT, H., Eds.), Innsbruck, Wien, Bozen: Studien Verlag, 2005.

[348] UTTAL, D.: Seeing the big picture: map use and the development of spatial cognition. *Developmental Science* 3 (2000) 3, pp. 247–264.

[349] VANDEVEER, R.; MENEFEE, M.: *Human Behavior in Organizations*. Upper Saddle River, New Jersey: Prentice Hall, 2005.

[350] VINSON, N.: Design guidelines for landmarks to support navigation in virtual environments. In: *Proc. of ACM Conf. on Human Factors in Computing Systems (CHI 1999), Pittsburgh, PA, USA*, 1999.

[351] VIVONI, E.; CAMILLI, R.: Real-time streaming of environmental field data. *Computers & Geosciences* 29 (2003), pp. 457–468.

[352] WAHLSTER, W.: SmartKom: Symmetric Multimodality in an Adaptive and Reusable Dialogue Shell. In: *Proc. of the Human Computer Interaction Status Conf. 2003, Berlin, Germany* (KRAHL, R.; GUNTHER, D., Eds.), 2003.

[353] WEIDEMANN, R.; GEIGER, W.; GRECEANU, C.; ZILLY, G.; SCHMID, E.; EBEL, R.: PaUla - Erste Komponenten für den Zugriff auf das UIS Baden-Württemberg mit dem Portal für mobile Umwelt-Assistenten. Wissenschaftliche Berichte, FZKA-6950 Phase IV, Forschungszentrum Karlsruhe GmbH, Karlsruhe, 2003.

[354] WEIR, C.; NOBEL, J.: A Window in your Pocket - Some Small Patterns for User Interfaces. In: *Proc. of 8th European Conf. on Pattern Languages of Programs (EuroPLoP 2003), Irsee, Germany*, Irsee: Universitätsverlag Konstanz, 2003.

[355] WEISS, D.: Chimer: Childrens Heritage Interactive Models for Evolving Repositories. Cultivate Interacitve 8, UKOLN, Bath, 2002.

[356] WEISS, S.: *Handheld Usability*. West Sussex: John Wiley & Sons, 2002.

[357] WEISS, S.: Paper Prototyping to Increase Usability of Applications. *Pocket PC magazine* May (2003).

[358] WERNER, S.; KRIEG-BRÜCKNER, B.; MALLOT, H.; SCHWEIZER, K.; FREKSA, C.: Spatial Cognition: The Role of Landmark, Route, and Survey Knowledge in Human and Robot Navigation. *Informatik aktuell* (1997), pp. 41–50.

[359] WESTBOMKE, J.; HAASE, M.; EBEL, R.; LEHNE, D.: Mobile Access to Environmental Information. In: *Environmental Online Communication* (SCHARL, A., Ed.), pp. 11–20, Berlin: Springer-Verlag, 2004.

[360] WHITE, S.; FEINER, S.; KOPYLEC, J.: Virtual Vouchers: Prototyping a Mobile Augmented Reality User Interface for Botanical Species Identification. In: *Proc. of the 3D User Interfaces (3DUI'06),Alexandria, VA, USA*, pp. 119–126, Washington, DC, USA: IEEE Computer Society, 2006.

[361] WIKIPEDIA: Wikipedia, The Free Encyclopedia. http://en.wikipedia.org, 2006.

[362] WILBANKS, T.; STERN, P.: New Tools for Environmental Protection: What We Know and Need to Know. In: *New Tools for Environmental Protection: Education, Information and Volutary Measures* (DIETZ, T.; STERN, P. C., Eds.), pp. 335–348, Washington: National Academy of Sciences, 2002.

[363] WILSON, N.: *Identifying Organisms with Computers: An Implementation of a Computerized Synoptic Identification System with Fungi as a Test Case*. Master's thesis, Computer and Information Science, University of California, 1994.

[364] WOHLERS, L.: Informelle Umweltbildung. In: *Umweltbildung: Theoretische Konzepte - empirische Erkenntnisse - praktische Erfahrungen* (BEYERSDORF, M.; MICHELSEN, G.; SIEBERT, H., Eds.), pp. 206–218, Neuwied: Luchterhand, 1998.

[365] WOODRUFF, A.; AOKI, P.; GRINTER, R.; HURST, A.; SZYMANSKI, M.; THORNTON, J.: Eavesdropping on Electronic Guidebooks: Observing Learning Resources in Shared Listening Environments. *Archives & Museum Informatics* (2002), pp. 21–30.

[366] WOODRUFF, A.; AOKI, P.; HURST, A.; SZYMANSKI, M.: Electronic Guidebooks and Visitor Attention. In: *Proc. of 6th Cultural Heritage Informatics (ICHIM '01), Milan, Italy*, pp. 623–637, Toronto: Archives & Museum Informatics, 2001.

[367] ZIPF, A.: User-Adaptive Maps for Location-Based Services (LBS) for Tourism. In: *Proc. of Int. Conf. Information and Communication Technologies in Tourism (ENTER 2002), Innsbruck, Austria* (WÖBER, K. W.; FREW, A.; HITZ, M., Eds.), 2002.

[368] ZIPF, A.; HÄUSSLER, J.: An Evaluation of the OpenLS Specifications for multi-modal mobile applications. In: *Journal of Geographic Information Sciences, Special Issue on LBS and UbiGIS, CPGIS, Berkeley, California*, 2004.